Lecture Notes in Computer Science 15402

Founding Editors

Gerhard Goos
Juris Hartmanis

Editorial Board Members

Elisa Bertino, *Purdue University, West Lafayette, IN, USA*
Wen Gao, *Peking University, Beijing, China*
Bernhard Steffen, *TU Dortmund University, Dortmund, Germany*
Moti Yung, *Columbia University, New York, NY, USA*

The series Lecture Notes in Computer Science (LNCS), including its subseries Lecture Notes in Artificial Intelligence (LNAI) and Lecture Notes in Bioinformatics (LNBI), has established itself as a medium for the publication of new developments in computer science and information technology research, teaching, and education.

LNCS enjoys close cooperation with the computer science R & D community, the series counts many renowned academics among its volume editors and paper authors, and collaborates with prestigious societies. Its mission is to serve this international community by providing an invaluable service, mainly focused on the publication of conference and workshop proceedings and postproceedings. LNCS commenced publication in 1973.

C. Aiswarya · Prabal Kumar Sen ·
Shashi Mohan Srivastava
Editors

Logic and Its Applications

11th Indian Conference, ICLA 2025
Kolkata, India, February 3–5, 2025
Proceedings

Editors
C. Aiswarya [iD]
Chennai Mathematical Institute
Siruseri, Tamil Nadu, India

Prabal Kumar Sen
University of Calcutta
Kolkata, West Bengal, India

Shashi Mohan Srivastava
RKMVERI
Belur, West Bengal, India

ISSN 0302-9743   ISSN 1611-3349 (electronic)
Lecture Notes in Computer Science
ISBN 978-3-031-89609-5   ISBN 978-3-031-89610-1 (eBook)
https://doi.org/10.1007/978-3-031-89610-1

© The Editor(s) (if applicable) and The Author(s), under exclusive license to Springer Nature Switzerland AG 2025

This work is subject to copyright. All rights are solely and exclusively licensed by the Publisher, whether the whole or part of the material is concerned, specifically the rights of translation, reprinting, reuse of illustrations, recitation, broadcasting, reproduction on microfilms or in any other physical way, and transmission or information storage and retrieval, electronic adaptation, computer software, or by similar or dissimilar methodology now known or hereafter developed.
The use of general descriptive names, registered names, trademarks, service marks, etc. in this publication does not imply, even in the absence of a specific statement, that such names are exempt from the relevant protective laws and regulations and therefore free for general use.
The publisher, the authors and the editors are safe to assume that the advice and information in this book are believed to be true and accurate at the date of publication. Neither the publisher nor the authors or the editors give a warranty, expressed or implied, with respect to the material contained herein or for any errors or omissions that may have been made. The publisher remains neutral with regard to jurisdictional claims in published maps and institutional affiliations.

This Springer imprint is published by the registered company Springer Nature Switzerland AG
The registered company address is: Gewerbestrasse 11, 6330 Cham, Switzerland

If disposing of this product, please recycle the paper.

Preface

This volume contains the proceedings of the 11th Indian Conference on Logic and Its Applications (ICLA). The conference was held as an in-person event during February 3–5, 2025, at the Indian Statistical Institute (ISI) Kolkata India.

The Indian Conference on Logic and its Applications (ICLA) is the primary conference of the Association for Logic in India (ALI). It is a forum for bringing together researchers from a variety of fields in which formal logic plays a significant and often foundational role: Mathematics, Computer Science, Philosophy, Linguistics and Cognitive Science. A special feature of ICLA is the inclusion of studies in systems of logic in the Indian tradition, as well as historical research on logic.

The list of topics in the scope of the conference was revised and expanded in this edition. The Program Committee was constituted to cover this wide range of topics. The reviewing followed a single-blind policy. All submitted papers received at least three reviews, and most submissions received four. Aside from reviews by the Program Committee (PC) members, in some cases there were reviews by external experts. The EasyChair system was used for submission and reviews; it proved to be quite convenient. We would like to express our deep appreciation to all the PC members for their efforts and support. We also thank all the external reviewers for their invaluable help.

For the first time, ICLA introduced a rebuttal phase as part of the reviewing process, where the initial reviews were shared with the authors giving them a chance to respond to questions and clarifications from the reviewers. This feature turned out to be particularly helpful in the PC discussions. Out of the 26 submissions, the program committee carefully selected 14 papers to be included in the proceedings.

The 11th edition of ICLA also introduced best paper and best student paper awards for the first time. The best paper award was shared by two papers – one by Benedikt Loewe and Han Xiao, and one by Abhishek De. The best student paper award was won by Pranshu Gaba and Arnab Sur.

ICLA 2025 included 5 invited talks, and 4 of these appear in the volume as full papers. We are immensely grateful to Su Gao, Janos Makowsky, Anil Nerode, Sophie Pinchinat and Nicholas Ramsey for kindly accepting our invitations.

In addition to the articles in the proceedings and the invited talks, ICLA 2025 also featured presentations of short abstracts and posters of already published work or work in progress. It also featured two panel discussions — on the importance of logic education and careers in logic, as well as on Gödel's incompleteness theorems. The conference was accompanied by two pre-conference workshops (on the Mīmāṁsā system of logical interpretation of imperatives organised by Manidipa Sanyal and Bama Srinivasan, and on Recent Developments in Arithmetic Theories and Applications organised by Kushraj Madnani and Georg Zetzsche) and a tutorial (on First-Order Modal Logic organised by R. Ramanujam), as well as a post-conference related event in the same city: the Asian Workshop on Philosophical Logic (AWPL) 2025.

Special thanks are due to ISI Kolkata, the organizing committee chaired by Sourav Chakraborty and Sujata Gosh, and all the volunteers, for making this edition of ICLA possible. Special thanks to the ALI council and the ALI executive committee, in particular for the Herculean efforts by Abhisekh Sankaran. We would also like to thank the generous sponsors of ICLA: ISI Kolkata, Indian Association for Research in Computing Science (IARCS) and DEST-SERB.

We are grateful to Springer, for agreeing to publish this volume in the LNCS series.

February 2025

C. Aiswarya
Prabal Kumar Sen
Shashi Mohan Srivastava

Organization

Program Committee Chairs

C. Aiswarya	Chennai Mathematical Institute, India
Prabal Kumar Sen	University of Calcutta, India
Shashi Mohan Srivastava	IACS Kolkata and RKMVERI Belur, India

Program Committee

Nikhil Balaji	IIT Delhi, India
Rupa Bandyopadhyay	Jadavpur University, India
Sanjukta Basu	Rabindra Bharati University, India
Sankha Basu	IIIT Delhi, India
Michael Benedikt	University of Oxford, UK
Kuntala Bhattacharya	Rabindra Bharati University Kolkata, India
Amita Chatterjee	Jadavpur University Kolkata, India
Madhumita Chatterjee	Jadavpur University Kolkata, India
Tran Chieu-Minh	National University of Singapore, Singapore
Deepak D'Souza	Indian Institute of Science, India
Anupam Das	University of Birmingham, UK
Hans Van Ditmarsch	University of Toulouse, CNRS, IRIT, France
Huimin Dong	TU Wien, Austria
Su Gao	Nankai University, China
Shibashis Guha	TIFR Mumbai, India
Petr Hliněný	Masaryk University, Czech Republic
Purbita Jana	Madras School of Economics, India
S. Krishna	IIT Bombay, India
Roman Kuznets	TU Wien, Austria
Moritz Lichter	RWTH Aachen, Germany
Khushraj Madnani	MPI-SWS Kaiserslautern, Germany
Amaldev Manuel	IIT Goa, India
Gayatri Panicker	Vellore Institute of Technology, India
Anand Pillay	University of Notre Dame, USA
R. Ramanujam	Azim Premji University, India
Nick Ramsey	University of Notre Dame, USA
Abhisekh Sankaran	TCS Research, India
Manidipa Sanyal	University of Calcutta, India

Sylvain Schmitz — Université Paris Cité, France
Chenwei Shi — Tsinghua University, China
Stephen G. Simpson — Pennsylvania State University and Vanderbilt University, USA
Slawomir Solecki — Cornell University, USA
Shichang Song — Beijing Jiaotong University, China
Bama Srinivasan — Anna University, India
Sergei Starchenko — University of Notre Dame, USA
Vaishnavi Sundararajan — IIT Delhi, India
Sourav Tarafder — St. Xavier's College, India
Ramanathan Thinniyam Srinivasan — Uppsala University, Sweden
Jouko Väänänen — University of Helsinki, Finland and University of Amsterdam, Netherlands

Local Organization

Organizing Chairs

Sourav Chakraborty — Indian Statistical Institute Kolkata, India
Sujata Ghosh — Indian Statistical Institute Chennai, India

Organizing Committee

Avijeet Ghosh — Indian Statistical Institute Kolkata, India
Rajarshi Ray — IACS Kolkata, India
Shashi Mohan Srivastava — IACS Kolkata and RKMVERI Belur, India

Additional Reviewers

Balabán, Jakub
Bhaskar, Ashwin
Block Gorman, Alexi
Chakraborty, Sourav
Felber, Stephan
Genco, Francesco Antonio
Härtel, Louis
Jedelský, Jan
Mukhopadhyay, Diganta
Neider, Daniel
Parikh, Rohit
Roy, Sayantan
Sano, Katsuhiko
Sen, Jayanta
Suresh, S. P.
Weber, Zach
Yamamoto, Kentaro
Égré, Paul

Contents

Invited Papers

On Extensions of Partial Isometries 3
   Mahmood Etedadialiabadi and Su Gao

The Specker-Blatter Theorem: An Application of Logic to Combinatorial
Counting 34
   Johann A. Makowsky

Propositional Dynamic Logic Formula Synthesis and Some Applications 38
   Sophie Pinchinat

NSOP1 as a Dividing Line 50
   Nicholas Ramsey

Contributed Papers

Asynchronous Transition System Games for Two Processes and Their
Analysis 69
   Bharat Adsul and Nehul Jain

Relational Companions of Logics 84
   Sankha S. Basu and Sayantan Roy

Bounded Henkin Quantifiers and the Exponential Time Hierarchy 97
   Abhishek De

Monotone Modal Logic Beyond Distributivity 111
   *Yiwen Ding, Krishna Manoorkar, Alessandra Palmigiano,
   and Ruoding Wang*

Recognizing Numbers 126
   Pranshu Gaba and Arnab Sur

There is Hope for Connexive Set Theories! 138
   Santiago Jockwich

A Semantics of Basic Modal Language via a Rough Set Framework 151
   Md. Aquil Khan and Ranjan

Modal and Intermediate Logics of Spiked Boolean Algebras 164
  Benedikt Löwe and Han Xiao

Equivalence of Deterministic Weighted Real-Time One-Counter Automata 176
  Prince Mathew, Vincent Penelle, Prakash Saivasan, and A. V. Sreejith

Passive Learning of Fuzzy Temporal Logic Rules from Finite Traces 190
  Sandip Paul and Bornali Paul

A Mīmāṃsā Inspired Framework Towards Temporal Reasoning in Large
Language Models 204
  Bama Srinivasan and Mohan Raj Vijayan

Measurement-Theoretic Foundations of Logic of Inexact Knowledge 218
  Satoru Suzuki

Craig Interpolation for Awareness Logics 233
  Kosuke Udatsu and Katsuhiko Sano

Knowable as Knowing How to Inquire 247
  Yiting Wang and Yanjing Wang

Author Index 263

Invited Papers

On Extensions of Partial Isometries

Mahmood Etedadialiabadi[1] and Su Gao[2(✉)]

[1] 1979 Milky Way, Verona, WI 53593, USA
[2] School of Mathematical Sciences and LPMC, Nankai University,
Tianjin 300071, China
sgao@nankai.edu.cn

Abstract. In this paper we define a notion of S-extension for a metric space and study minimality and coherence of S-extensions. We show that every S-extension can be identified with an algebraic object. We use this algebraic representation to give a complete characterization of all finite minimal S-extensions of a given finite metric space and a complete characterization of all minimal coherent S-extensions. We also define a notion of ultraextensive metric spaces and show that every countable metric space can be extended to a countable ultraextensive metric space. We also show that the isometry group of an infinite ultraextensive metric space has a dense locally finite subgroup, generalizing results in [8, 10, 13, 15]. We also study compact ultrametric spaces and show that every compact ultrametric space can be extended to a compact ultraextensive ultrametric space.

Keywords: Hrushovski property · extension property for partial automorphisms (EPPA) · partial isometry · s-extension · s-map · coherent · ultraextensive · ultrahomogeneous · locally finite · ultrametric

1 Introduction

The study in this paper was motivated by the following theorem of Solecki [13]:

Theorem 1 (Solecki [13]). *Every finite metric space X can be extended to a finite metric space Y such that every partial isometry of X extends to an isometry of Y.*

Alternative proofs have been presented in [7, 8, 10, 12]. Also, Vershik announced an unpublished proof in [17].

The main objective of this paper is to identify such extensions and related concepts with algebraic objects. Inspired by Solecki's theorem, we define the following notions. Given a metric space X, a distinguished point $a_0 \in X$ and a

---

The second author's research was partially supported by the National Natural Science Foundation of China (NSFC) grants 12250710128 and 12271263.

collection P of partial isometries of X such that $P = P^{-1}$ and $X \setminus \{a_0\} \subseteq P(a_0)$, a *P-type S-extension* is a pair (Y, ϕ), where Y is a metric space extending the metric space X and ϕ is a map from P into the set of all isometries of Y such that $\phi(p)$ extends p for all $p \in P$. When P is the set of all partial isometries of X, we call (Y, ϕ) an *S-extension* of X. Solecki's theorem can be restated as: Every finite metric space has a finite S-extension. We should also mention that similar notions in a more general context of finite relational structures have been studied by the authors in [2].

If X is a metric space and (Y, ϕ) is a P-type S-extension of X, then we say (Y, ϕ) is *minimal* (sometimes called *irreducible* in the literature) if for all $y \in Y$ there are partial isometries $p_1, \ldots, p_n \in P$ and $x \in X$ such that

$$y = \phi(p_1) \cdots \phi(p_n)(x).$$

This allows us to associate the point $y \in Y$ with $\phi(p_1) \cdots \phi(p_n) \in \phi(\mathbb{F}(P))$, where $\mathbb{F}(P)$ is the free group generated by elements of P, and therefore to view Y as an algebraic object.

Our first main result of the paper is to give an algebraic characterization of all finite minimal P-type S-extensions of a given finite metric space. Before even stating the result, we will give a direct constructive proof of Theorem 1. Recall that Solecki's proof in [13] uses a result of Herwig–Lascar [5], which is in turn a generalization of a celebrated result of Hrushovski [6] on extending partial isomorphisms of finite graphs. Our proof of Theorem 1 follows the ideas of Herwig–Lascar's proof and is essentially the same as the approaches in Rosendal's in [10] and Pestov's in [8]; but since we have a focus of characterizing all minimal S-extensions, our proof is somewhat different from them. Specifically, our proof is not as general as [10], and unlike [8], it does not need the full generality of Herwig–Lascar's result. Here we should mention that Hubička–Konečný–Nešetřil [7] provided a combinatorial proof of Theorem 1.

In a sense, our proof of Theorem 1 gives a "canonical" algebraic construction of a P-type S-extension from a parameter we call a *feasible prekernel*, which is a suitable normal subgroup of $\mathbb{F}(P)$. Given a feasible prekernel $N \trianglelefteq \mathbb{F}(P)$, we construct a canonical algebraic S-extension (Γ_N, Φ_N), and introduce a weight function w_N. The minimal S-extensions can then be characterized as follows.

Theorem 2. *Let (Y, ϕ) be a finite minimal P-type S-extension of X. Let $N = \ker(\phi)$ and $G = \Phi_N(\mathbb{F}(P))$. Then there is a G-invariant pseudometric ρ on Γ_N which is consistent with w_N such that (Y, ϕ) is isomorphic to $(\overline{\Gamma_N}^\rho, \overline{\Phi_N}^\rho)$.*

The next notion we study is that of coherence between P-type S-extensions. Here we introduce a notion of coherence between feasible prekernels, and our second main result demonstrates a correspondence between the two coherence notions.

Theorem 3. *Let $X_1 \subseteq X_2$ be finite metric spaces, (Y_1, ϕ_1) be a minimal P_1-type S-extension of X_1, and $P_1 \subseteq P_2$ where $P_2 = P_2^{-1}$ and $X_2 \setminus \{a_0\} \subseteq P_2(a_0)$. Then the following hold.*

(i) Let (Y_2, ϕ_2) be a P_2-type S-extension of X_2 that is coherent with (Y_1, ϕ_1). Then $N_2 = \ker(\phi_2)$ is a coherent extension of $N_1 = \ker(\phi_1)$.
(ii) Let $N_2 \trianglelefteq \mathbb{F}(P_2)$ be a coherent extension of $N_1 = \ker(\phi_1)$. Then letting $G_2 = \Phi_{N_2}(\mathbb{F}(P_2))$, there exists a G_2-invariant pseudometric ρ_2 on Γ_{N_2} which is consistent with w_{N_2}, such that $(Y_2, \phi_2) = (\overline{\Gamma_{N_2}}^{\rho_2}, \overline{\Phi_{N_2}}^{\rho_2})$ is coherent with (Y_1, ϕ_1).

Iterative coherent S-extensions lead to infinite metric spaces with striking properties. We introduce the notion of ultraextensive metric spaces and obtain some general results about them. Specifically, we call a metric space U *ultraextensive* if U is ultrahomogeneous, every finite $X \subseteq U$ has a finite S-extension (Y, ϕ) where $Y \subseteq U$, and if $X_1 \subseteq X_2 \subseteq U$ are finite and (Y_1, ϕ_1) is a finite minimal S-extension of X_1 with $Y_1 \subseteq U$, then there is a finite minimal S-extension (Y_2, ϕ_2) of X_2 such that $Y_2 \subseteq U$ and (Y_1, ϕ_1) and (Y_2, ϕ_2) are coherent.

Recall ultrahomogeneity means that any partial isometry can be extended to a full isometry of the entire space. Thus ultraextensiveness is a strengthening of ultrahomogeneity. It follows from our results that the universal Urysohn metric space $\mathbb{U}$ and the rational universal Urysohn space $\mathbb{QU}$ are ultraextensive. Another example of ultraextensive metric space is the countable random graph equipped with the path metric. Moreover, we will also establish the following results.

Theorem 4. *Every countable metric space can be extended to a countable ultraextensive metric space.*

Theorem 5. *If U is an ultraextensive metric space, then every countable subset $X \subseteq U$ can be extended to a countable ultraextensive $Y \subseteq U$.*

Theorem 6. *For any separable ultraextensive metric space U, $\mathrm{Iso}(U)$ contains a dense locally finite subgroup.*

For compact ultrametric spaces we prove the following.

Theorem 7. *Every compact ultrametric space can be extended to a compact ultraextensive ultrametric space. In particular, every compact ultrametric space has a compact ultrametric S-extension.*

The rest of the paper is organized as follows. In Sect. 2 we give some preliminaries of S-extensions, metrics on weighted graphs, and the profinite topology. In Sect. 3 we give the canonical algebraic construction of finite S-extensions in the style of Herwig–Lascar. In Sect. 4 we study finite minimal S-extensions and give a complete characterization of them. In Sect. 5 we study the notion of coherent S-extensions and give a complete characterization of coherent S-extensions along with several constructions and applications. In Sect. 6 we study ultraextensive spaces and establish the main results mentioned above. In Sect. 7 we study compact ultrametric spaces and show that they admit compact ultrametric S-extensions. In Sect. 8 we mention some open problems.

2 Preliminaries

2.1 S-Extensions

We fix some notations to be used in the rest of the paper. Let (X, d_X) and (Y, d_Y) be metric spaces. When there is no danger of confusion, we simply write X for (X, d_X) and Y for (Y, d_Y).

We say that Y is an *extension* of X if $X \subseteq Y$ and for all $x_1, x_2 \in X$, $d_Y(x_1, x_2) = d_X(x_1, x_2)$. Interchangeably, we use the same terminology when Y contains an isometric copy of X.

An *isometry* from X to Y is a bijection $\pi : X \to Y$ such that

$$d_Y(\pi(x_1), \pi(x_2)) = d_X(x_1, x_2)$$

for all $x_1, x_2 \in X$. An isometry from X to X is also called an *isometry of X*. The set of all isometries of X is denoted as $\text{Iso}(X)$. Under composition of maps, $\text{Iso}(X)$ becomes a group.

A *partial isometry* of X is an isometry between two finite subspaces of X. The set of all partial isometries of X is denoted as $\mathcal{P}(X)$. $\mathcal{P}(X)$ is not necessarily a group, but it is a groupoid. In particular, for each $p \in \mathcal{P}(X)$ we can speak of the inverse map p^{-1}, which is still a partial isometry.

If Y is an extension of X, then every partial isometry of X is also a partial isometry of Y. In symbols, we have $\mathcal{P}(X) \subseteq \mathcal{P}(Y)$ if $X \subseteq Y$.

If $p, q \in \mathcal{P}(X)$, we say that q *extends* p, and write $p \subseteq q$, if

$$\{(x, p(x)) : x \in \text{dom}(p)\} \subseteq \{(x, q(x)) : x \in \text{dom}(q)\}.$$

We let 1_X denote the identity isometry on X, i.e., $1_X(x) = x$ for all $x \in X$. Let $\mathcal{P}_X$ denote the set of all $p \in \mathcal{P}(X)$ such that $p \not\subseteq 1_X$. We refer to elements of $\mathcal{P}_X$ as *nonidentity partial isometries* of X.

The main concept we study in this paper is that of a P-type S-extension.

Definition 8. Let X be a finite metric space and fix a distinguished point $a_0 \in X$. Let $P \subseteq \mathcal{P}_X$ be such that $P = P^{-1}$ and $X \setminus \{a_0\} \subseteq P(a_0)$. A *$P$-type S-extension* of X is a pair (Y, ϕ), where $Y \supseteq X$ is an extension of X, and $\phi : P \to \text{Iso}(Y)$ such that $\phi(p)$ extends p for all $p \in P$. The map ϕ is called a *P-type S-map* for X.

When $P = \mathcal{P}_X$, we call (Y, ϕ) an *S-extension* of X and ϕ an *S-map* for X.

Note that an equivalent restatement of Solecki's theorem (Theorem 1) is that every finite metric space has a finite S-extension. It is well-known that the universal Urysohn space $\mathbb{U}$ is both universal (for all separable metric spaces) and ultrahomogeneous. These imply that every separable metric space has an S-extension (Y, ϕ) where Y is isometric with $\mathbb{U}$.

We will need the following notion of isomorphism between P-type S-extensions.

Definition 9. Let X be a metric space and (Y, ϕ) and (Z, ψ) be both P-type S-extensions of X. An *isomorphism* between (Y, ϕ) and (Z, ψ) is an isometry $\pi : Y \to Z$ such that $\psi(p) \circ \pi = \pi \circ \phi(p)$ for all $p \in P$. If there is an isomorphism between (Y, ϕ) and (Z, ψ), we say that (Y, ϕ) and (Z, ψ) are *isomorphic*, and write $(Y, \phi) \cong (Z, \psi)$.

2.2 Metrics on Weighted Graphs

We will study metric spaces derived from weighted graphs. A *weighted graph* is a pair (Γ, w), where $\Gamma = (V(\Gamma), E(\Gamma))$ is a (simple undirected) graph and $w : E(\Gamma) \to \mathbb{R}_+$ ($\mathbb{R}_+$ denotes the set of all positive real numbers). We call w the *weight function*. If w_1, w_2 are two weight functions on Γ, then we write $w_1 \leq w_2$ if $w_1(x, y) \leq w_2(x, y)$ for all $(x, y) \in E(\Gamma)$.

Given a weighted graph (Γ, w), let $L_w = \inf\{w(x, y) : (x, y) \in E(\Gamma)\}$ and $B_w = \sup\{w(x, y) : (x, y) \in E(\Gamma)\}$. Assuming $0 < L_w \leq B_w < \infty$, one can define a *path metric* d_w on $V(\Gamma)$ as follows: for any $x, y \in V(\Gamma)$, let

$$d_w(x, y) = \min\{B_w, \delta_w(x, y)\}$$

where

$$\delta_w(x, y) = \inf\left\{\sum_{i=1}^n w(x_i, x_{i+1}) : x_1 = x,\ x_{n+1} = y, \forall i \leq n\ (x_i, x_{i+1}) \in E(\Gamma)\right\}.$$

In particular, $\delta_w(x, y)$ is undefined when x and y are not connected by a path in Γ, in which case $d_w(x, y) = B_w$. It is easy to verify that d_w is indeed a metric. Also, if $(x, y) \in E(\Gamma)$, then $d_w(x, y) = \delta_w(x, y)$. We note the following simple fact about d_w without proof.

Lemma 10. *The following are equivalent:*

(i) For any $(x, y) \in E(\Gamma)$, $d_w(x, y) = w(x, y)$.
(ii) For any $(x, y) \in E(\Gamma)$ and any $x_1, \ldots, x_{n+1}$ where $x_1 = x$, $x_{n+1} = y$, and $(x_i, x_{i+1}) \in E(\Gamma)$ for all $i = 1, \ldots, n$, we have

$$w(x, y) \leq \sum_{i=1}^n w(x_i, x_{i+1}).$$

We introduce some new concepts about the consistency of metrics on weighted graphs.

Definition 11. Let (Γ, w) be a weighted graph and d be a metric on $V(\Gamma)$. We say that d is *consistent* with w if for all $(x, y) \in E(\Gamma)$, $d(x, y) = w(x, y)$. We say that w is *reduced* if d_w is consistent with w.

Lemma 12. *Let (Γ, w) be a connected weighted graph. Then there is a maximal reduced weight function w^* on Γ with $w^* \leq w$.*

Proof. For all $(x,y) \in E(\Gamma)$, define $w^*(x,y) = d_w(x,y) = \delta_w(x,y)$. Then $w^* \leq w$. To see that w^* is reduced we use Lemma 10 and consider $(x,y) \in E(\Gamma)$. Suppose $x_1, \ldots, x_{n+1} \in V(\Gamma)$ with $x_1 = x$, $x_{n+1} = y$, and $(x_i, x_{i+1}) \in E(\Gamma)$ for all $i = 1, \ldots, n$. Let $\epsilon > 0$. For each $i = 1, \ldots, n$, let $x_i^1 = x_i, x_i^2, \ldots, x_i^{k_i+1} = x_{i+1} \in V(\Gamma)$ with $(x_i^j, x_i^{j+1}) \in E(\Gamma)$ for all $j = 1, \ldots, k_i$ be such that

$$w^*(x_i, x_{i+1}) = d_w(x_i, x_{i+1}) \leq \sum_{j=1}^{k_i} w(x_i^j, x_i^{j+1}) \leq w^*(x_i, x_{i+1}) + \epsilon/n.$$

Then

$$w^*(x,y) = d_w(x,y) \leq \sum_{i=1}^n \sum_{j=1}^{k_i} w(x_i^j, x_i^{j+1}) \leq \sum_{i=1}^n w^*(x_i, x_{i+1}) + \epsilon.$$

Since ϵ is arbitrary, we have that

$$w^*(x,y) \leq \sum_{i=1}^n w^*(x_i, x_{i+1}).$$

Thus w^* is reduced. For the maximality of w^*, assume $u \leq w$ is a reduced weighted function. Then for all $(x,y) \in E(\Gamma)$, $u(x,y) = d_u(x,y) \leq d_w(x,y) = w^*(x,y)$. □

We can always turn a metric space into a weighted graph. If (X,d) is a metric space, for any $x,y \in X$ with $x \neq y$, we add an edge between x and y with weight $w_d(x,y) = d(x,y)$. Then (X, w_d) is a connected weighted graph and w_d is a reduced weight function.

We will also consider pseudometrics on weighted graphs.

Definition 13. Let (Γ, w) be a weighted graph and ρ be a pseudometric on $V(\Gamma)$. We say that ρ is *consistent* with w if for all $(x,y) \in E(\Gamma)$, $\rho(x,y) = w(x,y)$.

When a weight function w satisfies $B_w < \infty$ and $L_w = 0$, one can similarly define a *path pseudometric* d_w and the distance function δ_w the same way as above. The resulting path pseudometric is consistent with w.

Definition 14. Let (M, ρ) be a pseudometric space. An *isometry* of (M, ρ) is a map $\varphi : M \to M$ such that for all $x, y \in M$, $\rho(\varphi(x), \varphi(y)) = \rho(x,y)$. If G is a set of isometries of (M, ρ), we say that ρ is *G-invariant*.

Any pseudometric space (M, ρ) has a metric identification defined as follows. Let $\sim$ be an equivalence relation defined on M by $x \sim y$ iff $\rho(x,y) = 0$. For each $x \in M$, let $[x]_\sim$ denote the $\sim$-equivalence class of x. Then we can define $\overline{M} = \overline{M}^\rho = M/\sim$ and a metric $\overline{\rho}$ on $\overline{M}$ by $\overline{\rho}([x]_\sim, [y]_\sim) = \rho(x,y)$ for all $x, y \in M$. $(\overline{M}, \overline{\rho})$ is called the *metric identification* of (M, ρ). If φ is an isometry of (M, ρ), then we can define $\overline{\varphi} : \overline{M} \to \overline{M}$ by $\overline{\varphi}([x]_\sim) = [\varphi(x)]_\sim$ for all $x \in M$.

Then $\overline{\varphi}$ is an isometry of $(\overline{M}, \overline{\rho})$. Suppose G is a set of isometries of (M, ρ), then $\overline{G} = \overline{G}^\rho = \{\overline{\varphi} : \varphi \in G\}$ is a set of isometries of $(\overline{M}, \overline{\rho})$. We note that if G is a group, then $\overline{G}$ is also a group.

Let (Γ, w) be a weighted graph and ρ be a pseudometric on $V(\Gamma)$ consistent with w. Let $(\overline{\Gamma}, \overline{\rho})$ denote the metric identification of the pseudometric space (Γ, ρ). For $(x, y) \in E(\Gamma)$, define $\overline{w}([x]_\sim, [y]_\sim) = w(x, y)$. Then $(\overline{\Gamma}, \overline{w})$ is a weighted graph and $\overline{\rho}$ is consistent with $\overline{w}$. We will need this construction in the subsequent sections.

2.3 The Profinite Topology

One of the main tools we will be using is Ribes–Zalesskii theorem [9] on the profinite topology on an abstract group. Recall that if G is an abstract group, the *profinite topology* on G is the topology generated by all cosets of normal subgroups of finite index, that is, it has as a basis of open subsets all cosets of normal subgroups of finite index.

Theorem 15 (Ribes–Zalesskii [9]). *Let $\mathbb{F}$ be an abstract free group and $H_1, \ldots, H_n$ be finitely generated subgroups of $\mathbb{F}$. Then $H_1 \cdots H_n$ is closed in the profinite topology.*

A group G is said to have *property RZ* if for any finitely generated subgroups $H_1, \ldots, H_n$ of G, $H_1 \cdots H_n$ is closed in the profinite topology of G. All groups with property RZ are residually finite. We will also use the following theorem of Coulbois [1].

Theorem 16 (Coulbois [1]). *If G_1 and G_2 have property RZ, then so does the free product $G_1 * G_2$.*

Herwig–Lascar [5] used the Ribes–Zalesskii theorem in their study of the extension problems. This approach was explored further by Rosendal [10,11] to study extension problems for isometries, who showed that the Ribes–Zalesskii property for a group G is equivalent to an extension property for actions of G by isometries.

Definition 17. Let G be a group acting by isometries on a metric space (X, d_X). We say that the action is *finitely approximable* if for any finite $A \subseteq X$ and finite $F \subseteq G$ there is a finite metric space (Y, d_Y), on which G acts by isometries, and an isometry $\pi : A \to Y$ such that whenever $g \in F$ and $x, gx \in A$, then $\pi(gx) = g\pi(x)$.

Theorem 18 (Rosendal [10]). *The following are equivalent for a countable discrete group G:*

(1) G has property RZ;
(2) Any action of G by isometries on a metric space is finitely approximable.

3 Finite S-Extensions

In this section we give a direct constructive proof of Solecki's theorem (Theorem 1) following the ideas of Herwig–Lascar [5]. We will see in the following sections that the construction we present here is in some sense canonical.

For the rest of this section we fix a finite metric space X, a distinguished point $a_0 \in X$ and $P \subseteq \mathcal{P}_X$ where $P = P^{-1}$ and $X \setminus \{a_0\} \subseteq P(a_0)$. Recall that $\mathcal{P}_X$ is the set of all nonidentity partial isometries of X. Let $\mathbb{F}(P)$ be the free group generated by elements of P. For each $p \in P$, we identify the partial isometry $p^{-1} \in P$ with the formal inverse of p in $\mathbb{F}(P)$. Thus any nonidentity element of $\mathbb{F}(P)$ is a finite word of the form $p_1 \ldots p_n$ with $p_1, \ldots, p_n \in P$. We use 1 to denote the identity element of $\mathbb{F}(P)$. Of course, 1 can be identified with the identity isometry 1_X.

Let H be the set of all finite words $p_1 \cdots p_n$ with $p_1, \ldots, p_n \in P$ such that $p_1 \ldots p_n(a_0) = p_1(p_2(\cdots p_n(a_0) \cdots))$ is defined and $p_1 \ldots p_n(a_0) = a_0$. Since X is finite, H is a finitely generated subgroup of $\mathbb{F}(P)$.

Define $\Gamma = \mathbb{F}(P)/H$. We construct a weighted graph (Γ, w) as follows:

(1) for every $p, q \in P \cup \{1\}$ such that $p(a_0)$ and $q(a_0)$ are defined, there is an edge between pH and qH with $w(pH, qH) = d_X(p(a_0), q(a_0))$, and
(2) for every $g, g_1, g_2 \in \mathbb{F}(P)$, if there is an edge between $g_1 H$ and $g_2 H$, then there is an edge between $gg_1 H$ and $gg_2 H$ with $w(gg_1 H, gg_2 H) = w(g_1 H, g_2 H)$.

To see that w is well-defined, first note that if $w(g_1 H, g_2 H)$ is defined then there are $p, q \in P$ with $p(a_0)$ and $q(a_0)$ defined, and $g \in \mathbb{F}(P)$ such that $g_1 = gp$ and $g_2 = gq$. In this case, $w(g_1 H, g_2 H) = w(pH, qH) = d_X(p(a_0), q(a_0))$. Thus, to verify that w is well-defined, it suffices to make sure that if $p, q, r, s \in P$ and $d_X(p(a_0), q(a_0)) \neq d_X(r(a_0), s(a_0))$, then there does not exist $g \in \mathbb{F}(P)$ such that $gpH = rH$ and $gqH = sH$. Assume there is such a $g = p_1 \ldots p_n$. Then $r^{-1}gp, s^{-1}gq \in H$, and thus $r^{-1}p_1 \ldots p_n p(a_0) = a_0$ and $s^{-1}p_1 \ldots p_n q(a_0) = a_0$. It follows that

$$p_1 \ldots p_n(p(a_0)) = r(a_0) \text{ and } p_1 \ldots p_n(q(a_0)) = s(a_0).$$

Since all $p_1, \ldots, p_n$ are partial isometries, we have $d_X(p(a_0), q(a_0)) = d_X(r(a_0), s(a_0))$.

From the finiteness of P and the definition of w, it is clear that $L_w = \inf\{w(x,y) : (x,y) \in E(\Gamma)\} > 0$ and $B_w = \sup\{w(x,y) : (x,y) \in E(\Gamma)\} < \infty$. It follows that, equipped with the path metric d_w, Γ becomes a metric space. When there is no danger of confusion, we use Γ to denote the metric space (Γ, d_w).

We claim that Γ is essentially an extension of X. To see this, let $e : X \to \Gamma$ be defined by

$$e(a) = \begin{cases} H, & \text{if } a = a_0, \\ pH, & \text{where } p \in P \text{ and } p(a_0) = a, \text{ if } a \neq a_0. \end{cases}$$

To see that e is well-defined, first note that for any $a \neq a_0$ there is $p \in P$ with $p(a_0) = a$. If $p, q \in P$ with $p(a_0) = q(a_0)$, then $p^{-1}q \in H$ and therefore $pH = qH$. Thus e is well-defined. Furthermore, if $e(a) = pH = qH = e(b)$ and $p(a_0) = a$ and $q(a_0) = b$, then $p^{-1}q \in H$ and therefore $a = p(a_0) = q(a_0) = b$. This means that e is one-to-one. To see that e is an isometric embedding, we use the following lemmas.

Lemma 19. *Let (Y, ϕ) be a P-type S-extension of X. Then there is $\pi : \Gamma \to Y$ such that $d_Y(\pi(g_1 H), \pi(g_2 H)) = w(g_1 H, g_2 H)$ whenever $(g_1 H, g_2 H) \in E(\Gamma)$.*

Proof. We first expand $\phi : P \to \mathrm{Iso}(Y)$ to a map $\psi : \mathbb{F}(P) \to \mathrm{Iso}(Y)$ by

$$\psi(p_1 \ldots p_n) = \phi(p_1) \circ \cdots \circ \phi(p_n).$$

Define $\pi : \Gamma \to Y$ by $\pi(gH) = \psi(g)(a_0)$. It is easy to see that π is well-defined. Now, if $(g_1 H, g_2 H) \in E(\Gamma)$, then there exist $p, q \in P \cup \{1\}$ and $g \in \mathbb{F}(P)$ such that $g_1 = gp$, $g_2 = gq$, and $p(a_0), q(a_0)$ are defined. We have

$$\begin{aligned} d_Y(\pi(g_1 H), \pi(g_2 H)) &= d_Y(\pi(gpH), \pi(gqH)) = d_Y(\psi(g) \circ \phi(p)(a_0), \psi(g) \circ \phi(q)(a_0)) \\ &= d_Y(\phi(p)(a_0), \phi(q)(a_0)) = d_X(p(a_0), q(a_0)) \\ &= w(pH, qH) = w(gpH, gqH). \end{aligned}$$

□

Lemma 20. *w is reduced.*

Proof. We verify using Lemma 10 that for any $(g_1 H, g_2 H) \in E(\Gamma)$, $d_w(g_1 H, g_2 H) = w(g_1 H, g_2 H)$. Let $\gamma_1 = g_1, \gamma_2, \ldots, \gamma_{n+1} = g_2$ be elements of $\mathbb{F}(P)$ such that for all $i = 1, \ldots, n$, $(\gamma_i H, \gamma_{i+1} H) \in E(\Gamma)$. Let $Y = \mathbb{U}$ and (Y, ϕ) be a P-type S-extension of X. Let $\pi : \Gamma \to Y$ be given by Lemma 19. Then

$$\begin{aligned} w(g_1 H, g_2 H) &= d_Y(\pi(g_1 H), \pi(g_2 H)) \\ &\leq \sum_{i=1}^{n} d_Y(\pi(\gamma_i H), \pi(\gamma_{i+1} H)) = \sum_{i=1}^{n} w(\gamma_i H, \gamma_{i+1} H). \end{aligned}$$

□

Lemma 21. *e is an isometric embedding.*

Proof. Let $a, b \in X$ and $p, q \in P \cup \{1\}$ with $p(a_0) = a$ and $q(a_0) = b$. Then $(pH, qH) \in E(\Gamma)$. Since w is reduced, we have

$$d_w(e(a), e(b)) = d_w(pH, qH) = w(pH, qH) = d_X(p(a_0), q(a_0)) = d_X(a, b).$$

□

We identify X with

$$e(X) = \{pH : p \in P \cup \{1\} \text{ and } p(a_0) \text{ is defined}\} \subseteq \Gamma$$

and consider Γ an extension of X. For each $q \in P$, consider the partial map $\hat{q} : e(X) \to e(X)$ defined by $\hat{q}(pH) = qpH$ for all $pH, qpH \in e(X)$, i.e., whenever

$p(a_0)$ and $q(p(a_0))$ are defined; note that in this case there exists $r \in P$ such that $r(a_0) = q(p(a_0))$ and $rH = qpH$. Then it is straightforward to verify that for all $a, b \in X$, $q(a) = b$ iff $\hat{q}(e(a)) = e(b)$. Thus we may identify q with $\hat{q}$ on the domain $e(\text{dom}(q))$.

Define $\Phi : P \to \text{Iso}(\Gamma)$ by letting, for any $q \in P$,

$$\Phi(q)(gH) = qgH$$

for all $g \in \mathbb{F}(P)$. To see that $\Phi(q)$ is indeed an isometry of Γ, let $g_1, g_2 \in \mathbb{F}(P)$. From the definitions of w and δ_w, we get $\delta_w(g_1H, g_2H) = \delta_w(qg_1H, qg_2H)$ (including the case when one of these quantities is ∞). It follows that $d_w(g_1H, g_2H) = d_w(qg_1H, qg_2H)$.

Lemma 22. (Γ, Φ) *is a P-type S-extension of* X.

Proof. For any $q \in P$, $\Phi(q)$ is obviously an extension of $\hat{q}$. □

To construct a finite P-type S-extension of X our plan is to find a suitable normal subgroup N of finite index in $\mathbb{F}(P)$, and to use $\Gamma_N = \mathbb{F}(P)/NH$ as the underlying space of the P-type S-extension. Assuming such a normal subgroup $N \trianglelefteq \mathbb{F}(P)$ is found, we first turn Γ_N into a weighted graph (Γ_N, w_N) as follows:

(1) for every $p, q \in P \cup \{1\}$ with $p(a_0)$ and $q(a_0)$ defined, there is an edge between pNH and qNH with $w_N(pNH, qNH) = d_X(p(a_0), q(a_0))$, and
(2) for every $g, g_1, g_2 \in \mathbb{F}(P)$, if there is an edge between g_1NH and g_2NH, then there is an edge between gg_1NH and gg_2NH with

$$w_N(gg_1NH, gg_2NH) = w_N(g_1NH, g_2NH).$$

To guarantee that w_N is well-defined, we use a similar argument as before provided that the following condition holds:

(C1) For every $p, q, r, s \in P \cup \{1\}$ such that $d_X(p(a_0), q(a_0)) \neq d_X(r(a_0), s(a_0))$, there does not exist $g \in \mathbb{F}(P)$ such that $gpNH = rNH$ and $gqNH = sNH$, equivalently, $N \cap pHr^{-1}sHq^{-1} = \emptyset$.

To see the equivalence in the statement of (C1), suppose $gpNH = rNH$ and $gqNH = sNH$. Then by the normality of N we have $g \in rNHp^{-1} \cap sNHq^{-1}$, and thus $rNHp^{-1} \cap sNHq^{-1} \neq \emptyset$. It follows that $N \cap pHr^{-1}sHq^{-1}N \neq \emptyset$, or $N \cap pHr^{-1}sHq^{-1} \neq \emptyset$. All steps can be reversed to establish the backward implication.

Another similar argument as before shows that d_{w_N} is a metric on Γ_N. We again define

$$e_N(a) = \begin{cases} NH, & \text{if } a = a_0, \\ pNH, & \text{where } p \in P \text{ and } p(a_0) = a, \text{ if } a \neq a_0. \end{cases}$$

In order to guarantee that e_N is one-to-one, we argue similarly as before provided that the following condition holds for N:

(C2) For every $p, q \in P \cup \{1\}$, if $p(a_0)$ and $q(a_0)$ are defined and $p(a_0) \neq q(a_0)$, then $p^{-1}q \notin NH$, equivalently, $N \cap pHq^{-1} = \emptyset$.

Finally, to guarantee that e_N is an isometric embedding, we argue similarly as in the proof of Lemma 21 provided that w_N is reduced, which corresponds to the following condition:

(C3) For every $p, q, r_1, s_1, \ldots, r_n, s_n \in P \cup \{1\}$ such that

$$d_X(p(a_0), q(a_0)) > \sum_{i=1}^{n} d_X(r_i(a_0), s_i(a_0)),$$

there does not exist a path in Γ_N from pNH to qNH using translates of edges $(r_1NH, s_1NH), \ldots, (r_nNH, s_nNH)$ in the same order. That is, there do not exist $g_1, \ldots, g_n \in \mathbb{F}(P)$ such that

$$pNH = g_1 r_1 NH, \quad g_1 s_1 NH = g_2 r_2 NH,$$
$$\cdots\cdots$$
$$g_{n-1} s_{n-1} NH = g_n r_n NH, \quad g_n s_n NH = qH.$$

Equivalently, $N \cap pHr_1^{-1}s_1H \cdots Hr_n^{-1}s_n Hq^{-1} = \emptyset$.

To summarize, we need to find $N \trianglelefteq \mathbb{F}(P)$ of finite index so that (C1), (C2) and (C3) hold. Note that these correspond to finitely many conditions, and each condition is of the form $\gamma N \cap H_1 \cdots H_n = \emptyset$ where $\gamma \in \mathbb{F}(P)$, $H_1, \ldots, H_n$ are finitely generated subgroups of $\mathbb{F}(P)$, and $\gamma \notin H_1 \cdots H_n$. For example, the condition in (C1) can be rewritten as

$$(p^{-1}qs^{-1}r)N \cap H(r^{-1}sHs^{-1}r) = \emptyset.$$

Thus, by the Ribes–Zalesskii theorem, for each condition of the form $\gamma N \cap H_1 \cdots H_n = \emptyset$, where $\gamma \notin H_1 \cdots H_n$, there is a normal subgroup of finite index satisfying the condition. Taking the intersection of all these subgroups, we obtain still a normal subgroup of finite index to satisfy all conditions (C1), (C2) and (C3).

Now

$$e_N(X) = \{pNH : p \in P \text{ and} p(a_0) \text{is defined}\}.$$

Similar to the above, for each $q \in P$ we can define the partial map $\tilde{q} : e_N(X) \to e_N(X)$ by $\tilde{q}(pNH) = qpNH$. Then X is identified with $e_N(X)$ and q is identified with $\tilde{q}$ with domain $e(\mathrm{dom}(q))$. Define $\Phi_N : P \to \mathrm{Iso}(\Gamma_N)$ by

$$\Phi_N(q)(gNH) = qgNH.$$

Then it is obvious that $\Phi_N(q)$ extends $\tilde{q}$ for all $q \in P$.

We have thus established that (Γ_N, Φ_N) is a finite P-type S-extension of X.

Definition 23. Let X be a metric space and $P \subseteq \mathcal{P}_X$ be such that $X \setminus \{a_0\} \subseteq P(a_0)$. We say $N \trianglelefteq \mathbb{F}(P)$ is a *feasible P-type prekernel for X* if it satisfies (C1), (C2) and (C3). When there is no danger of confusion, we call N a *feasible prekernel*.

What we have shown in this section can be summarized as follows.

Lemma 24. *Let X be a metric space. If $N \trianglelefteq \mathbb{F}(P)$ is a feasible P-type prekernel for X, then (Γ_N, Φ_N) is a P-type S-extension of X.*

Thus Theorem 1 follows from the fact that for any finite metric space X there is a feasible P-type prekernel $N \trianglelefteq \mathbb{F}(P)$ that is of finite index in $\mathbb{F}(P)$. This in turn follows from property RZ of the free group $\mathbb{F}(P)$.

4 Minimal S-Extensions

In this section we give a complete characterization of all finite minimal P-type S-extensions of a given finite metric space. This is done by showing that the P-type S-extension we constructed in the previous section is canonical in several senses. We use the same notations from the previous section.

Throughout this section we still fix a finite metric space X, a distinguished point $a_0 \in X$ and $P \subseteq \mathcal{P}_X$ where $P = P^{-1}$ and $X \setminus \{a_0\} \subseteq P(a_0)$. We have constructed P-type S-extensions (Γ, Φ) and (Γ_N, Φ_N) for suitable $N \trianglelefteq \mathbb{F}(P)$. Here we first note that, as long as P is sufficiently rich, these P-type S-extensions do not depend on the choice of the point $a_0 \in X$. More explicitly, if $a'_0 \in X$ and $p_0(a_0) = a'_0$ for $p_0 \in P$, and if $X \setminus \{a'_0\} \subseteq P(a'_0)$, then we could similarly define $\Gamma' = \mathbb{F}(P)/H'$ and Φ'. It is easy to see that $H' = p_0 H p_0^{-1}$. Thus we may define a bijection $\pi : \Gamma \to \Gamma'$ by $\pi(gH) = gp_0^{-1}H'$ for all $g \in \mathbb{F}(P)$. It is straightforward to check that π is an isometry between Γ and Γ' such that $\pi(\Phi(q)(gH)) = \Phi'(q)(\pi(gH))$ for all $q \in P$ and $g \in \mathbb{F}(P)$. Thus π is indeed an isomorphism between the two P-type S-extensions. Similarly, when P is sufficiently rich, the finite P-type S-extension (Γ_N, Φ_N) does not depend on the choice of a_0 either.

Next we note that for any P-type S-extension (Y, ϕ) of X, the P-type S-map ϕ can be trivially extended to a map from all of $\mathbb{F}(P)$ to $\text{Iso}(Y)$ by letting

$$\hat{\phi}(p_1 \dots p_n) = \phi(p_1) \circ \dots \circ \phi(p_n)$$

for all $p_1, \dots, p_n \in \mathbb{F}(P)$. $\hat{\phi}$ is a semigroup homomorphism but not necessarily a group homomorphism. To turn it into a group homomorphism, we just need to make sure that $\phi(p^{-1}) = \phi(p)^{-1}$ for all $p \in P$, which is easy to arrange. In the rest of this paper, we will use ϕ to denote the extension $\hat{\phi}$, and thus regard ϕ as a map from $\mathbb{F}(P)$ to $\text{Iso}(Y)$. We will also tacitly assume that all the extended P-type S-maps $\phi : \mathbb{F}(P) \to \text{Iso}(Y)$ are indeed group homomorphisms and therefore their ranges are subgroups of $\text{Iso}(Y)$. We note that the extended P-type S-map $\Phi : \mathbb{F}(P) \to \text{Iso}(\Gamma_N)$ is already a group homomorphism.

The following lemma is one evidence of the canonicity of the construction (Γ_N, Φ_N).

Lemma 25. *Let $N \trianglelefteq \mathbb{F}(P)$ be a feasible P-type prekernel for X, that is, (Γ_N, Φ_N) is a P-type S-extension of X. Let $G = \Phi_N(\mathbb{F}(P)) \leq \mathrm{Iso}(\Gamma_N)$ and $N_G = \ker(\Phi_N)$. Then $N_G \trianglelefteq \mathbb{F}(P)$ is a normal subgroup and $NH = N_G H$. In particular, $(\Gamma_{N_G}, \Phi_{N_G}) = (\Gamma_N, \Phi_N)$.*

Note that the notion of "feasible prekernel" is justified by Lemma 25 which states that such a group N can be massaged into another normal subgroup N_G that produces the same S-extension of X and $N_G = \ker(\Phi_N) = \ker(\Phi_{N_G})$.

Proof. N_G is obviously a normal subgroup of $\mathbb{F}(P)$. We only need to verify $NH = N_G H$. Let $\gamma \in N_G$. Then $\Phi_N(\gamma) = 1$ and for all $g \in \mathbb{F}(P)$, $\gamma g NH = \Phi_N(\gamma)(gNH) = gNH$. In particular $\gamma NH = NH$, and so $\gamma \in NH$. This shows that $N_G \subseteq NH$ and so $N_G H \leq NH$. Conversely, suppose $\gamma \in N$. Then for all $g \in \mathbb{F}(P)$, we have $\Phi_N(\gamma)(gNH) = \gamma g NH = g(g^{-1}\gamma g)NH = gNH$. Thus $\gamma \in \ker(\Phi_N) = N_G$. This shows that $N \leq N_G$ and so $NH \leq N_G H$. □

Next we define minimality for P-type S-extensions.

Definition 26. *A P-type S-extension (Y, ϕ) of X is said to be minimal if for any $y \in Y$ there is $g \in \mathbb{F}(P)$ such that $y = \phi(g)(a_0)$.*

We state the following fact without proof.

Lemma 27. *Let (Y, ϕ) be a P-type S-extension of X. Then the following are equivalent:*

(i) (Y, ϕ) is minimal;
(ii) For any $y \in Y$ there exist $g \in \mathbb{F}(P)$ and $x \in X$ such that $y = \phi(g)(x)$.

Of course, the notion of minimality is motivated by the observation that if (Y, ϕ) is a P-type S-extension of X and let

$$Z = \{\phi(g)(x) : g \in \mathbb{F}(P), x \in X\},$$

then $Z \subseteq Y$ and for any $p \in P$ and $z \in Z$, $\phi(p)(z) \in Z$. Thus, by defining

$$\psi(p) = \phi(p) \upharpoonright Z$$

for all $p \in P$, we get another P-type S-extension (Z, ψ) of X which is a subextension of (Y, ϕ).

We also note that, if (Y, ϕ) is a P-type S-extension of X, then there are many ways to define proper superextensions of (Y, ϕ) by adding points to Y and defining metrics appropriately. Thus there is no hope to give a reasonable characterization of all finite P-type S-extensions of X. Below we concentrate on characterizing finite minimal P-type S-extensions of X. We will show that all finite minimal P-type S-extensions of X are derived from P-type S-extensions of the form (Γ_N, Φ_N).

Lemma 28. *Let (Y, ϕ) be a P-type S-extension of X. Let $N = \ker(\phi)$. Then N is a feasible P-type prekernel for X.*

Proof. Define $\Psi : \Gamma_N \to Y$ by $\Psi(gNH) = \phi(g)(a_0)$ for all $g \in \mathbb{F}(P)$. To see Ψ is well-defined, note that if $g_2^{-1}g_1 \in NH$, then for some $n \in N$ and $h \in H$, we have $\phi(g_2)^{-1}\phi(g_1)(a_0) = \phi(n)\phi(h)(a_0) = \phi(n)(a_0) = a_0$. Here we note that for any $n \in N$, $\phi(n)(a_0) = a_0$ since $n \in \ker(\phi)$, and for any $h \in H$, $\phi(h)(a_0) = a_0$ since $h(a_0) = a_0$ and $\phi(h)$ extends h. Thus $\phi(g_1)(a_0) = \phi(g_2)(a_0)$.

To verify (C1), let $p, q, r, s \in P \cup \{1\}$ and $g \in \mathbb{F}(P)$ be such that $p(a_0)$, $q(a_0)$, $r(a_0)$ and $s(a_0)$ are defined, $pNH = grNH$ and $qNH = gsNH$. Applying the map Ψ to these equations, we get $\phi(p)(a_0) = \phi(g)\phi(r)(a_0)$ and $\phi(q)(a_0) = \phi(g)\phi(s)(a_0)$. We need to show that $d_X(p(a_0), q(a_0)) = d_X(r(a_0), s(a_0))$. Since $\phi(g)$ is an isometry of Y, we have

$$\begin{aligned} d_X(p(a_0), q(a_0)) &= d_Y(\phi(p)(a_0), \phi(q)(a_0)) \\ &= d_Y(\phi(g)\phi(r)(a_0), \phi(g)\phi(s)(a_0)) \\ &= d_Y(\phi(r)(a_0), \phi(s)(a_0)) = d_X(r(a_0), s(a_0)). \end{aligned}$$

To verify (C2), let $p, q \in P \cup \{1\}$ be such that $p(a_0), q(a_0)$ are defined and $pNH = qNH$. We get

$$p(a_0) = \phi(p)(a_0) = \Psi(pNH) = \Psi(qNH) = \phi(q)(a_0) = q(a_0).$$

Finally, to verify (C3), let $p, q, r_1, s_1, \ldots, r_n, s_n \in P \cup \{1\}$ and $g_1, \ldots, g_n \in \mathbb{F}(P)$ be such that $p(a_0), q(a_0), r_1(a_0), s_1(a_0), \ldots, r_n(a_0), s_n(a_0)$ are all defined, and

$$pNH = g_1 r_1 NH, \quad g_1 s_1 NH = g_2 r_2 NH,$$
$$\cdots\cdots$$
$$g_{n-1} s_{n-1} NH = g_n r_n NH, \quad g_n s_n NH = qH.$$

Applying Ψ to all these equations, we get

$$p(a_0) = \phi(g_1)(r_1(a_0)), \quad \phi(g_1)(s_1(a_0)) = \phi(g_2)(r_2(a_0)),$$
$$\cdots\cdots$$
$$\phi(g_{n-1})(s_{n-1}(a_0)) = \phi(g_n)(r_n(a_0)), \quad \phi(g_n)(s_n(a_0)) = q(a_0).$$

It follows that

$$d_X(p(a_0), q(a_0)) = d_Y(p(a_0), q(a_0)) \leq \sum_{i=1}^n d_Y(r_i(a_0), s_i(a_0)) = \sum_{i=1}^n d_X(r_i(a_0), s_i(a_0)).$$

□

Thus, for any finite P-type S-extension (Y, ϕ), we have that $N = \ker(\phi)$ is a normal subgroup of $\mathbb{F}(P)$ of finite index and that N is a feasible prekernel. Furthermore, we will be able to carry out the construction of (Γ_N, Φ_N) as in Section 3 as a P-type S-extension of X based on the weighted graph (Γ_N, w_N). In particular, the weight function w_N, the path metric d_{w_N}, the isometric embedding e_N, etc. are all well-defined.

Theorem 29. *Let (Y, ϕ) be a finite minimal P-type S-extension of X. Let $N = \ker(\phi)$ and $G = \Phi_N(\mathbb{F}(P))$. Then there is a G-invariant pseudometric ρ on Γ_N which is consistent with w_N such that (Y, ϕ) is isomorphic to $(\overline{\Gamma_N}, \overline{\Phi_N})$.*

Proof. We again define $\Psi : \Gamma_N \to Y$ by $\Psi(gNH) = \phi(g)(a_0)$ for all $g \in \mathbb{F}(P)$. As in the proof of Lemma 28, Ψ is well-defined. Since ϕ is minimal, Ψ is onto.

We define a pseudometric ρ on Γ_N by

$$\rho(g_1NH, g_2NH) = d_Y(\Psi(g_1NH), \Psi(g_2NH)) = d_Y(\phi(g_1)(a_0), \phi(g_2)(a_0)).$$

It is easy to verify that ρ is indeed a pseudometric on Γ_N.

Recall that $\Phi_N : P \to \text{Iso}(\Gamma_N)$ is defined by

$$\Phi_N(p)(gNH) = pgNH$$

for all $p \in P$ and $g \in \mathbb{F}(P)$. From previous section, the $\text{Iso}(\Gamma_N)$ refers to the group of isometries for the metric space (Γ_N, d_{w_N}). Here we claim that the maps $gNH \mapsto pgNH$ are also isometries of the pseudometric space (Γ_N, ρ). To see this, we only need to check

$$\rho(pg_1NH, pg_2NH) = d_Y(\phi(pg_1)(a_0), \phi(pg_2)(a_0))$$
$$= d_Y(\phi(p)\phi(g_1)(a_0), \phi(p)\phi(g_2)(a_0)) = \rho(g_1NH, g_2NH).$$

Extending Φ_N to a group homomorphism from $\mathbb{F}(P)$ to $\text{Iso}(\Gamma_N)$, the group of all isometries of the pseudometric space (Γ_N, ρ), it follows that ρ is G-invariant.

To verify that ρ is consistent with w_N, we consider an edge in the weighted graph (Γ_N, w_N), which is of the form $(gpNH, gqNH)$ where $g \in \mathbb{F}(P)$ and $p, q \in P \cup \{1\}$ are such that $p(a_0)$ and $q(a_0)$ are defined. Note that $w_N(gpNH, gqNH) = d_X(p(a_0), q(a_0))$. We have

$$\rho(gpNH, gqNH) = d_Y(\phi(g)\phi(p)(a_0), \phi(g)\phi(q)(a_0))$$
$$= d_Y(\phi(p)(a_0), \phi(q)(a_0))$$
$$= d_X(p(a_0), q(a_0)) = w_N(gpNH, gqNH).$$

We can now consider the metric identification of the pseudometric space (Γ_N, ρ), which is denoted by $(\overline{\Gamma_N}, \overline{\rho})$. Since ρ is consistent with w_N, so is $\overline{\rho}$. Since ρ is G-invariant, for each $\varphi \in G$ we can define an isometry $\overline{\varphi} \in \overline{G}$ for $(\overline{\Gamma_N}, \overline{\rho})$. Thus it makes sense to define $\overline{\Phi_N} : P \to \text{Iso}(\overline{\Gamma_N})$ by $\overline{\Phi_N}(p) = \overline{\Phi_N(p)}$.

Finally, let $\pi : \overline{\Gamma_N} \to Y$ be defined as $\pi([gNH]_\sim) = \phi(g)(a_0)$. Then π is an isometry between the metric spaces $(\overline{\Gamma_N}, \overline{\rho})$ and (Y, d_Y). To complete the proof of the theorem, we only need to verify that for any $p \in P$, $\pi \circ \overline{\Phi_N}(p) = \phi(p) \circ \pi$. We have

$$[\pi \circ \overline{\Phi_N}(p)]([gNH]_\sim) = \pi[\overline{\Phi_N}(p)([gNH]_\sim)] = \pi([pgNH]_\sim)$$
$$= \phi(pg)(a_0) = \phi(p)\phi(g)(a_0)$$
$$= \phi(p)[\pi([gNH]_\sim)] = [\phi(p) \circ \pi]([gNH]_\sim).$$

$\square$

Theorem 30. *Let $N \trianglelefteq \mathbb{F}(P)$ be a feasible P-type prekernel that is of finite index. Let $G = \Phi_N(\mathbb{F}(P))$. Let ρ be a G-invariant pseudometric on Γ_N which is consistent with the weight function w_N. Then $(\overline{\Gamma_N}, \overline{\Phi_N})$ is a finite minimal P-type S-extension of X.*

Proof. Consider the metric identification $(\overline{\Gamma_N}, \overline{\rho})$. Since ρ is G-invariant, $\overline{G}$ is a set of isometries for $\overline{\Gamma_N}$. Since ρ is consistent with w_N, so is $\overline{\rho}$. Define $\overline{e_N} : X \to \overline{\Gamma_N}$ by $\overline{e_N}(a) = [e_N(a)]_\sim$. Then $\overline{e_N}$ is an isometric embedding from X into $\overline{\Gamma_N}$. Note that $\overline{e_N}(a_0) = [NH]_\sim$. Thus we can identify a_0 with $[NH]_\sim$. It follows from similar arguments as before that $(\overline{\Gamma_N}, \overline{\Phi_N})$ is a P-type S-extension of X. To see that it is minimal, we just note that for any $g \in \mathbb{F}(P)$, $\overline{\Phi_N}(g)([NH]_\sim) = [gNH]_\sim$. □

We summarize the characterization of all finite minimal P-type S-extensions of X in the following theorem.

Theorem 31. *The following are equivalent:*

(i) (Y, ϕ) is a finite minimal P-type S-extension of X;
(ii) There exists a feasible P-type prekernel $N \trianglelefteq \mathbb{F}(P)$ of finite index, and, letting $G = \Phi_N(\mathbb{F}(P))$, there exists a G-invariant pseudometric ρ on Γ_N which is consistent with w_N, such that (Y, ϕ) is isomorphic to $(\overline{\Gamma_N}, \overline{\Phi_N})$;
(iii) For $N = \ker(\phi)$ and $G = \Phi_N(\mathbb{F}(P))$, there exists a G-invariant pseudometric ρ on Γ_N which is consistent with w_N, such that (Y, ϕ) is isomorphic to $(\overline{\Gamma_N}, \overline{\Phi_N})$.

5 Coherent S-Extensions

5.1 Coherent S-Extensions and Strongly Coherent S-Extensions

In this section we study a notion of coherence for (P-type) S-extensions. The terminology has been used for a different notion in Siniora–Solecki [15] which refers to a slightly stronger condition. We call their notion of coherence *strongly coherent*.

Definition 32 (Solecki). Let X be a metric space. An S-extension (Y, ϕ) of X is *strongly coherent* if for every triple (p, q, r) of partial isometries of X such that $p \circ q = r$, we have $\phi(p) \circ \phi(q) = \phi(r)$.

The following is a strengthening of Solecki's theorem on existence of finite S-extensions.

Theorem 33 (Solecki [10,14,15]). *Let X be a finite metric space. Then X has a finite strongly coherent S-extension (Y, ϕ).*

The following notion of coherence is slightly weaker than the notion of strongly coherent but is sufficient for our study of ultraextensive metric spaces in subsequent sections.

Definition 34. Let $X_1 \subseteq X_2$ be metric spaces and (Y_i, ϕ_i) be a P_i-type S-extension of X_i for $i = 1, 2$ where $P_1 \subseteq P_2$. We say that (Y_1, ϕ_1) and (Y_2, ϕ_2) are *coherent* if

(i) Y_2 extends Y_1,
(ii) $\phi_2(p)$ extends $\phi_1(p)$ for all $p \in P_1 \subseteq P_2$, and

(iii) letting $K_i = \phi_i(\mathbb{F}(P_i)) \leq \mathrm{Iso}(Y_i)$ for $i = 1, 2$, and letting $\kappa : K_1 \to K_2$ be such that $\kappa(\phi_1(p)) = \phi_2(p)$ for all $p \in P_1$, then κ has a unique extension to a group isomorphic embedding from K_1 into K_2.

The following lemma makes it precise that the notion of strong coherence is a stronger notion than coherence.

Lemma 35. *Let $X_1 \subseteq X_2$ be finite metric spaces and (Y_1, ϕ_1) be a P_1-type S-extension of X_1. Let $P_2 \supseteq P_1$ be such that $P_2 = P_2^{-1}$ and $X_2 \setminus \{a_0\} \subseteq P_2(a_0)$. Suppose (Y_2, ϕ) is a strongly coherent S-extension of $X_2 \cup Y_1$. Then there is ϕ_2 such that (Y_2, ϕ_2) is a P_2-type S-extension of X_2 which is coherent with (Y_1, ϕ_1).*

Proof. Let (Y_2, ϕ) be a strongly coherent S-extension of $X_2 \cup Y_1$. Define $\phi_2 : P_2 \to \mathrm{Iso}(Y_2)$ by

$$\phi_2(p) = \begin{cases} \phi(\phi_1(p)), & \text{if } p \in P_1, \\ \phi(p), & \text{if } p \in P_2 \setminus P_1. \end{cases}$$

Then (Y_2, ϕ_2) is obviously a P_2-type S-extension of X_2. The construction also guarantees that $Y_1 \subseteq Y_2$ and that $\phi_1(p) \subseteq \phi_2(p)$ for $p \in P_1$. Since (Y_2, ϕ) is a strongly coherent S-extension of Y_1, ϕ restricted to $\mathrm{Iso}(Y_1)$ is a group isomorphism embedding from $\mathrm{Iso}(Y_1)$ into $\mathrm{Iso}(Y_2)$. When further restricted to $K_1 = \phi_1(\mathbb{F}(P_1))$, it gives a group isomorphic embedding into $K_2 = \phi_2(\mathbb{F}(P_2))$. □

5.2 A Characterization of Coherent S-Extensions

Although the existence of coherent S-extensions follows from results of Solecki [10,14,15] by Lemma 35, we introduce a notion of coherent extensions for groups and utilize this notion to give a characterization of all possible minimal coherent S-extensions.

Let $X_1 \subseteq X_2$ be finite metric spaces, (Y_1, ϕ_1) be a minimal P_1-type S-extension of X_1, and $P_1 \subseteq P_2 \subseteq \mathcal{P}_{X_2}$ where $P_2 = P_2^{-1}$ and $X_2 \setminus \{a_0\} \subseteq P_2(a_0)$. Let (Y_2, ϕ_2) be a minimal P_2-type S-extension of X_2 that is coherent with (Y_1, ϕ_1). Next, we characterize all such coherent S-extensions.

Definition 36. Let $X_1 \subseteq X_2$ be finite metric spaces, (Y_1, ϕ_1) be a P_1-type S-extension of X_1 and $P_1 \subseteq P_2 \subseteq \mathcal{P}_{X_2}$ where $P_2 = P_2^{-1}$ and $X_2 \setminus \{a_0\} \subseteq P_2(a_0)$. Let $N_1 = \ker(\phi_1)$. We say $N_2 \trianglelefteq \mathbb{F}(P_2)$ is a *coherent extension* of N_1 if it is a feasible P_2-type prekernel for X_2 and satisfies the following conditions:

(D1) $N_1 = N_2 \cap \mathbb{F}(P_1)$;

(D2) For every $g, h, k, l \in \mathbb{F}(P_1)$ such that

$$d_{Y_1}(\phi_1(g)(a_0), \phi_1(h)(a_0)) \neq d_{Y_1}(\phi_1(k)(a_0), \phi_1(l)(a_0)),$$

we have $N_2 \cap gH_2k^{-1}lH_2h^{-1} = \emptyset$;

(D3) For every $g, h \in \mathbb{F}(P_1)$ and $p, q \in P_2$ with both $p(a_0)$ and $q(a_0)$ defined, if

$$d_{Y_1}(\phi_1(g)(a_0), \phi_1(h)(a_0)) \neq d_{X_2}(p(a_0), q(a_0)),$$

we have $N_2 \cap gH_2p^{-1}qH_2h^{-1} = \emptyset$.

Note that since N_2 is a feasible prekernel, letting $G_2 = \Phi_{N_2}(\mathbb{F}(P_2))$, for any G_2-invariant pseudometric ρ_2 on Γ_{N_2} which is consistent with w_{N_2}, $(\overline{\Gamma_{N_2}}, \overline{\Phi_{N_2}})$ is a minimal S-extension of X_2.

Theorem 37. *Let $X_1 \subseteq X_2$ be finite metric spaces, (Y_1, ϕ_1) be a minimal P_1-type S-extension of X_1, and $P_1 \subseteq P_2 \subseteq \mathcal{P}_{X_2}$ where $P_2 = P_2^{-1}$ and $X_2 \setminus \{a_0\} \subseteq P_2(a_0)$.*

(i) Let (Y_2, ϕ_2) be a P_2-type S-extension of X_2 that is coherent with (Y_1, ϕ_1). Then $N_2 = \ker(\phi_2)$ is a coherent extension of $N_1 = \ker(\phi_1)$.

(ii) Let $N_2 \trianglelefteq \mathbb{F}(P_2)$ be a coherent extension of $N_1 = \ker(\phi_1)$. Then letting $G_2 = \Phi_{N_2}(\mathbb{F}(P_2))$, there exists a G_2-invariant pseudometric ρ_2 on Γ_{N_2} which is consistent with w_{N_2}, such that $(Y_2, \phi_2) = (\overline{\Gamma_{N_2}}, \overline{\Phi_{N_2}})$ is coherent with (Y_1, ϕ_1).

Proof. We first prove (i). Let (Y_2, ϕ_2) be a P_2-type S-extension of X_2 that is coherent with (Y_1, ϕ_1). Let $N_2 = \ker(\phi_2)$. Then by Theorem 29, there is a pseudometric ρ_2 on Γ_{N_2} such that (Y_2, ϕ_2) is isomorphic to $(\overline{\Gamma_{N_2}}, \overline{\Phi_{N_2}})$. By Lemma 28, N_2 is a feasible prekernel. Next we show that N_2 is a coherent extension of N_1.

For (D1), note that since (Y_2, ϕ_2) is coherent with (Y_1, ϕ_1), we have $\phi_1(\mathbb{F}(P_1)) \cong \phi_2(\mathbb{F}(P_1))$ via the map $\phi_1(g) \mapsto \phi_2(g)$. Thus,
$$N_2 \cap \mathbb{F}(P_1) = \ker(\phi_2) \cap \mathbb{F}(P_1) = \ker(\phi_1) = N_1.$$

For (D2), we need to verify that if for $g, h, k, l \in \mathbb{F}(P_1)$
$$d_{Y_1}(\phi_1(g)(a_0), \phi_1(h)(a_0)) \neq d_{Y_1}(\phi_1(k)(a_0), \phi_1(l)(a_0)),$$
then $N_2 \cap gH_2k^{-1}lH_2h^{-1} = \emptyset$. Toward a contradiction, assume there is $n \in N_2 \cap gH_2k^{-1}lH_2h^{-1}$. Then there are $\eta, \eta' \in H_2$ with $k\eta^{-1}g^{-1}n = l\eta'h^{-1}$, which implies
$$\phi_2(k\eta^{-1}g^{-1}n) = \phi_2(l\eta'h^{-1}).$$
From the definitions of H_2 and of N_2, if we apply the left-hand-side element to $\phi_2(g)(a_0) = \phi_1(g)(a_0)$, the resulting value is $\phi_2(k)(a_0) = \phi_1(k)(a_0)$. Similarly, if we apply the right-hand-side element to $\phi_2(h)(a_0) = \phi_1(h)(a_0)$, the resulting value is $\phi_2(l)(a_0) = \phi_1(l)(a_0)$. Thus, both sides of the equation represent the same partial isometry of Y_1 with $\phi_1(g)(a_0)$ and $\phi_1(h)(a_0)$ in its domain and with $\phi_1(k)(a_0)$ and $\phi_1(l)(a_0)$ in its range. We conclude that $d_{Y_1}(\phi_1(g)(a_0), \phi_1(h)(a_0)) = d_{Y_1}(\phi_1(k)(a_0), \phi_1(l)(a_0))$, a contradiction.

The argument for (D3) is similar. This finishes the proof of (i).

For (ii), let $G_1 = \Phi_{N_1}(\mathbb{F}(P_1))$ and, by Theorem 29, let ρ_1 be a G_1-invariant pseudometric that is consistent with w_{N_1} such that $(\overline{\Gamma_{N_1}}, \overline{\Phi_{N_1}}) \cong (Y_1, \phi_1)$. For notational simplicity we assume $(Y_1, \phi_1) = (\overline{\Gamma_{N_1}}, \overline{\Phi_{N_1}})$. Let $N_2 \trianglelefteq \mathbb{F}(P_2)$ be a coherent extension of $N_1 = \ker(\phi_1)$. Since N_2 is a feasible prekernel, one can define Γ_{N_2}, w_{N_2}, and Φ_{N_2} as before.

Define a map $\pi : \Gamma_{N_1} \to \Gamma_{N_2}$ by letting $\pi(gN_1H_1) = gN_2H_2$ for all $g \in \mathbb{F}(P_1)$. To see π is well-defined, note that if $gN_1H_1 = g'N_1H_1$, then $g^{-1}g' \in N_1H_1 \leq N_2H_2$, and therefore $gN_2H_2 = g'N_2H_2$.

Recall that w_{N_1} is defined for pairs (gpN_1H_1, gqN_1H_1) where $g \in \mathbb{F}(P_1)$ and $p, q \in P_1$ with $p(a_0)$ and $q(a_0)$ defined, and its value is $d_{X_1}(p(a_0), q(a_0))$. Let $\pi(w_{N_1})$ on $\pi(\Gamma_{N_1})$ be the push-forward weight function, that is,

$$\pi(w_{N_1})(gpN_2H_2, gqN_2H_2) = w_{N_1}(gpN_1H_1, gqN_1H_1).$$

Note that (D2) implies that $\pi(w_{N_1})$ is well-defined. Also note that w_{N_2} coincides with $\pi(w_{N_1})$ on $\pi(\Gamma_{N_1})$. In fact, w_{N_2} is defined in the same way on the image of such pairs under π, that is, on pairs of the form (gpN_2H_2, gqN_2H_2) for $g \in \mathbb{F}(P_1) \subseteq \mathbb{F}(P_2)$ and $p, q \in P_1 \subseteq P_2$.

Recall that Φ_{N_2} is defined by $\Phi_{N_2}(p)(gN_2H_2) = pgN_2H_2$ for all $g \in \mathbb{F}(P_2)$ and $p \in P_2$, and is extended to a group homomorphism from $\mathbb{F}(P_2)$ to the symmetric group of Γ_{N_2}. Let $G_2 = \Phi_{N_2}(\mathbb{F}(P_2))$.

We are now ready to define a G_2-invariant pseudometric ρ_2 on Γ_{N_2} that is consistent with w_{N_2} and satisfies $\rho_2 \upharpoonright \pi(\Gamma_{N_1}) = \pi(\rho_1)$. Here $\pi(\rho_1)$ is the pseudometric on $\pi(\Gamma_{N_1})$ defined by $\pi(\rho_1)(gN_2H_2, hN_2H_2) = \rho_1(gN_1H_1, hN_1H_1)$ for $g, h \in \mathbb{F}(P_1)$. Note that (D2) implies that $\pi(\rho_1)$ is well-defined.

We define ρ_2 as follows. First, for $g, h \in \mathbb{F}(P_1)$, define

$$\rho_2(gN_2H_2, hN_2H_2) = \pi(\rho_1)(gN_2H_2, hN_2H_2)$$
$$= \rho_1(gN_1H_1, hN_1H_1) = d_{Y_1}(\phi_1(g)(a_0), \phi_1(h)(a_0)).$$

Next, for $p, q \in P_2$ with $p(a_0)$ and $q(a_0)$ defined, and for $\gamma \in \mathbb{F}(P_2)$, define

$$\rho_2(\gamma p N_2H_2, \gamma q N_2H_2) = w_{N_2}(\gamma p N_2H_2, \gamma q H_2N_2) = d_{X_2}(p(a_0), q(a_0)).$$

To see that these do not conflict with each other, note that (D3) implies that for $g, h \in \mathbb{F}(P_1)$ and $p, q \in P_2$ with both $p(a_0)$ and $q(a_0)$ defined, if

$$d_{Y_1}(\phi_1(g)(a_0), \phi(h)(a_0)) \neq d_{X_2}(p(a_0), q(a_0)),$$

then there is no $\gamma \in \mathbb{F}(P_2)$ with $\gamma p N_2H_2 = gN_2H_2$ and $\gamma q N_2H_2 = hN_2H_2$. We continue to define ρ_2 so that if $g, h \in \mathbb{F}(P_2)$ and $\rho_2(gN_2H_2, hN_2H_2)$ is already defined, then we define

$$\rho_2(\gamma g N_2H_2, \gamma h N_2H_2) = \rho_2(gN_2H_2, hN_2H_2)$$

for any $\gamma \in \mathbb{F}(P_2)$. To see that this does not create a conflict among the existing definitions of ρ_2 values, note that condition (D2) implies that for any $g, h, k, l \in \mathbb{F}(P_1)$, if there is $\gamma \in \mathbb{F}(P_2)$ such that $\gamma g N_2H_2 = kN_2H_2$ and $\gamma h N_2H_2 = lN_2H_2$, then

$$d_{Y_1}(\phi_1(g)(a_0), \phi_1(h)(a_0)) = d_{Y_1}(\phi_1(k)(a_0), \phi_1(l)(a_0)).$$

To complete the definition of ρ_2, we consider the existing values of ρ_2 as a weight function and define ρ_2 to be the path pseudometric. Since the weight function is G_2-invariant, it follows from the definition of the path pseudometric that the resulting ρ_2 is also G_2-invariant.

Since for every $g, h \in \mathbb{F}(P_1)$ we have

$$\rho_2(gN_2H_2, hN_2H_2) = \pi(\rho_1)(gN_2H_2, hN_2H_2) = \rho_1(gN_1H_1, hN_1H_1),$$

we also have that $\overline{\rho_2} \upharpoonright \pi(\overline{\Gamma_{N_1}}) = \pi(\overline{\rho_1})$. Note that (D2) with $l, k = 1$ implies that the induced map $\overline{\pi} : \overline{\Gamma_{N_1}} = \overline{\Gamma_{N_1}}^{\overline{\rho_1}} \to \overline{\Gamma_{N_2}} = \overline{\Gamma_{N_2}}^{\overline{\rho_2}}$ is an isometric embedding. Thus $\overline{\Gamma_{N_1}} \cong \pi(\overline{\Gamma_{N_1}})$ is a subspace of $\overline{\Gamma_{N_2}}$.

Letting $Y_2 = \overline{\Gamma_{N_2}}$ and $\phi_2 = \overline{\Phi_{N_2}}$. We have that (Y_2, ϕ_2) is a P_2-type S-extension of X_2 and $Y_1 \subseteq Y_2$ via the isomorphism of (Y_1, ϕ_1) with $(\overline{\Gamma_{N_1}}, \overline{\Phi_{N_1}})$. To see the coherence of (Y_2, ϕ_2) with (Y_1, ϕ_1), let $p \in P_1$. Then $\Phi_{N_1}(p)(gN_1H_1) = pgN_1H_1$ for all $g \in \mathbb{F}(P_1)$ and $\Phi_{N_2}(p)(gN_2H_2) = pgN_2H_2$ for all $g \in \mathbb{F}(P_2)$. Via the induced embedding $\overline{\pi} : \overline{\Gamma_{N_1}} \to \overline{\Gamma_{N_2}}$ and the isomorphism of (Y_1, ϕ_1) with $(\overline{\Gamma_{N_1}}, \overline{\Phi_{N_1}})$, and because $\overline{\rho_2} \upharpoonright \pi(\overline{\Gamma_{N_1}}) = \pi(\overline{\rho_1})$, we have $\phi_1(p) \subseteq \phi_2(p)$. Finally, it is clear that the map $\phi_1(p) \mapsto \phi_2(p)$ for all $p \in P_1$ generates a group isomorphic embedding from G_1 to G_2. □

5.3 A Construction of Coherent Prekernel Extensions

The existence of finite coherent S-extensions via Lemma 35 and Theorem 37 imply the existence of coherent prekernel extensions. In this subsection we provide a direct construction of coherent prekernel extensions of finite index.

Lemma 38. *Let $X_1 \subseteq X_2$ be finite metric spaces, (Y_1, ϕ_1) be a P_1-type S-extension of X_1 and $P_1 \subseteq P_2 \subseteq \mathcal{P}_{X_2}$ where $P_2 = P_2^{-1}$ and $X_2 \setminus \{a_0\} \subseteq P_2(a_0)$. Then there exists a P_2-type S-map $\phi_\mathbb{U} : \mathbb{F}(P_2) \to \mathrm{Iso}(\mathbb{U})$ such that $(\mathbb{U}, \phi_\mathbb{U})$ is a P_2-type S-extension of X_2 which is coherent with (Y_1, ϕ_1).*

Proof. Following Uspenskij's proof in [16], which uses the Katetov construction of $\mathbb{U}$ to show that the isometry group of every Polish space can be embedded into $\mathrm{Iso}(\mathbb{U})$ (see also Sects. 1.2 and 2.5 of [4] for details), we obtain an isometric embedding $i : Y_1 \to \mathbb{U}$ and a group isomorphic embedding $j : \mathrm{Iso}(Y_1) \to \mathrm{Iso}(\mathbb{U})$ such that for every $\varphi \in \mathrm{Iso}(Y_1)$, $j(\varphi) \supseteq \varphi$. In addition, from the ultrahomogeneity of $\mathbb{U}$ we obtain an isometric copy of X_2 in $\mathbb{U}$ as a superset of X_1. Now for each $p \in P_2$, let $\phi_\mathbb{U}(p) \in \mathrm{Iso}(\mathbb{U})$ be an extension of p guaranteed to exist by the ultrahomogeneity of $\mathbb{U}$ such that if $p \in P_1$, then $\phi_\mathbb{U}(p) = j(\phi_1(p))$. Then $\phi_\mathbb{U}$ is as required. □

Proposition 39. *Suppose $X_1 \subseteq X_2$ are finite metric spaces and $P_1 \subseteq P_2 \subseteq \mathcal{P}_{X_2}$ where $P_2 = P_2^{-1}$ and $X_2 \setminus \{a_0\} \subseteq P_2(a_0)$. Let (Y_1, ϕ_1) be a finite P_1-type S-extension of X_1 and $N_1 = \ker(\phi_1)$. Then, there exists a coherent extension, $N_2 \trianglelefteq \mathbb{F}(P_2)$, of N_1 of finite index.*

Proof. By Lemma 28, $N_1 = \ker(\phi_1)$ is a feasible prekernel. We may define w_{N_1} and Φ_{N_1} and let $G_1 = \Phi_{N_1}(\mathbb{F}(P_1))$ and $\Gamma_{N_1} = \mathbb{F}(P_1)/N_1H_1$. Since (Y_1, ϕ_1) is a minimal P_1-type S-extension of X_1, by Theorem 29, there is a G_1-invariant pseudometric ρ_1 on Γ_{N_1} such that it is consistent with w_{N_1}, Y_1 is isometric to $(\overline{\Gamma_{N_1}}, \overline{\rho_1})$ and (Y_1, ϕ_1) is isomorphic to $(\overline{\Gamma_{N_1}}, \overline{\Phi_{N_1}})$.

Since $P_1 \subseteq P_2$, we have that all of N_1, H_1 and $\mathbb{F}(P_1)$ are subgroups of $\mathbb{F}(P_2)$. We will find a coherent extension, $N_2 \trianglelefteq \mathbb{F}(P_2)$, of N_1 of finite index.

Let $\mathcal{G} = G_1 * \mathbb{F}(P_2 \setminus P_1)$ be the free product of G_1 with $\mathbb{F}(P_2 \setminus P_1)$. We define a group homomorphism $\psi : \mathbb{F}(P_2) \to \mathcal{G}$ by letting

$$\psi(p) = \begin{cases} \Phi_{N_1}(p), & \text{if } p \in P_1, \\ p, & \text{otherwise} \end{cases}$$

for all $p \in P_2$. Since H_2 is a finitely generated subgroup of $\mathbb{F}(P_2)$, $\psi(H_2)$ is a finitely generated subgroup of $\mathcal{G}$. We will find $M \trianglelefteq \mathcal{G}$ of finite index and set $N_2 = \psi^{-1}(M)$. To guarantee that N_2 is a coherent extension of N_1, we need M to satisfy the following corresponding conditions:

(R1) For every $p, q, r, s \in P_2 \cup \{1\}$ such that $d_{X_2}(p(a_0), q(a_0)) \neq d_{X_2}(r(a_0), s(a_0))$, we have $M \cap \psi(p)\psi(H_2)\psi(r)^{-1}\psi(s)\psi(H_2)\psi(q)^{-1} = \emptyset$;

(R2) For every $p, q \in P_2 \cup \{1\}$, if $p(a_0)$ and $q(a_0)$ are defined and $p(a_0) \neq q(a_0)$, we have $M \cap \psi(p)\psi(H_2)\psi(q)^{-1} = \emptyset$;

(R3) For every $p, q, r_1, s_1, \ldots, r_n, s_n \in P_2 \cup \{1\}$ such that

$$d_{X_2}(p(a_0), q(a_0)) > \sum_{i=1}^{n} d_{X_2}(r_i(a_0), s_i(a_0)),$$

we have

$$M \cap \psi(p)\psi(H_2)\psi(r_1)^{-1}\psi(s_1)\psi(H_2)\cdots\psi(H_2)\psi(r_n)^{-1}\psi(s_n)\psi(H_2)\psi(q)^{-1} = \emptyset;$$

(S1) $M \cap G_1 = \{1\}$;

(S2) For every $g, h, k, l \in \mathbb{F}(P_1)$ such that

$$d_{Y_1}(\phi_1(g)(a_0), \phi_1(h)(a_0)) \neq d_{Y_1}(\phi_1(k)(a_0), \phi_1(l)(a_0)),$$

we have $M \cap \psi(g)\psi(H_2)\psi(k)^{-1}\psi(l)\psi(H_2)\psi(h)^{-1} = \emptyset$;

(S3) For every $g, h \in \mathbb{F}(P_1)$ and $p, q \in P_2$ with both $p(a_0)$ and $q(a_0)$ defined, if

$$d_{Y_1}(\phi_1(g)(a_0), \phi_1(h)(a_0)) \neq d_{X_2}(p(a_0), q(a_0)),$$

we have $M \cap \psi(g)\psi(H_2)\psi(p)^{-1}\psi(q)\psi(H_2)\psi(h)^{-1} = \emptyset$.

To see that (S1) implies (D1), note that $N_2 \cap \mathbb{F}(P_1) = \psi^{-1}(M) \cap (\psi^{-1}(G_1) \cap \mathbb{F}(P_1)) = (\psi^{-1}(M) \cap \psi^{-1}(G_1)) \cap \mathbb{F}(P_1) = \psi^{-1}(M \cap G_1) \cap \mathbb{F}(P_1) = \ker(\psi) \cap \mathbb{F}(P_1) = N_1$. The other conditions for M obviously imply the corresponding conditions for N_2. Note also that each of conditions (R1), (R2) and (R3) is a finite collection of conditions of the form $\gamma M \cap L_1 \cdots L_n = \emptyset$ for $\gamma \in \mathcal{G}$ and finitely generated subgroups $L_1, \ldots, L_n$ (in fact each L_i is a conjugate of $\psi(H_2)$) with $\gamma \notin L_1 \cdots L_n$. Since G_1 is finite, condition (S1) is also a finite collection of conditions of the form $\gamma M \cap \{1\} = \emptyset$ for nonidentity $\gamma \in G_1$. Conditions (S2) and (S3) appear to be about infinitely many elements in $\mathbb{F}(P_1)$. However, since $G_1 = \psi(\mathbb{F}(P_1))$ is finite, they all end up being about finitely many elements of

G_1, and so each of (S2) and (S3) is still a finite collection of conditions of the form $\gamma M \cap L_1 \cdots L_n = \emptyset$ for finitely generated subgroups $L_1, \ldots, L_n$. We verify that in each case, $\gamma \notin L_1 \cdots L_n$.

Using the P_2-type S-map $\phi_\mathbb{U}$ from Lemma 38, we note that for any $g \in \mathbb{F}(P_2)$, if $\psi(g) = 1$, then $\phi_\mathbb{U}(g) = 1$. This follows from the definition of ψ and of $\phi_\mathbb{U}$.

For (R1), we need to verify that $1 \notin \psi(p)\psi(H_2)\psi(r)^{-1}\psi(s)\psi(H_2)\psi(q)^{-1}$. Toward a contradiction, if $1 \in \psi(p)\psi(H_2)\psi(r)^{-1}\psi(s)\psi(H_2)\psi(q)^{-1}$, then there exist $\eta_1, \eta_2 \in H_2$ such that $\psi(p\eta_1 r^{-1} s\eta_2 q^{-1}) = 1$. Let $\alpha = s\eta_2 q^{-1}$ and $\beta = r\eta_1^{-1}p^{-1}$. Then $\psi(\alpha) = \psi(\beta)$ and therefore $\phi_\mathbb{U}(\alpha) = \phi_\mathbb{U}(\beta)$. Since $\eta_1, \eta_2 \in H_2$, $\phi_\mathbb{U}(\alpha)(q(a_0)) = s(a_0)$ and $\phi_\mathbb{U}(\beta)(p(a_0)) = r(a_0)$. Now since $\phi_\mathbb{U}(\alpha) = \phi_\mathbb{U}(\beta)$ is an isometry, we should have $d_{X_2}(p(a_0), q(a_0)) = d_{X_2}(r(a_0), s(a_0))$.

For (R2), similar argument shows that if $\alpha = p\eta_1 q^{-1}$, then $\phi_\mathbb{U}(\alpha)(q(a_0)) = p(a_0)$. Now since $\psi(\alpha) = 1$, $\phi_\mathbb{U}(\alpha) = 1$ and therefore $q(a_0) = p(a_0)$.

For (R3), if

$$1 \in \psi(p)\psi(H_2)\psi(r_1)^{-1}\psi(s_1)\psi(H_2)\cdots\psi(H_2)\psi(r_n)^{-1}\psi(s_n)\psi(H_2)\psi(q)^{-1}$$

then for some $h_1, \ldots, h_{n+1} \in H_2$ we have

$$1 = \psi(p)\psi(h_1)\psi(r_1)^{-1}\psi(s_1)\psi(h_2)\cdots\psi(h_n)\psi(r_n)^{-1}\psi(s_n)\psi(h_{n+1})\psi(q)^{-1}.$$

Consider the sequence

$$b_0 = \phi_\mathbb{U}(p)(a_0) = p(a_0),$$
$$b_1 = \phi_\mathbb{U}(ph_1 r_1^{-1} s_1)(a_0),$$
$$b_2 = \phi_\mathbb{U}(ph_1 r_1^{-1} s_1 h_2 r_2^{-1} s_2)(a_0),$$
$$\cdots\cdots$$
$$b_n = \phi_\mathbb{U}(ph_1 r_1^{-1} s_1 h_2 \cdots r_n^{-1} s_n)(a_0) = \phi_\mathbb{U}(qh_{n+1}^{-1})(a_0) = q(a_0).$$

We have

$$d_\mathbb{U}(b_0, b_1) = d_\mathbb{U}(\phi_\mathbb{U}(r_1 h_1^{-1} p^{-1})(b_0), \phi_\mathbb{U}(r_1 h_1^{-1} p^{-1})(b_1))$$
$$= d_\mathbb{U}(r_1(a_0), s_1(a_0)) = d_{X_2}(r_1(a_0), s_1(a_0)),$$

and similarly $d_\mathbb{U}(b_1, b_2) = d_{X_2}(r_2(a_0), s_2(a_0))$, $\ldots$, $d_\mathbb{U}(b_{n-1}, b_n) = d_{X_2}(r_n(a_0), s_n(a_0))$. Thus

$$d_{X_2}(p(a_0), q(a_0)) \leq \sum_{i=1}^n d_{X_2}(b_{i-1}, b_i) = \sum_{i=1}^n d_{X_2}(r_i(a_0), s_i(a_0)).$$

For (S2), we need to verify that if

$$d_{Y_1}(\phi_1(g)(a_0), \phi_1(h)(a_0)) \neq d_{Y_1}(\phi_1(k)(a_0), \phi_1(l)(a_0)),$$

then $1 \notin \psi(g)\psi(H_2)\psi(k)^{-1}\psi(l)\psi(H_2)\psi(h)^{-1}$. Toward a contradiction, assume $1 \in \psi(g)\psi(H_2)\psi(k)^{-1}\psi(l)\psi(H_2)\psi(h)^{-1}$. Then there are $\eta, \eta' \in H_2$ with

$$\psi(g)\psi(\eta)\psi(k)^{-1} = \psi(h)\psi(\eta')\psi(l)^{-1}.$$

From the definitions of H_2 and of ψ, if we apply the left-hand-side element to $\psi(k)(a_0) = \phi_1(k)(a_0)$, the resulting value is $\psi(g)(a_0) = \phi_1(g)(a_0)$. Similarly, if we apply the right-hand-side element to $\psi(l)(a_0) = \phi_1(l)(a_0)$, the resulting value is $\psi(h)(a_0) = \phi_1(h)(a_0)$. Thus, both sides of the equation represent the same partial isometry of Y_1 with $\phi_1(k)(a_0)$ and $\phi_1(l)(a_0)$ in its domain and with $\phi_1(g)(a_0)$ and $\phi_1(h)(a_0)$ in its range. We conclude that

$$d_{Y_1}(\phi_1(g)(a_0), \phi_1(h)(a_0)) = d_{Y_1}(\phi_1(k)(a_0), \phi_1(l)(a_0)),$$

a contradiction.

The argument for (S3) is similar.

Now by Coulbois' theorem (Theorem 16), the group $\mathcal{G} = G_1 * \mathbb{F}(P_2 \setminus P_1)$ has property RZ. Thus, there exists $M \trianglelefteq \mathcal{G}$ of finite index such that all conditions (R1)–(S3) are satisfied. Consequently, $N_2 = \psi^{-1}(M) \trianglelefteq \mathbb{F}(P_2)$ is a coherent extension of N_1 of finite index. □

5.4 Extending Isometry Groups

In this subsection we apply the algebraic method from the preceding subsection to obtain a construction of S-extensions with prescribed isometry groups. Since we deal with only finite isometry groups, it suffices to consider groups extended by one more generator.

Theorem 40. *Let X_1 be a finite metric space and (Y_1, ϕ_1) be a finite minimal P_1-type S-extension of X_1. Let $G_1 = \phi_1(\mathbb{F}(P_1))$ and $G_2 = \langle G_1, k \rangle$ be an overgroup of G_1 with one element $k \notin G_1$. Then there exist a finite metric space $X_2 \supseteq X_1$, $l \in \mathcal{P}_{X_2}$ and a P_2-type S-extension (Y_2, ϕ_2) of X_2, where $P_2 = P_1 \cup \{l, l^{-1}\}$, such that (Y_1, ϕ_1) and (Y_2, ϕ_2) are coherent and $G_2 \cong \phi_2(\mathbb{F}(P_2))$.*

Proof. By Theorem 29, Y_1 is isometric to $(\overline{\Gamma_{N_1}}, \overline{\Phi_{N_1}})$ where $N_1 = \ker(\phi_1)$. Let $X_2 = Y_1 \cup \{a\}$ be the one point extension of Y_1 where $d_{X_2}(a, b) = \text{diam}(Y_1)$ for every $b \in Y_1$. Let $P_2 = P_1 \cup \{l, l^{-1}\}$ where $l = \{(a_0, a)\}$ is the partial isometry that sends a_0 to a. We define a homomorphism $\phi_2 : \mathbb{F}(P_2) \to G_2$ such that $\phi_2 \upharpoonright_{\mathbb{F}(P_1)} = \phi_1$ and $\phi_2(l) = k$. Let $N_2 = \ker(\phi_2)$. We claim there exists a G_2-invariant pseudometric ρ_2 on Γ_{N_2} which is consistent with w_{N_2}, such that $(Y_2, \phi_2) \cong (\overline{\Gamma_{N_2}}, \overline{\Phi_{N_2}})$ is as desired. Note that because of the definition of X_2 and P_2, $H_2 = H_1$. By Theorem 37, it suffices to show that N_2 is a coherent extension of N_1.

(C1) For every $p, q, r, s \in P_2 \cup \{1\}$ such that $d_{X_2}(p(a_0), q(a_0)) \neq d_{X_2}(r(a_0), s(a_0))$, we have $N_2 \cap pH_1 r^{-1} s H_1 q^{-1} = \emptyset$. If p, q, r, s are different from l, then since $(\overline{\Gamma_{N_1}}, \overline{\Phi_{N_1}})$ is a P_1-type S-extension of X_1, by (C0) we have $N_2 \cap pH_1 r^{-1} s H_1 q^{-1} = N_1 \cap pH_1 r^{-1} s H_1 q^{-1} = \emptyset$. If one of p, q, r, s is equal to l, then we have $\phi_2(ph_1 r^{-1} sh_2 q^{-1}) = 1$ for some $h_1, h_2 \in H_1$. Since l appears exactly once in $ph_1 r^{-1} sh_2 q^{-1}$, this means $\phi_2(l) = k \in G_1$, which is a contradiction. Other cases are obvious.

(C2) For every $p, q \in P_2 \cup \{1\}$, if $p(a_0)$ and $q(a_0)$ are defined and $p(a_0) \neq q(a_0)$, we have $N_2 \cap pH_1 q^{-1} = \emptyset$. This is similar to (C1).

(C3) For every $p, q, r_1, s_1, \ldots, r_n, s_n \in P_2 \cup \{1\}$ such that

$$d_{X_2}(p(a_0), q(a_0)) > \sum_{i=1}^{n} d_{X_2}(r_i(a_0), s_i(a_0)),$$

we have $N_2 \cap pH_1 r_1^{-1} s_1 H_1 \cdots H_1 r_n^{-1} s_n H_1 q^{-1} = \emptyset$. This is also similar to (C1).

(D1) $N_1 = N_2 \cap \mathbb{F}(P_1)$. If $g \in N_2 \cap \mathbb{F}(P_1)$ then $\phi_1(g) = \phi_2(g) = 1$. Therefore, $g \in N_1$.

(D2) For every $g, h, k, l \in \mathbb{F}(P_1)$ such that

$$d_{Y_1}(\phi_1(g)(a_0), \phi_1(h)(a_0)) \neq d_{Y_1}(\phi_1(k)(a_0), \phi_1(l)(a_0)),$$

we have $N_2 \cap gH_1 k^{-1} l H_1 h^{-1} = \emptyset$. This is a direct consequence of (D1).

(D3) For every $g, h \in \mathbb{F}(P_1)$ and $p, q \in P_2$ with both $p(a_0)$ and $q(a_0)$ defined, if

$$d_{Y_1}(\phi_1(g)(a_0), \phi_1(h)(a_0)) \neq d_{X_2}(p(a_0), q(a_0)),$$

we have $N_2 \cap gH_1 p^{-1} q H_1 h^{-1} = \emptyset$. This is a direct consequence of (D1).

□

We remark that it is possible to give a combinatorial proof of Theorem 40. This is done in [3] Lemma 5.1, which was in turn motivated by a result of Rosendal (Lemma 16 of [11]). As in [3], Theorem 40 can be used to show that the Hall's universal locally finite group can be embedded as a dense subgroup of the isometry group of the Urysohn space.

6 Ultraextensive Metric Spaces

In this section we study ultraextensive metric spaces.

Definition 41. A metric space U is *ultraextensive* if

(i) U is ultrahomogeneous, i.e., there is a ϕ such that (U, ϕ) is an S-extension of U;
(ii) Every finite $X \subseteq U$ has a finite S-extension (Y, ϕ) where $Y \subseteq U$;
(iii) If $X_1 \subseteq X_2 \subseteq U$ are finite and (Y_1, ϕ_1) is a finite minimal S-extension of X_1 with $Y_1 \subseteq U$, then there is a finite minimal S-extension (Y_2, ϕ_2) of X_2 such that $Y_2 \subseteq U$ and (Y_1, ϕ_1) and (Y_2, ϕ_2) are coherent.

Motivated by Hrushovski [6], Solecki [13] and Vershik [17], Pestov in [8] introduced a notion of *Hrushovski–Solecki–Vershik property*, which is correspondent to the first two clauses of the above definition. He used the notion to study the nonexistence of uniform and coarse embeddings from the universal Urysohn metric space into reflexive Banach spaces. He also gave a proof of Solecki's theorem (Theorem 1) using Herwig–Lascar's theorem [5].

Recall that the random graph is the Fraïssé limit of the class of all finite graphs. We equip it with the path metric and turn it into a metric space, which is denoted by $\mathcal{R}$.

Proposition 42. *The Urysohn space $\mathbb{U}$, the rational Urysohn space $\mathbb{QU}$ and the random graph $\mathcal{R}$ are ultraextensive.*

Proof. The ultraextensiveness for $\mathbb{U}$ follows directly from its universality and ultrahomogeneity, and from Theorem 37.

The space $\mathbb{QU}$ is also ultrahomogeneous and universal for all finite metric spaces with rational distances. From our proof of Theorem 1 it is clear that if X is a finite metric space with rational distances, then there is a finite S-extension (Y, ϕ) of X where the distances of Y are finite sums of the distances in X, and therefore also rational. This implies clause (ii) of the definition of ultraextensiveness for $\mathbb{QU}$. The same observation applies to the proof of Theorem 37. Namely, in every construction of the proof of Theorem 37 we used the path (pseudo)metric to define new distances. Thus the distances in Y_2 are finite sums of distances in $Y_1 \cup X_2$. Therefore, if distances in X_1, X_2, Y_1 are rational, then we can find Y_2 with rational distances. Together with the ultrahomogeneity and universality of $\mathbb{QU}$, this implies clause (iii) of the definition of ultraextensiveness for $\mathbb{QU}$.

Note that the random graph $\mathcal{R}$ as a metric space has only distances 0, 1 and 2. In fact, two distinct vertices have distance 1 if and only if they are connected with an edge. If we endow every finite graph with such a metric, namely, two distinct vertices have distance 1 if they are connected with an edge, and have distance 2 otherwise, then $\mathcal{R}$ as a metric space is ultrahomogeneous and universal for this class of finite metric spaces. Then clause (ii) of the definition of ultraextensiveness for $\mathcal{R}$ follows from this universality of $\mathcal{R}$ and from Hrushovski's theorem [6]. Finally, in Theorem 37, if X_1, X_2, Y_1 are finite metric spaces coming from graphs, then they have distances 0, 1 and 2, and our constructions give that the distances in Y_2 are natural numbers. Now if we redefine every distance ≥ 3 to be 2 in Y_2, then any isometry of Y_2 continues to be an isometry in this new metric, and from ultrahomogeneity and universality we again obtain clause (iii) of the definition of ultraextensiveness for $\mathcal{R}$. □

Theorem 43. *Every countable metric space can be extended to a countable ultraextensive metric space.*

Proof. Let X be a countable metric space. Write X as an increasing union of finite metric spaces F_n for $n = 1, 2, \ldots$. For $n \geq 1$, inductively define increasing sequences of finite metric spaces X_n, Y_n and Z_n as follows. Let $X_1 = F_1$ and (Y_1, ϕ_1) be a finite minimal S-extension of $X_1 = F_1$. We define $Y_1 \subseteq Z_1$ such that for every $D \subseteq D' \subseteq Y_1$ and a minimal S-extension of D, (E, ϕ), where $E \subseteq Y_1$, there exists a minimal S-extension of D', (E', ϕ'), where $E' \subseteq Z_1$ and (E, ϕ) and (E', ϕ') are coherent. Note that this is possible since there are only finitely many triples (D, D', E) and for any such triple by Theorem 37 we can fix a coherent extension E'. Finally, to construct Z_1, we add $E' \setminus E$ to Y_1 for all E' corresponding to the triple (D, D', E) such that the union of the new points $(E' \setminus E)$ and $E \subseteq Y_1$ is an isometric copy of E'. Then, this new set with the path metric is Z_1. Let X_2 be the metric space that is obtained by adding $F_2 \setminus F_1$ to

Z_1 such that the union of $(F_2 \setminus F_1)$ and F_1 is isometric to F_2 and the distance between points in $F_2 \setminus F_1$ and $Z_1 \setminus F_1$ comes from the path metric.

In general, assume finite $Y_{n-1} \subseteq Z_{n-1} \subseteq X_n$ has been defined. Apply Theorem 37 to find (Y_n, ϕ_n) a finite minimal S-extension of $X_n \supseteq X_{n-1}$ that is coherent with (Y_{n-1}, ϕ_{n-1}). We use a similar construction to the construction of Z_1 from Y_1 to define $Z_n \supset Y_n$. Note that Z_n has the property that every minimal S-extension in Y_n (that is , $D \subseteq E \subseteq Y_n$ where (E, ϕ) is a minimal S-extension of D) has a coherent minimal S-extension in Z_n for every $D \subseteq D' \subseteq Y_n$. Let X_{n+1} be the metric space that is obtained by adding $F_{n+1} \setminus F_n$ to Z_n such that the union of $(F_{n+1} \setminus F_n)$ and F_n is isometric to F_{n+1} and the distance between points in $F_2 \setminus F_1$ and $Z_1 \setminus F_1$ comes from the path metric.

Let Y be the union of the increasing sequence Y_n. We verify that Y is ultraextensive. To verify Definition 41 (i), let $p \in \mathcal{P}_Y$. Then there is $n \geq 1$ such that $p \in \mathcal{P}_{X_n}$. Let n_p be the least such n. Then for all $m \geq n_p$, $p \subseteq \phi_m(p) \subseteq \phi_{m+1}(p)$ by the coherence of (Y_m, ϕ_m) and (Y_{m+1}, ϕ_{m+1}). Define $\phi(p) = \bigcup_{m \geq n_p} \phi_m(p)$. Then $\phi(p)$ is an isometry of Y that extends p.

For Definition 41 (ii), let $F \subseteq Y$ be finite. Then there is n such that $F \subseteq X_n$, and it follows that $(Y_n, \phi_n \upharpoonright \mathcal{P}_F)$ is an S-extension of F.

Finally, for Definition 41 (iii), let $F \subseteq F' \subset Y$ be finite and assume that (E, ψ) is a finite minimal S-extension of F with $E \subseteq Y$. Then, there is a natural number n such that $E \subseteq Y_n$. By the construction of Z_n, there exists a minimal S-extension of F', (E', ϕ') (corresponding to the triple (F, F', E)), such that $E' \subseteq Z_n \subseteq Y$ and that (E', ϕ') is coherent with (E, ϕ). □

Theorem 44. *Let U be an ultraextensive metric space and $X \subseteq U$ be a countable subset. Then there exists a countable ultraextensive subset $Y \subseteq U$ with $X \subseteq Y$.*

Proof. The proof is similar to that of Theorem 43. The differences are that in the construction Y_n and Z_n are obtained by applying clauses (ii) and (iii) of the definition of ultraextensive metric space for U; and $X_{n+1} = F_{n+1} \cup Z_n$. □

Pestov [8] showed that $\mathrm{Iso}(\mathbb{U})$ contains a countable dense locally finite subgroup. Solecki strengthened this result by showing that $\mathrm{Iso}(\mathbb{QU})$ contains a countable dense locally finite subgroup. Rosendal [10] presented a different proof of the result by Solecki. Here we note that such dense locally finite subgroups are present in the isometry group of every separable ultraextensive space.

Theorem 45. *For every separable ultraextensive metric space U, $\mathrm{Iso}(U)$ contains a dense locally finite subgroup.*

Proof. Note that $\mathrm{Iso}(U)$ has a countable dense subset D. Let $X \subseteq U$ be a countable dense subset with the property that for all $x \in X$ and $\varphi \in D$, $\varphi(x) \in X$. Apply Theorem 44 to obtain a countable ultraextensive $Y \subseteq U$ with $X \subseteq Y$. Then $\mathrm{Iso}(Y)$ is dense in $\mathrm{Iso}(U)$.

It suffices to show that $\mathrm{Iso}(Y)$ contains a dense locally finite subgroup. As in the proof of Theorem 44 we can write Y as an increasing union $\bigcup_n Y_n$. We also have group isomorphic embeddings from each $\mathrm{Iso}(Y_n)$ to $\mathrm{Iso}(Y_{n+1})$. Let $G = \bigcup_n \mathrm{Iso}(Y_n)$. Then it is clear that G is locally finite and G is dense in $\mathrm{Iso}(Y)$. □

7 Compact Ultrametric Spaces

In this section we show that every compact ultrametric space can be extended to a compact ultraextensive ultrametric space. We first study finite ultrametric spaces and show that the notions of homogeneity, ultrahomogeneity, and ultraextensiveness coincide on finite ultrametric spaces.

We will use the following fact about homogeneity for every minimal S-extension.

Lemma 46. *Let X be a metric space and (Y, ϕ) be a minimal S-extension of X. Then Y is homogeneous.*

Proof. Let $y_1, y_2 \in Y$. Since (Y, ϕ) is minimal, there are $g_1, g_2 \in \mathbb{F}(\mathcal{P}_X)$ such that $\phi(g_1)(a_0) = y_1$ and $\phi(g_2)(a_0) = y_2$. Hence, $\phi(g_1 g_2^{-1})(y_2) = y_1$. Since $\phi(g_1 g_2^{-1})$ is an isometry of Y, Y is homogeneous. □

Theorem 47. *Let Y be a finite ultrametric space. Then the following are equivalent:*

(i) Y is homogeneous;
(ii) Y is ultrahomogeneous;
(iii) Y is ultraextensive.

Proof. (i)⇒(ii): Let $D(Y) = \{d(x,y) : x \neq y \in Y\}$. We prove this by induction on $|D(Y)|$. If $|D(Y)| = 1$ then Y is clearly ultrahomogeneous. Suppose $|D(Y)| > 1$ and let r be the least element of $D(Y)$. For each $x \in Y$ let $B_r(x) = \{y \in Y : d(x,y) \leq r\} = \{x\} \cup \{y \in Y : d(x,y) = r\}$. Then for any $x, y \in Y$, either $B_r(x) = B_r(y)$ or $B_r(x) \cap B_r(y) = \emptyset$. In the latter case, we also have that for any $z_1 \in B_r(x)$ and $z_2 \in B_r(y)$, $d(z_1, z_2) = d(x, y)$. Let $Y_1 = \{B_r(x) : x \in Y\}$. Then Y_1 is a partition of Y. For disjoint $B_r(x)$ and $B_r(y)$, we define $d_1(B_r(x), B_r(y)) = d(x,y)$. It is easy to check that (Y_1, d_1) is again an ultrametric space, and $D(Y_1) = D(Y) \setminus \{r\}$. If $\varphi \in \mathrm{Iso}(Y)$, then φ induces an isometry φ_1 of Y_1, where $\varphi_1(B_r(x)) = B_r(\varphi(x))$. Since Y is homogeneous, so is Y_1, and by the inductive hypothesis, Y_1 is ultrahomogeneous. Now suppose $p : A \to B$ is a partial isometry of Y. It induces a partial isometry $p_1 : \{B_r(a) : a \in A\} \to \{B_r(b) : b \in B\}$ of Y_1. Thus there is an isometry $\varphi_1 \in \mathrm{Iso}(Y_1)$ extending p_1. Note that for any $x, y \in Y$, $B_r(x)$ is isometric to $B_r(y)$ by the homogeneity of Y, and each $B_r(x)$ is ultrahomogeneous. Now for each $B_r(x) \in Y_1$, we define an isometry from $B_r(x)$ to $\varphi_1(B_r(x))$ as follows. If $B_r(x) \cap A = \emptyset$, then we arbitrarily fix an isometry from $B_r(x)$ to $\varphi_1(B_r(x))$. If $B_r(x) \cap A \neq \emptyset$, then $|B_r(x) \cap A| = |\varphi_1(B_r(x)) \cap B|$, and we fix an isometry from $B_r(x)$ to $\varphi_1(B_r(x))$ that sends each $a \in B_r(x) \cap A$ to $p(a) \in \varphi_1(B_r(x)) \cap B$. Putting all of these isometries together, we obtain an isometry of Y extending p. Thus Y is ultrahomogeneous.

(ii)⇒(iii): We use a similar induction as in the above proof. If $|D(Y)| = 1$ then Y is clearly ultraextensive. Assume $|D(Y)| > 1$ and let r be the least element of Y. Define Y_1 similarly as above. Then by the inductive hypothesis Y_1 is ultraextensive. For any $x \in Y$, $B_r(x)$ is also ultraextensive. Arbitrarily fix an $x \in Y$ and let $Y_2 = B_r(x)$. Consider $Y_1 \times Y_2$ and define a metric d' by

$$d'((B_r(y_1), z_1), (B_r(y_2), z_2)) = \max\{d_1(B_r(y_1), B_r(y_2)), d(z_1, z_2)\}.$$

Then (Y, d) is isometric to $(Y_1 \times Y_2, d')$. Thus we will view Y as $Y_1 \times Y_2$. Enumerate the elements of Y_1 by $b_1 = B_r(y_1), \ldots, b_m = B_r(y_m)$. We show that Y is ultraextensive.

Since Y is finite and ultrahomogeneous, it is enough to show that for every minimal S-extension (Y_0, ϕ_0) of X where $X, Y_0 \subseteq Y$ there is a group embedding $\pi : \text{Iso}(Y_0) \to \text{Iso}(Y)$ such that $\pi(g) \upharpoonright Y_0 = g$.

Since (Y_0, ϕ_0) is a minimal S-extension, by Lemma 46, Y_0 is homogeneous and therefore ultrahomogeneous by the previous argument. It follows that the non-empty intersections of Y_0 with $b_i = B_r(y_i)$ are isometric. That is, if $Y_0 \cap B_r(y_i) \neq \emptyset$ and $Y_0 \cap B_r(y_j) \neq \emptyset$, then $Y_0 \cap B_r(y_i)$ and $Y_0 \cap B_r(y_j)$ are isometric. Arbitrarily fix such a non-empty intersection Y_{02}. Let $Y_{01} = \{B_r(x) : x \in Y_0\}$. Then Y_0 is isometric to $Y_{01} \times Y_{02}$ as a subset of $Y_1 \times Y_2$. Now, for every $g \in \text{Iso}(Y_0)$, g induces an isometry of Y_{01}, which we denote by $\phi_{01}(g)$. Furthermore, for every $g \in \text{Iso}(Y_0)$ and every $1 \leq i, j \leq m$ such that $\phi_{01}(g)(b_i) = b_j$, g induces an isometry of Y_{02}, which we denote by $\phi(i,j)(g)$. More precisely, if $g(b_i, z_1) = (b_j, z_2)$ for some $1 \leq i, j \leq m$ and $z_1, z_2 \in Y_{02}$, then $\phi(i,j)(g)(z_1) = z_2$. Since Y_1 and Y_2 are ultraextensive, there are group embeddings $\pi_{01} : \text{Iso}(Y_{01}) \to \text{Iso}(Y_1)$ and $\pi_{02} : \text{Iso}(Y_{02}) \to \text{Iso}(Y_2)$ such that $\pi_{01}(g) \upharpoonright Y_{01} = g$ and $\pi_{02}(g) \upharpoonright Y_{02} = g$. Let $\pi : \text{Iso}(Y_0) \to \text{Iso}(Y)$ be such that for $b_i \in Y_1$ and $z \in Y_2$ where $\phi_{01}(g)(b_i) = b_j$ we have

$$\pi(g)(b_i, z) = (\pi_{01}(\phi_{01}(g))(b_i), \pi_{02}(\phi(i,j)(g))(z)).$$

Then, π is as desired. That is, π is a group embedding and if $g(b_i, z) = (b_j, z')$, then $\pi(g)(b_i, z) = (b_j, z')$. Therefore, Y is ultraextensive.

(iii)⇒(i) is obvious. □

In view of Theorem 47 it is easy to construct finite ultrahomogeneous or ultraextensive ultrametric spaces.

Definition 48. Let (Γ, w) be a connected (undirected) weighted graph. The *maximum path metric* on Γ is the metric defined by

$$d(x, y) = \inf\{\max\{w(y_i, y_{i+1}) : i = 1, \ldots, n\} : y_1 = x, y_{n+1} = y \text{ and }$$
$$(y_i, y_{i+1}) \text{ is an edge in } \Gamma \text{ for all } i = 1, \ldots, n\}.$$

If (Γ, w) is a connected finite weighted graph, then it is easy to see that Γ with the maximum path metric is an ultrametric space.

Proposition 49. *Let X be a finite ultrametric space. Then X can be extended to a finite ultraextensive ultrametric space Y. Furthermore, there is such Y so that the set of distances in X and Y are the same.*

Proof. By Theorem 47 it suffices to construct an extension of X that is homogeneous. We use the same notation as in the proof of Theorem 47. Our proof will be by induction on $|D(X)|$. If $|D(X)| = 1$ then X is already homogeneous. Assume $|D(X)| > 1$ and let r be the lease element of $D(X)$. Define $X_1 = \{B_r(x) : x \in X\}$ and d_1 on X_1. Then $|D(X_1)| = |D(X)| - 1$. By the inductive hypothesis, X_1 can be extended to a homogenous Y_1 with the same distances as in X_1. Now each $B_r(x)$ is a homogeneous space with every pair of points having distance r. Let $N = \max\{|B_r(x)| : x \in X\}$ and let $x_0 \in X$ be such that $|B_r(x_0)| = N$. Then $X_2 = B_r(x_0)$ is a homogeneous extension of each of $B_r(x)$. It follows that $Y_1 \times X_2$ is a homogeneous ultrametric space extending X. □

Lemma 50. *Let $\epsilon > 0$. Let $X_1 \subseteq X_2$ be finite ultrametric spaces such that X_1 is an ϵ-net in X_2. Let (Y_1, ϕ_1) be a finite minimal S-extension of X_1 such that Y_1 is an ultrametric space with the same distances as in X_1. Then there is a minimal S-extension (Y_2, ϕ_2) of X_2 such that Y_2 is an ultrametric space, (Y_2, ϕ_2) is coherent with (Y_1, ϕ_1), and Y_1 is an ϵ-net in Y_2.*

Proof. By Lemma 46 and Theorem 47, Y_1 is homogeneous and therefore ultraextensive. It is enough to construct an S-extension (Y_2, ϕ_2) of X_2 that satisfies the prescribed conditions, as a minimal S-extension can always be extracted from an S-extension. Since X_1 is an ϵ-net in X_2, we have that the set of $B_{<\epsilon}(x) = \{y \in X_2 : d_{X_2}(x,y) < \epsilon\}$, when x varies over X_1, is a partition of X_2. Let $B_{<\epsilon} = X_2/\sim$ where $\sim$ identifies all points of X_1 and the metric on $B_{<\epsilon}$ is the maximum path metric. Then $B_{<\epsilon}$ extends $B_{<\epsilon}(x)$ for every $x \in X_1$ and therefore X_2 corresponds to a subset of the product $X_1 \times B_{<\epsilon}$. Now Y_1 is a homogeneous extension of X_1 with the same distances as X_1. In particular any distance between distinct points in Y_1 is $\geq \epsilon$. We can let $Y_2 = Y_1 \times B$ where B is a homogeneous extension of $B_{<\epsilon}$ with the same set of distances as $B_{<\epsilon}$. Then Y_2 is obviously homogeneous, and therefore ultrahomogeneous. It is clear that Y_2 is an ultrametric space, and that for every $y_2 \in Y_2$ there is $y_1 \in Y_1$ with $d_{Y_2}(y_2, y_1) < \epsilon$. We define a group homomorphism $\phi_2 : \mathbb{F}(\mathcal{P}_{X_2}) \to \mathrm{Iso}(Y_2)$ such that for every $p \in \mathcal{P}_{X_1}$

$$\phi_2(p)(y_1, y_2) = (\phi_1(p)(y_1), y_2)$$

and for every $p \in \mathcal{P}_{X_2} \setminus \mathcal{P}_{X_1}$ let $\phi_2(p)$ be an isometry of Y_2 such that $p \subseteq \phi_2(p)$. Note that since Y_2 is ultrahomogeneous, it is possible to find $\phi_2(p)$ as required for every $p \in \mathcal{P}_{X_2} \setminus \mathcal{P}_{X_1}$. It is clear that (Y_2, ϕ_2) is coherent with (Y_1, ϕ_1). □

Theorem 51. *Every compact ultrametric space can be extended to a compact ultraextensive ultrametric space. In particular, every compact ultrametric space has a compact ultrametric S-extension.*

Proof. Let $\{X_k\}_{k=1}^\infty$ be an increasing sequence of finite subsets of X such that for each k, X_k is a $\frac{1}{2^k}$-net. Then by Proposition 49 and Lemma 50, there is a sequence of S-extensions $\{(Y_k, \phi_k)\}_{k=1}^\infty$ such that $\{Y_k\}_{k=1}^\infty$ is an increasing sequence of finite ultraextensive ultrametric spaces, (Y_k, ϕ_k) is an S-extension of X_k, (Y_{k+1}, ϕ_{k+1}) is coherent with (Y_k, ϕ_k), and Y_k is a $\frac{1}{2^k}$-net in Y_{k+1}. Let Y be the completion of $\bigcup_{k=1}^{\infty} Y_k$. Then, Y is clearly an ultrametric space; Y is compact since $\bigcup_{k=1}^{\infty} Y_k$ is totally bounded. In fact, Y_k is a $\frac{1}{2^k}$-net in Y. Since each Y_k is ultraextensive, so is $\bigcup_{k=1}^{\infty} Y_k$. We show that Y is ultraextensive.

We first show that Y is ultrahomogeneous. For this, let $p : A \to B$ be a partial isometry of Y. Let $\frac{1}{2^k}$ be less than the smallest non-zero distance between points of A. Since Y_k is a $\frac{1}{2^k}$-net in Y, there are $A_k, B_k \subseteq Y_k$ and $p_k : A_k \to B_k$ such that points in A are approximated by points in A_k, points in B are approximated by points in B_k, and consequently p_k is also a partial isometry. Each p_k can be extended via $\phi_n(p_k)$ for $n > k$ to $\bigcup_{n > k} \phi_n(p_k)$, an isometry of $\bigcup_{k=1}^{\infty} Y_k$, and then uniquely to an isometry P_k of Y. Since $\mathrm{Iso}(Y)$ is compact, the collection of P_k has an accumulation point φ, which is an isometry of Y. Since each P_k approximates p with an error less than $\frac{1}{2^k}$, it follows that $\varphi \supseteq p$. This shows that any partial isometry of Y can be extended to an isometry of Y. In particular, it also shows that any partial isometry of X can be extended to an isometry of Y, thus there is a suitable ϕ such that (Y, ϕ) is an S-extension of X.

For the remaining properties of ultraextensiveness, it suffices to show that any finite subset of Y can be extended to a finite homogeneous, and therefore ultraextensive, subset of Y. For this, let $A \subseteq Y$ be finite and let $\frac{1}{2^k}$ be less than the smallest non-zero distance between points in A. Since Y_k is a $\frac{1}{2^k}$-net in Y, there is a set $A_k \subseteq Y_k$ such that for each $a \in A$ there is a unique point $a_k \in A_k$ such that $d(a, a_k) < \frac{1}{2^k}$. Consider the set $Z_k = (Y_k \setminus A_k) \cup A$. It is easy to see that the map $\pi : Z_k \to Y_k$ defined by $\pi(a) = a_k$ for $a \in A$ and $\pi(y) = y$ otherwise is an isometry. Thus Z_k is a finite homogenenous subset of Y extending A. □

8 Open Problems

One general problem is to determine if a certain class of finite metric spaces admit finite S-extensions in the same class. For example, we do not know if the class of finite Euclidean metric spaces has this property.

Question 1. *Let $X \subseteq \mathbb{R}^n$ be a finite subset. Does X have a finite S-extension (Y, ϕ) with $Y \subseteq \mathbb{R}^m$ for some $m \geq n$?*

As stated in Theorem 45, we know that the isometry group of every ultraextensive metric space has a dense locally finite subgroup. It is of interest to know if the isomorphism group of other well-known mathematical objects have the same property. In particular, the following questions are open.

Question 2. *Does the homeomorphism group of the Hilbert cube have a dense locally finite subgroup?*

Question 3. *Does the linear isometry group of the Gurarij space have a dense locally finite subgroup?*

References

1. Coulbois, T.: Free product, profinite topology and finitely generated subgroups. Internat. J. Algebra Comput. **11**(2), 171–184 (2001)
2. Etedadialiabadi, M., Gao, S.: On extensions of partial isomorphisms. J. Symbolic Logic **87**(1), 416–435 (2022)
3. Etedadialiabadi, M., Gao, S., Le Maître, F., Melleray, J.: Dense locally finite subgroups of automorphism groups of ultraextensive spaces. Adv. Math. **391**, 107966 (2021)
4. Gao, S.: Invariant descriptive set theory, Pure and Applied Mathematics, A Series of Monographs and Textbooks, 293. CRC Press (2009)
5. Herwig, B., Lascar, D.: Extending partial automorphisms and the profinite topology on free groups. Trans. Am. Math. Soc. **352**, 1985–2021 (1999)
6. Hrushovski, E.: Extending partial isomorphisms of graphs. Combinatorica **12**, 411–416 (1992)
7. Hubička, J., Konečný, M., Nešetřil, J.: A combinatorial proof of the extension property for partial isometries. Comment. Math. Univ. Carol. **60**(1), 39–47 (2019)
8. Pestov, V.: A theorem of Hrushovski-Solecki-Vershik applied to uniform and coarse embeddings of the Urysohn metric space. Topology Appl. **155**, 1561–1575 (2008)
9. Ribes, L., Zalesskii, P.A.: On the profinite topology on a free group. Bull. Lond. Math. Soc. **25**, 37–43 (1993)
10. Rosendal, C.: Finitely approximable groupsand actions part I: the Ribes-Zalesskii property. J. Symbolic Logic **76**(4), 1297–1306 (2011)
11. Rosendal, C.: Finitely approximable groups and actions part II: generic representations. J. Symbolic Logic **76**(4), 1307–1321 (2011)
12. Sabok, M.: Automatic continuity for isometry groups. J. Inst. Math. Jussieu 1–30 (2017)
13. Solecki, S.: Extending partial isometries. Israel J. Math. **150**, 315–332 (2005)
14. Solecki, S.: Notes on a strengthening of the Herwig–Lascar Extension Theorem, unpublished notes (2009). https://faculty.math.illinois.edu/~ssolecki/papers/HervLascfin.pdf
15. Siniora, D., Solecki, S.: Coherent extension of partial automorphisms, free amalgamation, and automorphism groups. J. Symbolic Logic **85**(1), 199–223 (2020)
16. Uspenskij, V.V.: On the group of isometries of the Urysohn universal metric space. Comment. Math. Univ. Carol. **31**(1), 181–182 (1990)
17. Vershik, A.M.: Globalization of the partial isometries of metric spaces and local approximation of the group of isometries of Urysohn space. Topology Appl. **155**(14), 1618–1626 (2008)

The Specker-Blatter Theorem: An Application of Logic to Combinatorial Counting

Johann A. Makowsky[✉]

Department of Computer Science, Technion-IIT, Haifa, Israel
janos@cs.technion.ac.il
http://janos.cs.technion.ac.il

Abstract. Over forty years ago, E.P. Specker and C. Blatter discovered an astonishing meta-theorem about combinatorial counting functions definable in Monadic Second Order Logic (MSOL). In this talk we discuss extensions and limit of this theorem and its wide-ranging applications.

Keywords: Combinatorial counting · Monadic Second Logic · Integer sequences

This talk is dedicated to the memory of Ernst P. Specker (1920–2011) and Christian Blatter (1935–2021)[1]. They were both among my first teachers at ETHZ (the Swiss Federal Institute of Technology in Zurich). Blatter taught me Real Analysis and impressed me for his expository skills. Specker taught me Mathematical Logic, mathematical and intellectual curiosity, and uncompromised honesty.
The talk is based on recent joint work with Yuval Filmus, Eldar Fischer and Vsevolod Rakita, [5,7,8].

1 Motivation

While teaching an introductory topology course in the late 1970 s, Ernst Specker asked the students to compute the sequence $t(n)$ listed as sequence A000798 of OEIS[2], which counts of the number of finite topologies on the set $[n] = \{1, 2, \ldots, n\}$. At the time when Specker asked the question, two papers gave different results for $t(5)$:

$$t(5) = 7181 \text{ and } t(5) = 6942.$$

---

[1] https://en.wikipedia.org/wiki/Ernst_Specker, https://mathshistory.st-andrews.ac.uk/Biographies/Specker/, See also [4]. https://de.wikipedia.org/wiki/Christian_Blatter.

[2] The On-Line Encyclopedia of Integer Sequences, https://oeis.org/.

The question arose how to prove for at least one of them that it was wrong? Even today there is no explicit formula known for $t(n)$ nor is there a known recurrence relations which allows one to compute $t(n)$ precisely.

C. Blatter and E.P. Specker [1,2] devised a method how to compute for every positive integer m, the number

$$a(n) = t(n) \pmod{m}.$$

They computed $a(5) = 2$ and concluded that $t(5) = 7181$ was wrong.

2 MC-Finite Sequences of Integers

A sequence of natural numbers $s(n)$ is *C-finite* if it satisfies a linear recurrence relation with constant coefficients. $s(n)$ is MC-finite if it satisfies a linear recurrence relation with constant coefficients modula m for each m separately. A typical example of an C-finite sequence is the sequence $f(n)$ of Fibonacci numbers. A typical example of an MC-finite sequence which is not C-finite is the sequence $B(n)$ of Bell numbers. The Bell number $B(n)$ counts the number of partitions of the set $[n]$ of the numbers $\{1, 2, \ldots, n\}$. Let $Eq(n)$ be number of equivalence relations on $[n]$. Clearly, $B(n) = Eq(n)$.

In [9] G. Pfeiffer discusses counting other transitive relations besides $Eq(n)$, in particular, partial orders $PO(n)$, quasi-orders (aka preorders) $QO(n)$ and just transitive relations $Tr(n)$. Using a growth argument one can see that none of these functions is C-finite. It will follow directly from the Specker-Blatter Theorem below that $PO(n), QO(n), Tr(n)$ are MC-finite. However, to the best of our knowledge, this has not been stated explicitly in the literature, although it must have been noticed by many. This may be due to the fact that no precise formulas are known for these functions, nor are there useful recurrence relations. The Specker-Blatter Theorem establishes MC-finiteness even in the absence of explicit formulas.

Growth arguments do not suffice to show that an integer sequence is not MC-finite. The Catalan numbers $C(n)$ are not MC-finite, but its proof does not generalize. This and other examples are given in [5].

3 The Original Specker-Blatter Theorem

Let ϕ_E be the formula in First Order Logic FOL which says that $E(x, y)$ is an equvalence relation. $E(n)$ now can be written as

$$Eq(n) = |\{E \subseteq [n]^2 : ([n], E) \models \phi_E\}|.$$

$PO(n), QO(n), Tr(n)$ can be written in a similar way.

The original Specker-Blatter Theorem from 1981, [1,2,10], gives a general criterion for certain integer sequences to be MC-finite. Let $R = \{R_1, \ldots, r_m\}$ be a finite set of relation symbols of arity ρ_i respectively and ϕ be a formula of Monadic Second Order Logic MSOL without free variables using relation symbols from R.

Let $Sp_\phi(n)$ be the number of ways we can interpret the relation symbols in R on $[n]$ such that the resulting structures where A_i is the interpretation of R_i satisfies ϕ. Formally

$$Sp_\phi(n) = |\{A_i \subseteq [n]^{\rho_i}, i \leq m : ([n], A_1, \ldots, A_m) \models \phi\}|$$

Theorem 1. *Let R be a finite set of* **binary relations** *and ϕ be a formula of MSOL using relation symbols in R. Then the sequence $Sp_\phi(n)$ is MC-finite.*

4 Why Binary Relations only

The methods of C. Blatter and E.P. Specker summarized in [10] do not work for relations of higher arity. However, no counterexamples are given. In 2002, E. Fischer [6] gave a counter example for a FOL-definable class for one *quaternary* relation. Finally, in 2023, more than 40 years after [1], we have shown in [8]:

Theorem 2. *There is an* FOL-*sentence with* **only one** ternary *relation symbol ϕ_1, such that $Sp_{\phi_1}(n)$ is not an MC-sequence.*

5 MC-Finiteness of Integer Sequences

In [8] we also extend the Specker-Blatter Theorem and in [5] we investigate the applicability of our extensions to selected integer sequences listed in OEIS. There are two methods to show that certain integer sequences are MC-finite, the model theoretic method of the extended Specker-Blatter Theorem, and a method based on analyzing recurrence relations which define the sequence, [3]. The latter was discovered only very recently. Both methods are extensively discussed in [5] and they are shown to be incomparable. Finally, in [7], we establish MC-finiteness of sequences counting the number of finite topologies subject to additional restrictions.

Acknowledgements. I would like to thank the organizers of ICLA 2025 for inviting me.

References

1. Blatter, C., Specker, E.: Le nombre de structures finies d'une théorie à charactère fini, pp. 41–44. Sciences Mathématiques, Fonds Nationale de la recherche Scientifique, Bruxelles pp (1981)
2. Blatter, C., Specker, E.: Recurrence relations for the number of labeled structures on a finite set. In: Logic and Machines: Decision Problems and Complexity: Proceedings of the Symposium "Rekursive Kombinatorik" held from May 23–28, 1983 at the Institut für Mathematische Logik und Grundlagenforschung der Universität Münster/Westfalen. pp. 43–61. Springer (1984)

3. Cadilhac, M., Mazowiecki, F., Paperman, C., Pilipczuk, M., Sénizergues, G.: On polynomial recursive sequences. Theor. Comput. Syst. **68**(4), 593–614 (2024)
4. Engeler, E., Hungerbühler, N., Makowsky, J.A.: Remembering ernst Specker (1920–2011). Elem. Math. **67**(3), 89–115 (2012)
5. Filmus, Y., Fischer, E., Makowsky, J.A., Rakita, V.: MC-finiteness of restricted set partition functions. J. Integer Sequences **26**(2), 3 (2023)
6. Fischer, E.: The Specker-Blatter theorem does not hold for quaternary relations. J. Comb. Theor. Ser. A **103**, 121–136 (2003)
7. Fischer, E., Makowsky, J.A.: Counting finite topologies. Enumerative Combinatorics and Applications **ECA 4:4 (2024)**, #S2R27 (2024)
8. Fischer, E., Makowsky, J.A.: Extensions and limits of the Specker–Blatter theorem. J. Symbolic Logic 1–29 (2024)
9. Pfeiffer, G.: Counting transitive relations. J. Integer Seq. **7**(2), 3 (2004)
10. Specker, E.: Application of logic and combinatorics to enumeration problems. In: Ernst Specker Selecta, pp. 324–350. Springer (1990)

Propositional Dynamic Logic Formula Synthesis and Some Applications

Sophie Pinchinat[✉]

IRISA Laboratory, University of Rennes, Rennes, France
sophie.pinchinat@irisa.fr
https://people.irisa.fr/Sophie.Pinchinat

Abstract. Logic in Computer Science plays important roles ranging from formal methods to Human-Computer Interaction, including Software and Hardware Engineering, etc. Regarding the former broad application domain of formal methods, a tremendous amount of work concern verification, namely satisfiability checking (for model synthesis) and model checking (for counterexample/witness synthesis). I want to discuss a fairly novel question, called *Formula Synthesis Problem*, that I believe is extremely natural to address, and yet has not received enough attention in logic. In its most general form, the Formula Synthesis Problem consists in deciding whether some formula in a given set is satisfied by a given model (and output one if any). Obviously, if the input set of formulas is finite, this amounts to model checking. On the contrary, if this set is infinite, say obtained by some tree-grammar for formulas, then the answer becomes extraordinary challenging.

As far as I am aware of, only in [17], the authors (part of which I am) have addressed the problem for the first time, in the particular case of the logic Propositional Dynamic Logic (PDL) extended with shuffle (PDL$^{\parallel}$), a deeply-studied logic in the literature. The obtained results regarding this instance of the Formula Synthesis Problem, called SYNTHPDL$^{\parallel}$, were published in a strongly AI-tainted conference, since it is AAAI 2022.

I wish hereby to let these results be known by a broader audience, and in particular by the community of logic and formal methods. This, all the more than the contribution on SYNTHPDL$^{\parallel}$ opens up connections to other problems in other Computer Science fields such as planning and security.

Keywords: Formula synthesis Problem · Propositional Dynamic Logic · Formal languages · Tree Grammars · Attack trees · Hierarchical Planning

1 Introduction

Model checking is a fundamental computational problem in logic which asks to decide if a given structure (aka model, interpretation) satisfies a given formula. This problem is at the core of automated system verification and has been extensively studied for various classes of models and logics [3].

In [17], the authors propose a generalization of the model checking problem, called *Formula Synthesis Problem*, that takes as input a (possibly infinite) set of formulas, and asks to find a formula in this set that is satisfied in the model (or to say that there is none). In this work, sets of formulas are specified by tree-grammars; these are context-free rewriting systems that generate terms over a ranked alphabet, and formulas in particular. In other words, the grammars specify recipes for generating the input set of formulas. For the particular case where the grammar generates a single formula, we recover the model checking problem against the logic this formula belongs to. The work [17] focuses on the particular case where the input grammar generates formulas in Propositional Dynamic Logic (PDL) introduced by Fisher and Ladner in [6] where, additionally, the *programs* in the formulas may contain the shuffle operator, sometimes called *interleaving*. This extension of PDL, noted $PDL^{\parallel}$ here, has already been considered in the past [1,10]. These Dynamic Logics were introduced to formally reason about programs whose state can change, in contrast with classical "static" logics which are evaluated in a fixed state (i.e., evaluation of the variables). Such formulas are interpreted over *transition systems* whose edges are labeled by actions. For instance, the PDL formula $\langle (a)^*; b \rangle p$ says that some execution of the nondeterministic program "execute a some finite amount of times, and then execute a" results in a state that satisfies the atomic property b.

From a broad perspective, the results I will present pave the way for a huge research program that investigates the Formula Synthesis Problem, a significant generalization of the model checking problem, itself a cornerstone problem in computational logic, to other logics besides $PDL^{\parallel}$. The relevance of the Formula Synthesis Problem lies in its ability to handle model checking against some kind of infinitary logics, or against extension of some logic with fixpoints. Besides, the original contribution [17] can actually handle the logic $FPDL^{\parallel}$, the extension of PDL with fixpoints (see for instance [12]). Variants of the problem, that I do not consider here, where the question is to decide whether every generated formula holds in the input model or not. Note that it trivially reduces to the original problem if the logic is closed under negation.

Because $PDL^{\parallel}$ formulas contain programs, the Formula Synthesis Problem turns out to have remarkable applications. This is the case for settings where hierarchical description is central, such as Attack Tree Synthesis in Security [7,18,21] and Hierarchical Task Network problems in Planning [8]. Those settings, arising from two different communities, can be seen as a single: both involve compound tasks that are recursively decomposed into sequences of subtasks, until elementary steps are reached. Grammars that generated $PDL^{\parallel}$ formulas whose *programs*[1] can represent such decompositions of compound tasks. Taking the particular example of Hierarchical Task Network problems in Planning, compound tasks are decomposed into planning scenarios, transition systems represent planning domains, tree-grammars replace methods in task-networks, and the Formula Synthesis Problem replaces the Hierarchical Task Network Planning Problem.

---

[1] A key feature in formulas.

2 Term Grammars and PDL$^{\parallel}$

We first set useful notations for standard notions in formal language theory, and second define the syntax and semantics of the logic PDL$^{\parallel}$ and the subsequent Formula Synthesis Problem.

2.1 Preliminary Notions

We use standard notations for binary relations over set S. Given two binary relations R_1 and R_2, we let $R_1 \circ R_2 := \{(s,s') : \exists s''.(s,s'') \in R_1 \text{ and } (s'',s') \in R_2\}$, and $R^* := \bigcup_{n \geq 0} R^n$ with R^n defined as follows: $R^0 = \{(s,s) : s \in S\}$, $R^{n+1} = R^n \circ R$ for $n > 0$.

Let $\Sigma = \{a, a', a_1, \ldots\}$ be a finite alphabet. A *word* over Σ is a finite sequence $\alpha = a_1 a_2 \ldots a_n$ of elements of Σ. We denote by ϵ the empty word. We recall classic formal language operations. Let $\mathcal{L}_1$ and $\mathcal{L}_2$ be two languages over Σ. We define the *concatenation* of two languages $\mathcal{L}_1$ and $\mathcal{L}_2$ by $\mathcal{L}_1 \circ \mathcal{L}_2 := \{\alpha\beta \mid \alpha \in \mathcal{L}_1 \text{ and } \beta \in \mathcal{L}_2\}$. Next, we recall the *shuffle* of two languages as the language obtained by shuffling (or *interleaving*) each word of $\mathcal{L}_1$ with each word of $\mathcal{L}_2$. Formally, the *shuffle* $\alpha \parallel \beta$ of two words α and β is defined by induction as follows: $\epsilon \parallel \alpha = \alpha \parallel \epsilon = \{\alpha\}$ and $(a\alpha) \parallel (a'\beta) = a \circ (\alpha \parallel a'\beta) \cup a' \circ (a\alpha \parallel \beta)$. For example, $ab \parallel cd \ni abcd, acbd, cadb$, etc. The *shuffle* of two languages $\mathcal{L}_1$ and $\mathcal{L}_2$ is then defined by $\mathcal{L}_1 \parallel \mathcal{L}_2 := \bigcup_{\alpha \in \mathcal{L}_1, \beta \in \mathcal{L}_2} \alpha \parallel \beta$.

We use regular tree-grammars to describe sets of terms. We now recall the basic notions (the reader may refer to [4, Subsection 2.1, p. 51] for more details). A *ranked alphabet* is a set Σ and associated ranks (arities) of the symbols. The rank of $f \in \Sigma$ is denoted $rk(f)$, *constants* are symbols of rank 0. The rank of standard function symbols are obvious, e.g., the rank of $\wedge$ is 2, and the rank of $\top$ is 0.

The set of *terms*, also known as *trees*, over Σ is the smallest set of expressions satisfying: all constants are terms, and if $t_1, \cdots, t_k$ are terms and $rk(f) = k$ then $f(t_1, \cdots, t_k)$ is a term. A *(term) language* over Σ is a set of terms over Σ.

A *regular tree-grammar* over ranked alphabet Σ is a tuple $G = (N, \mathfrak{S}, \Sigma, \mathcal{R})$ where:

- N is a finite non-empty set of *nonterminals*,
- $\mathfrak{S} \in N$ is the *axiom*,
- Σ is the set of *terminals*, and
- $\mathcal{R}$ a finite set of *production rules* of the form $\mathbf{X} \to t$ where $\mathbf{X} \in N$ and t is a term over $\Sigma \cup N$, with convention that nonterminals have rank 0.

For terms t_1, t_2 over the ranked alphabet $\Sigma \cup N$, write $t_1 \Rightarrow t_2$ if t_2 is the result of replacing a nonterminal $\mathbf{X}$ in t_1 by the right hand side of some rule $\mathbf{X} \to t$. Write $\Rightarrow^*$ for the reflexive and transitive closure of $\Rightarrow$. The *language generated by* G is $L(G) := \{t \text{ over } \Sigma \mid \mathfrak{S} \Rightarrow^* t\}$. The *size* of a grammar G is the sum of the length of all the rules in $\mathcal{R}$. A (term) language is *regular* if it is the language generated by some regular tree-grammar G.

We recall that, given two regular tree-grammars G_1 and G_2, one can build in polynomial time a regular tree-grammar G that generates the intersection of the languages generated by G_1 and G_2 (see [4]). Also, given a regular tree-grammar G one can decide in polynomial time if the language it generates is non-empty, and in this case one can return in polynomial time a term generated by G (see [4] for details).

2.2 The Logic PDL$^{\|}$

Fix PROP, a set of *atomic propositions* and ACT of a set *atomic programs*, both finite, and define. Formulas of PDL$^{\|}$ as follows:

$$\varphi ::= p \mid \top \mid \neg\varphi \mid \varphi_1 \wedge \varphi_2 \mid \langle \rho \rangle \varphi$$
$$\rho ::= a \mid \varphi? \mid \rho_1 + \rho_2 \mid \rho_1; \rho_2 \mid \rho* \mid \rho_1 \| \rho_2$$

where $p \in$ PROP and $a \in$ ACT. Note that we write $\rho*$ instead of ρ^* since we reserve the superscript for the semantics (see Table 1). Expressions φ are PDL$^{\|}$ *formulas* forming PDL$^{\|}$, and expressions ρ are *programs* forming *Prog*.

We write PDL for the fragment of PDL$^{\|}$-formulas that do not contain the shuffle symbol $\|$, that will correspond to the logic originally defined by [6].

One interprets PDL$^{\|}$ formulas over *transition systems (TS)* over PROP and ACT. A transition system is a structure $\mathcal{S} = (S, \lambda, \theta)$ composed of a finite nonempty set S of *states*, a labeling of states by propositions $\lambda :$ PROP $\to 2^S$, and a transition relation $\theta \subseteq S \times$ ACT $\times S$.

In the following, we let $\tau :$ ACT $\to 2^{S \times S}$ be defined by

$$\tau(a) := \{(s, s') \in S \times S \mid (s, a, s') \in \theta\} \tag{1}$$

In a TS $\mathcal{S} = (S, \lambda, \theta)$, a *pure transition* is a pair (s, s'), with $(s, a, s') \in \theta$ for some $a \in$ ACT, and a *stuttering* transition is a pair (s, s') where $s \in S$. A transition of $\mathcal{S}$ is either a pure or a stuttering transition, and we gather both kinds of transtions in the set $\mathcal{T}_\mathcal{S}$, or simply $\mathcal{T}$ as it will be clear from the context; we use notations $\tau, \tau', \ldots$ for its typical elements of $\mathcal{T}$. We then denote by $\mathcal{T}^*$ the set of finite sequences over $\mathcal{T}$. Given a sequence of transitions $\tau_1 \ldots \tau_n \in \mathcal{T}^*$, we let the first state of this sequence be $\texttt{fst}(\tau_1 \ldots \tau_n) = s$ if $\tau_1 = (s, s')$ for some s', and we define the last state, $\texttt{lst}(\tau_1 \ldots \tau_n)$, in a similar way.

There are particular sequences of transitions $\tau_1\tau_2 \ldots \tau_n$ called *paths*, where $\texttt{lst}(\tau_i) = \texttt{fst}(\tau_{i+1})$, for every $1 \leq i < n$. Those correspond to the classic notion of path in a transition system. We say that a path $\tau_1\tau_2 \ldots \tau_n$ *starts from state* s whenever $\texttt{fst}(\tau_1) = s$, and we write $\Pi^s_\mathcal{S} \subseteq \mathcal{T}^*$ for the set of paths that start from state s.

Semantics of PDL$^{\|}$ is obtained by extending by simultaneous induction the labeling function λ and relation τ $\lambda_\mathcal{S} :$ PDL$^{\|} \to 2^S$ and $\tau_\mathcal{S} : Prog \to 2^{\mathcal{T}^*}$ as described in Table 1, where we have dropped the $\mathcal{S}$ subscript for readability.

Table 1. PDL semantics.

τ(Programs)	λ(Formulas)
$\tau(\rho_1 + \rho_2) = \tau(\rho_1) \cup \tau(\rho_2)$	$\lambda(\top) = S$
$\tau(\rho_1; \rho_2) = \tau(\rho_1) \circ \tau(\rho_2)$	$\lambda(p)$ is defined from $\mathcal{S} = (S, \lambda, \theta)$
$\tau(\rho*) = (\tau(\rho))^*$	$\lambda(\neg\varphi) = S \setminus \lambda(\varphi)$
$\tau(\rho_1 \parallel \rho_2) = \tau(\rho_1) \parallel \tau(\rho_2)$	$\lambda(\varphi_1 \wedge \varphi_2) = \lambda(\varphi_1) \cap \lambda(\varphi_2)$
$\tau(\varphi?) = \{(s,s) \in S^2 \mid s \in \lambda(\varphi)\}$	$\lambda(\langle\rho\rangle\varphi) = \{s \in S \mid \exists \pi \in \varPi\mathcal{S}_s \cap \tau(\rho), \mathtt{1st}(\pi) \in \lambda(\varphi)\}$

Write $\mathcal{S}, s \models \varphi$ if $s \in \lambda(\varphi)$, and say that a program ρ is *executable* in $\mathcal{S}$ from s if $\mathcal{S}, s \models \langle\rho\rangle\top$, namely, according to the semantics above, whenever there exists a path π in $\mathcal{S}$ starting from s and whose end state satisfies φ.

The set of PDL$^{\parallel}$ formulas can be viewed as terms over the ranked alphabet

$$\Sigma := \text{Act} \cup \text{Prop} \cup \{\top, *, ?, \neg, ;, +, \wedge, \langle\ \rangle, \parallel\}$$

with the following arities of symbols: symbols in Act and Prop and the symbol $\top$ have arity 0, the symbols $*$, $?$, and $\neg$ have arity 1, and the remaining ones have arity 2, but we keep standard notations: term $\langle\ \rangle(t,t')$ is rather written $\langle t \rangle t'$, and so on.

For example, terms that describe arbitrary PDL$^{\parallel}$ formulas can be generated by the regular tree-grammar $G = (\{\mathfrak{S}, \mathbf{P}\}, \mathfrak{S}, \Sigma, \mathcal{R})$ where the axiom symbol $\mathfrak{S}$ derives the PDL$^{\parallel}$ formulas and the nonterminal $\mathbf{P}$ derives the PDL$^{\parallel}$ programs, and where the set $\mathcal{R}$ of rules is:

$$\begin{array}{ll} \mathfrak{S} \rightarrow p & \mathbf{Y} \rightarrow a \\ \mathfrak{S} \rightarrow \top & \mathbf{Y} \rightarrow \mathfrak{S}? \\ \mathfrak{S} \rightarrow \neg\mathfrak{S} & \mathbf{Y} \rightarrow \mathbf{Y} + \mathbf{Y} \\ \mathfrak{S} \rightarrow \mathfrak{S} \wedge \mathfrak{S} & \mathbf{Y} \rightarrow \mathbf{Y}; \mathbf{Y} \\ \mathfrak{S} \rightarrow \langle\mathbf{Y}\rangle\mathfrak{S} & \mathbf{Y} \rightarrow \mathbf{Y}* \\ & \mathbf{Y} \rightarrow \mathbf{Y} \parallel \mathbf{Y} \end{array}$$

3 The Formula Synthesis Problem

In [17], the Formula Synthesis Problem (for PDL$^{\parallel}$), written SYNTHPDL$^{\parallel}$, is formally defined as follows.

Definition 1 (SYNTHPDL$^{\parallel}$)**.**

> **Input:** *a finite set* Prop *of atomic propositions, a finite set* Act *of atomic programs, a TS* $\mathcal{S}$ *(over* Prop *and* Act*), a state* $s \in S$*, a regular tree-grammar G generating a set of* PDL$^{\parallel}$ *formulas.*
> **Output:** *does there exists φ generated by G such that $\mathcal{S}, s \models \varphi$? (and find one if there is.)*

As model checking against PDL$^{\parallel}$ is PSPACE-complete [10, Proposition 2.10], while model checking against PDL is P-complete [11,14], we obtain lower bounds for SYNTHPDL$^{\parallel}$ and SYNTHPDL respectively. Actually, these lower bounds are misleading since, as we will see, those problems are much more difficult to solve.

Theorem 1 ([17]). SYNTHPDL$^{\parallel}$ *is recursively enumerable (RE) complete (and therefore undecidable), even when restricted to sets of formulas expressing executability of programs with shuffle and concatenation only.*

The problem is clearly RE with an enumerator that produces all formulas generated by the grammar and that filters out those that do not hold in the distinguished transition system state - recall that model checking against PDL$^{\parallel}$ is decidable [10]. For the RE-hardness, the proof technique exhibits a reduction from the emptiness of the intersection of two context-free languages specified by context-free grammars (CFGs), known to be RE-hard [2]: the problem is, given two CFGs H_1 and H_2 over some alphabet Γ in Chomsky-normal-form with the same terminal alphabet Γ, decide if $L(H_1) \cap L(H_2) \neq \emptyset$.

Remark that the proposed reduction to show the undecidability of SYNTHPDL$^{\parallel}$ (Theorem 1) yields, first, a recursive input grammar that generates formulas with (a single) shuffle operator, and second, that the obtained input TS has infinitely many paths. Consequently, natural subproblems arise from considering restrictions on the input of the general SYNTHPDL$^{\parallel}$ problem that turn out to become decidable.

We let SYNTHPDL be the subproblem of SYNTHPDL$^{\parallel}$ where the input regular tree-grammar generates a set of PDL formulas (no shuffle in programs), that are terms over $\Sigma[\text{PDL}]$.

Theorem 2 ([17]). SYNTHPDL *is* EXPTIME-*complete.*

We give a proof sketch of Theorem 2: Given an instance of SYNTHPDL, say $(\mathcal{S}, s, G)$, it turns out that one can effectively compute a tree-grammar $G_{\mathcal{S},s}$, of size exponential in the size of $\mathcal{S}$, that generates exactly the PDL-formulas φ such $\mathcal{S}, s \models \varphi$. It is then sufficient to check emptiness of the grammar resulting from the "intersection" of $G_{\mathcal{S},s}$ with the input grammar G. This can be done in polynomial time [4].

EXPTIME-hardness of SYNTHPDL comes from a reduction from the non universality problem for *(bottom up) nondeterministic finite tree automata*, known to be EXPTIME-complete [15, p. 91].

Another remarkable decidable case is the one of *non-recursive grammars*. A regular tree grammar is *non-recursive* [16] whenever every nonterminal occurs at most once in every derivation; as a consequence a non-recursive grammar generates a finite language. Remark that the grammar constructed in the undecidability proof for SYNTHPDL$^{\parallel}$ (Theorem 1) is recursive as soon as one of the two word-grammars is.

Theorem 3 ([17]). SYNTHPDL$^{\parallel}$ *with non-recursive grammars is in* EXPSPACE.

The proof idea is as follows. Non-recursiveness of the grammar entails an exponential number (in the size of the grammar G) of derivable formulas, where each formula is of size $O(2^{poly(|G|)})$. One can then design the following algorithm: Generate all those formulas, and model check against each of them (recall that model checking against PDL$^{\parallel}$ is PSPACE[10]). As each formula of size $O(2^{poly(|G|)})$, the algorithm requires an exponential amount of memory.

The upper bound decreases significantly when only PDL formulas are allowed, based on an alternating algorithm that runs in polynomial time.

Theorem 4 ([17]). SYNTHPDL *with non-recursive grammars is in* PSPACE.

Due to lack of space we only mention two extra results in [17] regarding linear PDL grammars, namely grammars where each rule has at most one nonterminal in its right-hand side. In this case, the size of the formulas generated by the grammar G is in $O(|G|)$. As a result, SYNTHPDL restricted to linear grammars is PSPACE-complete, and it can be shown that SYNTHPDL restricted to non-recursive linear grammars is NP-complete.

We now consider a restriction of the input TS of SYNTHPDL$^{\parallel}$.

A DAG^2-like TS $\mathcal{S} = (S, \lambda, \theta)$ essentially has finitely many paths. This allows us to retrieve decidability for arbitrary PDL$^{\parallel}$ tree-grammars.

Theorem 5 ([17]). SYNTHPDL$^{\parallel}$ *restricted to DAG-like transition systems is in* 2-EXPTIME.

Indeed, when the TS is a DAG, the semantics of programs is bounded, i.e., the image, namely $\Pi_{\mathcal{S}} \cap \tau(\rho)$, of every program ρ is finite. One can then effectively build a regular tree-grammar that generates the PDL$^{\parallel}$ formulas that hold in some fixed state and thus design an algorithm similar to the one for PDL grammars (Theorem 2). The obtained algorithm is in time doubly-exponential in the size of the $\mathcal{S}$ and polynomial-time in the size of G. It is still open whether the problem is 2-EXPTIME-hard, and if any other kind of restriction on the input TS also yields decidability.

We now turn to two applications of the Formula Synthesis Problem for the logic PDL$^{\parallel}$.

4 Security: Attack Tree Automated Generation

A notable work that concerns hierarchical decomposition is the *attack tree synthesis problem* in Security. Attack trees (AT) are acknowledged graphical models for reasoning in Risk Analysis [20]. Their automated generation (i.e., synthesis) is considered as a Holy Grail by security experts so as to avoid their tedious and error-prone manual design. This synthesis problem has gained interest in the formal method community in the past decade [7,18,21].

---

[2] Directed Acyclic Graph.

In a nutshell, ATs are finite terms over alphabet[3]

$$\Sigma := \{a\}_{a \in \text{ACT}} \cup \{\text{OR}, \text{SAND}, \text{AND}\}$$

where each a ranging over ACT (a finite set) has rank 0 and all other have rank 2 – actually, AND is rather an unranked symbol, and we will get back to this point later.

We give here a simple example of an AT with the intuitive semantics behind such a tree. Consider the AT depicted in Fig. 1, where constants a, b, c, d are atomic actions an attacker can perform.

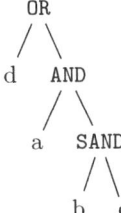

Fig. 1. An attack tree.

The AT of Fig. 1 describes several ways of attacking by executing either of the following sequence of actions among $\{d, abc, bac, bca\}$ – we call this set the semantics of the AT and its elements are the *attack plans of the AT*. In the world of formal languages (over alphabet ACT), the semantics of leaf ATs a is the singleton with the single-letter word a, and the semantics for compound ATs are compositional by interpreting function symbols OR, SAND and AND as union, concatenation and shuffle, respectively.

On this basis, one can read an AT as a hierarchical description of how to achieve the attack of the system: in order to achieve attack specified by term OR(d, AND(a, SAND(b, c))), one can perform (basic) attack/action d or achieve the subattack specified by the subtree AND(a, SAND(b, c)). Achieving the latter attack AND(a, SAND(b, c)) consists in performing (basic) attack/action a and in achieving subattack SAND(b, c). In turn, the latter subattack SAND(b, c) is achieved by performing (basic) attack b and sequentially performing (basic) attack c.

In practice, experts do build ATs manually – which is very tedious and error-prone – by reasoning on the system they want to analyze. Once the AT is built, it provides the recipe of how to attack the system, and helps the experts to select efficient (e.g., low cost) countermeasures, as well as to identify attacker profiles in e.g., forensic [19].

An alternative way, which nowadays attracts growing interest, is to automatically generate the recipe from observations of the system. This is the problem

---

[3] We use here notations considered standard in the community, as explained in [22].

formally addressed in [18]: given some observation, like bac of the attacked system, is there an AT that provides the recipe? As is, the question is not very well-formed as one can always define a trivial AT, namely $\text{SAND}(b, \text{SAND}(a, c))$ for the case of observation bac. Such an AT does not provide any additional information. Rather, one would like to recognize/identify from this sequence of actions some know-hows or an attacker profile that knowledge databases[4] gather along time.

In essence, knowledge databases are informal (textual) sets of grammatical rules for $\text{PDL}^{\|}$-like programs, since function symbols OR, SAND and AND occurring in ATs correspond to the $+$, $;$ and $\|$ program operators in $\text{SYNTHPDL}^{\|}$, respectively. In particular, the authors of [18] introduce the notion of *library*, a mere tree-grammar G as in $\text{SYNTHPDL}^{\|}$, that becomes an input to the (formal) AT synthesis problem – it should be clear that libraries arising in practice are non-recursive.

Definition 2 (Attack Tree Synthesis Problem).

> **Input:** *a finite sequence α of actions ranging over a finite set* ACT *and a* library, *namely a regular non-recursive tree-grammar over ranked alphabet* $\{a\}_{a \in \text{ACT}} \cup \{OR, SAND, AND\}$.
> **Output:** *does there exists a term t generated by G such that sequence α is an attack plan of t? (and find one if there is.)*

As noticed in [17], it is not hard to see that the attack tree synthesis problem of [18] polynomially reduces to $\text{SYNTHPDL}^{\|}$:

- the new tree-grammar G' has all the rules of G and an extra one of the form $\mathfrak{S}' \to \langle \mathfrak{S} \rangle \text{end}$, where $\mathfrak{S}'$ is the axiom symbol of grammar G, while $\mathfrak{S}$ is the axiom symbol of input grammar G, and
- the action sequence α becomes a single-path TS whose transitions are labeled by the actions of α in order, with an extra last transition labeled by $\#$ and the last reached state is the only state labeled by atomic proposition end.

By Theorem 4, one gets the following.

Corollary 1. *The Attack Tree Synthesis Problem is in* EXPSPACE. *Additionally, if the library is a linear (non-recursive) grammar, the Attack Tree Synthesis Problem is in* NP.

The EXPSPACE upper bound of the corollary above is in contrast with the NP-complete complexity obtained in [18]. This is due to the strong restriction on the input TS that is single-path. Besides, the proof technique in [18] is utterly different from the approach in [17]: it amounts to a slight adaptation of the classic CYK dynamic-programming syntactic analysis algorithm[5] to take the shuffle

---

[4] Such as the *Common Vulnerabilities and Exposures* (CVE), a dictionary of publicly disclosed cybersecurity vulnerabilities and exposures https://cve.mitre.org/cve/.
[5] [13].

operation into account. Because of this very peculiar approach, the authors of [18] left open the problem where, instead of considering a single input sequence of actions α, one considered a finite amount of those. Then [17] claim that the reduction of this generalized Attack Tree Synthesis Problem to SYNTHPDL$^{\|}$ is feasible, but actually one needs a stronger machinery: instead of using term-grammars, or equivalently tree-automata, one has to rely on tree-automata with constraints between brothers (see their definition in [4, Section 4.3])[6].

5 Hierarchical Planning

We here turn to the automated planning community (whose foundations can be found in [9]), with a focus on a branch of this field that concerns *Hierarchical Task Networks* (HTN) Planning [8].

An HTN involves task-network methods that decompose compound tasks into a combination of subtasks, just as attack trees decompose attack into sub-attacks.

The *HTN* Planning Problem (also known as the *task problem*) consists in synthesising (if any) a plan in an input planning domain that is obtained by applying methods from the input HTN.

While HTN can be very complex objects involving quantitative constraints and yielding a zoo of complexity results [5], when restricting to the propositional setting of HTN where all methods are linearly ordered task decomposition (L) or unordered task decomposition (U)[7], task-network methods can be looked at as tree-grammars, and a similar approach to the one for attack trees above applies.

[[17]] "The task problem on HTNP^{L+U} instances, i.e., propositional instance only consisting of L and U methods, polynomially reduces to SYNTHPDL$^{\|}$."

A careful inspection of the proof of the undecidability of the task problem for arbitrary HTN instances (see in [5]) reveals that undecidability already holds for the HTNP^{L+U} instances. Besides, undecidability result for SYNTHPDL$^{\|}$ (Theorem 1) becomes a mere corollary and the proof in [17, Theorem 1] borrows ideas from [5]. While it is possible to obtain complexity results for HTNP^{L+U} Planning Problem from the ones of SYNTHPDL$^{\|}$, it should be noticed that in the HTN setting, the planning domain is symbolic, while the input transition system in SYNTHPDL$^{\|}$ is explicit. Also, since task-network methods in general HTN Planning problems may rely on arbitrary partial orders between subtasks, the path through SYNTHPDL$^{\|}$ is hopeless for general HTN problems: indeed, PDL$^{\|}$ program based on sequential and shuffle compositions cannot specify arbitrary partial orders.

---

[6] I am currently writing an article on that topic.
[7] Also known as *TOTD* and *UTD* respectively in the literature, see [8, Section 2.2.1].

References

1. Abrahamson, K.R.: Decidability and expressiveness of logics of processes. University of Washington (1980)
2. Aho, A.V., Hopcroft, J.E., Ullman, J.D.: The Design and Analysis of Computer Algorithms. Addison-Wesley (1974)
3. Clarke, E.M., Henzinger, T.A., Veith, H., Bloem, R., et al.: Handbook of model checking, vol. 10. Springer (2018)
4. Comon, H., et al.: Tree automata techniques and applications (2008). https://inria.hal.science/hal-03367725
5. Erol, K., Hendler, J.A., Nau, D.S.: Complexity results for HTN planning. Ann. Math. Artif. Intell. **18**(1), 69–93 (1996)
6. Fischer, M.J., Ladner, R.E.: Propositional dynamic logic of regular programs. J. Comput. Syst. Sci. **18**(2), 194–211 (1979). https://doi.org/10.1016/0022-0000(79)90046-1
7. Gadyatskaya, O., Jhawar, R., Mauw, S., Trujillo-Rasua, R., Willemse, T.A.C.: Refinement-aware generation of attack trees. In: STM, LNCS, vol. 10547, pp. 164–179. Springer (2017)
8. Georgievski, I., Aiello, M.: HTN planning: overview, comparison, and beyond. Artif. Intell. **222**, 124–156 (2015)
9. Ghallab, M., Nau, D., Traverso, P.: Automated Planning: theory and practice. Elsevier (2004)
10. Göller, S.: Computational complexity of propositional dynamic logics. Ph.D. thesis, University of Leipzig (2008). https://d-nb.info/99245168X
11. Harel, D., Kozen, D., Tiuryn, J.: Dynamic Logic. MIT Press (2000)
12. Harel, D., Kozen, D., Tiuryn, J.: Dynamic logic. ACM SIGACT News **32**(1), 66–69 (2001)
13. Kasami, T.: An efficient recognition and syntax-analysis algorithm for context-free languages. Coordinated Science Laboratory Report no. R-257 (1966)
14. Lange, M.: Model checking propositional dynamic logic with all extras. J. Appl. Log. **4**(1), 39–49 (2006)
15. Löding, C.: Basics on tree automata. In: D'Souza, D., Shankar, P. (eds.) Modern Applications of Automata Theory, IISc Research Monographs Series, vol. 2, pp. 79–110. World Scientific (2012). https://doi.org/10.1142/9789814271059_0003
16. Nederhof, M.J., Satta, G.: Parsing non-recursive cfgs. In: Proceedings of the 40th Annual Meeting of the Association for Computational Linguistics, pp. 112–119 (2002)
17. Pinchinat, S., Rubin, S., Schwarzentruber, F.: Formula synthesis in propositional dynamic logic with shuffle. In: Thirty-Sixth AAAI Conference on Artificial Intelligence, AAAI 2022, Thirty-Fourth Conference on Innovative Applications of Artificial Intelligence, IAAI 2022, The Twelveth Symposium on Educational Advances in Artificial Intelligence, EAAI 2022 Virtual Event, February 22 - March 1, 2022, pp. 9902–9909. AAAI Press (2022). https://doi.org/10.1609/AAAI.V36I9.21227
18. Pinchinat, S., Schwarzentruber, F., Lê Cong, S.: Library-based attack tree synthesis. In: Eades III, H., Gadyatskaya, O. (eds.) GraMSec 2020. LNCS, vol. 12419, pp. 24–44. Springer, Cham (2020). https://doi.org/10.1007/978-3-030-62230-5_2
19. Poolsapassit, N., Ray, I.: Investigating computer attacks using attack trees. In: IFIP International Conference on Digital Forensics, pp. 331–343. Springer (2007)
20. Schneier, B.: Attack trees. Dr. Dobb's J. **24**(12), 21–29 (1999)

21. Vigo, R., Nielson, F., Nielson, H.R.: Automated generation of attack trees. In: 2014 IEEE 27th Computer Security Foundations Symposium, pp. 337–350. IEEE (2014)
22. Wideł, W., Audinot, M., Fila, B., Pinchinat, S.: Beyond 2014: formal methods for attack tree-based security modeling. ACM Comput. Surv. (CSUR) **52**(4), 1–36 (2019)

NSOP1 as a Dividing Line

Nicholas Ramsey(✉)

University of Notre Dame, Notre Dame, IN 46556, USA
sramsey5@nd.edu

Abstract. The strong order property 1 (SOP_1) is a model-theoretic tree property introduced by Džamonja and Shelah. We give an exposition of a family of results around $NSOP_1$ theories, explaining why the $NSOP_1/SOP_1$ divide is a meaningful dividing line among first-order theories. This involves summarizing work on Kim-independence and on the interpretability order, as well as aspects of Mutchnik's work on SOP_1 and SOP_2.

The aim of this article is to explain a certain circle of ideas around the model-theoretic tree property called SOP_1 and situate it in a broader context. Our main thesis is that the property SOP_1 is a *dividing line*, that is, a fundamental marker of complexity in the space of first-order theories. The $NSOP_1$ theories, i.e. those theories that do not have the strong order property 1, may be seen as tame: there is a well-developed structure theory for them, centered on a notion of independence called *Kim-independence*. Additionally, there is a compelling non-structure theory for the theories that do have SOP_1. They are known to be maximally complicated in Keisler's order and, assuming some set theory, to characterize maximality in the related ⊴*-order, both of which compare the complexity of first-order theories in terms of the difficulty of producing saturated models of them.

The system of results concerning the $SOP_1/NSOP_1$ divide grew out of an earlier theory for simple theories, which has been a core topic of research in both pure and applied model theory for the past few decades. The $NSOP_1$ theories properly contain the simple theories and encompass many new examples of intrinsic interest, allowing for the analysis of several concrete mathematical structures coming from algebra and combinatorics. These theorems, together with complementary theorems about theories without the tree property of the second kind, indicate that, the theory for simple theories should be regarded as a first instance of what ought to be seen as an *area* within model theory in its own right: the field of 'model-theoretic tree properties.' This corner of model-theoretic research has its own unifying set of questions, as well as a coherent set of methods and techniques. We hope to show that the theory of the $SOP_1/NSOP_1$ divide can provide an illuminating case study for what these methods look like and what shape the general theory of model-theoretic tree properties might take in future research.

This article is structured as follows. In Sect. 1, we describe Shelah's 'dividing lines' philosophy and explain the key results that establish Shelah's notion of simplicity as a dividing line. The simple theories are defined to be the theories that do not have the *tree property*, which is the simplest instance of a model-theoretic tree property. We then turn, in Sect. 2, to a discussion of the property SOP_1, and describe the core developments in the structure theory for $NSOP_1$ theories, emphasizing the analogy with simple theories. In Sect. 3, we explain how TP_1, the *tree property of the first kind*, makes it harder to produce saturated models. This is made precise in terms of the notion of the interpretability order $\trianglelefteq^*$. Finally, we explain how the structure theory and nonstructure theory can be brought together in a single dividing line, by way of Mutchnik's theorem, which shows that SOP_1 and TP_1 coincide. We conclude with a discussion of what this network of results has to say about the future development of 'model-theoretic tree properties' as an area of model-theoretic research.

In what follows, we follow standard model-theoretic conventions and notation. We write T for a complete first-order theory and $\mathbb{M} \models T$ denotes a monster model of T. When we assert the existence of tuples, sequences, models, etc. we will always assume they are small— that is, of size smaller than the degree of saturation of $\mathbb{M}$— and contained in $\mathbb{M}$ without explicitly remarking so. We will not notationally distinguish between elements and tuples.

1 Background and Context

1.1 Classification Theory and the Dividing Lines Philosophy

Shelah's book *Classification Theory* [28] was dedicated to understanding when models of a complete first-order theory can be characterized by cardinal invariants. The simplest examples in which a characterization like this is possible are the strongly minimal theories, like the theory of $\mathbb{Q}$-vector spaces or the theory of algebraically closed fields of characteristic zero. The models of these theories can be classified by a single number— the dimension in the case of a vector space, the transcendence degree over $\mathbb{Q}$ in the case of ACF_0. Towards understanding when this kind of classification is possible, Shelah introduced an abstract notion of independence, called non-forking independence, and showed that, in certain contexts, this notion of independence allows one to identify geometries associated to a given theory, whose dimensions determine a model. In the other direction, Shelah proved that, outside of these contexts, theories have too many models for classification by cardinal invariants to be possible. For example, the theories RCF of real closed fields and DLO of dense linear orders without endpoints have 2^κ models of size κ for every uncountable cardinal κ. Shelah found that, at least for countable complete first-order theories, there is a small list of properties that separate out the theories with classifiable models from those with too many models to classify.

In delineating the distinction between the contexts where classification is possible and those where it is not, Shelah discovered *dividing lines*. The dividing lines are properties of a first-order theory that mark significant jumps in complexity. Dividing lines split first-order theories into tame and wild: to be a dividing line, one should be able to prove *structure theorems* for theories without the property, and *non-structure theorems* for the theories that have the property. Dividing lines should also admit both *inside* characterizations in terms of syntactic properties of a formula or family of formulas, and *outside* characterizations in terms of models. These features of dividing lines suggest that these properties cut mathematical reality at its joints; they are meaningful ways of attributing simplicity or complexity to first-order theories. Although the first dividing lines were isolated in connection to understanding classification by cardinal invariants, the idea that, in trying to understand first-order theories, one should search for dividing lines has emerged as an important methodological principle within model theory. This has shaped model-theoretic research well beyond the motivating classification problem.

1.2 Simplicity Theory

In this subsection, we will give a brief sketch of the main results concerning simple theories. Simplicity is an important dividing line and it is the most basic example of a 'model-theoretic tree property.' The theory developed to understand it serves as the paradigm for understanding model-theoretic tree properties more generally. Simplicity theory is now a large and well-established part of model theory, so we will only scratch the surface but more details can be found in [1, 13, 21, 32].

The following definition is due to Shelah [29].

Definition 1. *A formula $\varphi(x;y)$ is said to have the* tree property *if there is a collection of tuples $(a_\eta)_{\eta \in \omega^{<\omega}}$ and some $k < \omega$ satisfying the following conditions:*

1. *(Paths are consistent) For all $\eta \in \omega^\omega$, $\{\varphi(x; a_{\eta|i}) : i < \omega\}$ is consistent.*
2. *(Siblings are k-inconsistent) For all $\eta \in \omega^{<\omega}$, $\{\varphi(x; a_{\eta^\frown \langle i \rangle}) : i < \omega\}$ is k-inconsistent.*

The theory T is called simple *if no formula has the tree property modulo T.*

Here we give an example of a simple theory and an example of a theory with the tree property:

Example 1. We will give an example and a non-example:

1. The theory T_{rg} of the random graph is the theory in the language $L = \{R\}$ consisting of a single binary relation which asserts that R is a symmetric irreflexive relation, there are infinitely many vertices, and, for all $n \geq 1$,

$$(\forall v_1, \ldots, v_{2n})(\exists w) \bigwedge_{i=1}^{n} R(w, v_i) \wedge \neg R(w, v_{n+i}).$$

The theory T_{rg} is the theory of the Fraïssé limit of all finite graphs and thus is $\aleph_0$-categorical and eliminates quantifiers. Using quantifier elimination, one can show that if $\langle a_i : i \in \omega \rangle$ is a sequence and $\varphi(x; y)$ is an L-formula such that $\{\varphi(x; a_i) : i < \omega\}$ is k-inconsistent, then $\varphi(x; a_i)$ must entail that x is

equal to some element of a_i not contained in the tuples $a_{i'}$ for $i' \neq i$. Thus one cannot find a tree of tuples as in Definition 1 of height greater than the number of variables in the tuple x, and T_{rg} is simple.
2. DLO denotes the theory of dense linear orders without endpoints in the language $L = \{\leq\}$. In $(\mathbb{Q}, <)$, one can choose pairs $(a_\eta, b_\eta)_{\eta \in \omega^{<\omega}}$ such that $a_\eta < b_\eta$ for all η and, moreover, $\eta \trianglelefteq \nu$ implies that the interval (a_ν, b_ν) is contained in the interval (a_η, b_η) and $i \neq j$ implies that the intervals $(a_{\eta^\frown \langle i \rangle}, b_{\eta^\frown \langle i \rangle})$ and $(a_{\eta^\frown \langle j \rangle}, b_{\eta^\frown \langle j \rangle})$ are disjoint. Now consider the formula $\varphi(x; y, z) = (y < x < z)$. Note that if $\eta \in \omega^\omega$, then $\{\varphi(x; a_{\eta|i}, b_{\eta|i}) : i < \omega\}$ is consistent, since this asserts that x is contained in a nested sequence of intervals. On the other hand, if $\eta \perp \nu$, then $\{\varphi(x; a_\eta, b_\eta), \varphi(x; a_\nu, b_\nu)\}$ is inconsistent, since this asserts x is contained in two disjoint intervals. Therefore, DLO is not a simple theory.

Simplicity theory has its origins in three distinct lines of inquiry within model theory. Two of these grew out of an analysis of concrete examples. Situating pseudo-finite fields in a broader model-theoretic context, Hrushovski initiated the study of bounded PAC substructures of a strongly minimal set in [14]. He developed a great deal of simplicity in finite rank in order to apply it to understand definability in these structures. In a similar vein, a certain amount of simplicity theory was developed by Cherlin and Hrushovski [3] in the course of the analysis of smoothly approximable homogeneous structures. In these cases, the development of the theory was necessitated by the treatment of concrete structures or classes of structures. From a totally different direction, Shelah [29] introduced the saturation spectrum of a theory, and proved that the simple theories can be characterized set-theoretically in terms of the behavior of the saturation spectrum. These parallel developments were eventually consolidated into a theory by Kim's dissertation [18,19] and work by Kim and Pillay [22], who showed that the theory of non-forking independence could unify and explain the fundamental phenomena at play in understanding simple theories.

Definition 2. *Suppose C is a set of parameters.*

1. *We say a formula $\varphi(x; b)$ divides over C if there is a C-indiscernible sequence $\langle b_i : i < \omega \rangle$ such that $\{\varphi(x; b_i) : i < \omega\}$ is inconsistent.*
2. *We say a formula $\varphi(x; b)$ forks over C if there are finitely many formulas $(\psi_i(x; d_i))_{i<k}$, each of which divides over C, such that $\varphi(x; b) \vdash \bigvee_{i<k} \psi_i(x; d_i)$.*
3. *We stipulate that the tuple a is non-forking independent from the tuple b over C, written $a \downarrow^f_C b$ if $\mathrm{tp}(a/Cb)$ contains no formula that forks over C.*

One thinks of a set defined by a formula that divides as being small: it has a family of conjugates which are almost disjoint. The sets defined by the forking formulas, then, are exactly the sets in the ideal generated by the dividing formulas. The notion of independence $a \downarrow^f_C b$ means informally that you cannot use the parameters bC to trap the tuple a into a small set, where *small* is determined relative to the base C.

Example 2. In the theory of $\mathbb{Q}$-vector spaces, $A \underset{C}{\overset{f}{\downarrow}} B$ means the vectors in the set A are linearly independent from those in B over the vectors in C. In the theory ACF_0, it means that A and B are algebraically independent over C.

For a somewhat more transparent example, we can consider forking in the theory of an equivalence relation:

Example 3. Let the theory T denote the theory in the language $L = \{E\}$, consisting of a single binary relation, which asserts that E is an equivalence relation with infinitely many classes, all of which are infinite. The theory T is also the theory of the Fraïssé limit of finite equivalence relations and is therefore $\aleph_0$-categorical and has elimination of quantifiers.

Choose an element b in any model of T. Then the formula $E(x, b)$ divides over $\emptyset$. To see this, choose an indiscernible sequence $\langle b_i : i < \omega \rangle$ with $b_0 = b$ such that b_i and b_j are in different E-classes, whenever $i \neq j$. Then $\{E(x, b_i) : i < \omega\}$ is 2-inconsistent.

More generally, we have $A \underset{C}{\overset{f}{\downarrow}} B$ if and only if $A \cap B \subseteq C$ and the intersection of the sets of equivalence classes represented by elements of A and by elements of B are contained in the set of equivalence classes represented by elements of C.

In order to understand dividing in simple theories, a crucial technical tool is the notion of a Morley sequence, which is an indiscernible sequence in which each term is independent from its predecessors.

Definition 3. *Suppose C is a set of parameters. We say that $I = \langle b_i : i < \omega \rangle$ is a Morley sequence over C if I is a C-indiscernible sequence and $b_i \underset{C}{\overset{f}{\downarrow}} b_{<i}$ for all $i < \omega$.*

Example 4. Let T denote the theory of an equivalence relation with infinitely many classes, all of which are infinite, from Example 3. Letting $M \models T$ be the unique countable model, pick an element $b \in M$. By quantifier elimination, one sees that, up to an automorphism of M, there are exactly 3 indiscernible sequences $I = \langle b_i : i < \omega \rangle$ starting with b:

1. The constant sequence, in which $b_i = b$ for all i.
2. A non-constant equivalent sequence, in which $M \models E(b_i, b_j)$ and $b_i \neq b_j$ for all $i \neq j$.
3. An inequivalent sequence, in which $M \models \neg E(b_i, b_j)$ for all $i \neq j$.

It is clear that the constant sequence in (1) is not Morley (over $\emptyset$) since b_1 realizes the formula $x = b_0$, and $x = b_0$ divides (witnessed by any non-constant indiscernible sequence starting with b). Moreover, the non-constant equivalent sequence is not Morley since b_1 realizes the formula $E(x, b_0)$ and we saw in Example 3 that $E(x, b)$ divides. Thus (3) is the Morley sequence.

Notice that when we showed in Example 3 that the formula $E(x, b)$ divides over $\emptyset$, we took the witnessing indiscernible sequence to be the Morley sequence in (3). Notice that if I is of type (1) or (2), then $\{E(x; b_i) : i < \omega\}$ is consistent, hence this sequence will not witness the dividing.

The most important fact about Morley sequences in simple theories is *Kim's Lemma* [19] which says that whether or not a formula divides is entirely determined by its behavior with respect to Morley sequences.

Lemma 1 (Kim's Lemma). *Suppose T is a simple theory. The following are equivalent for a formula $\varphi(x;b)$ and set of parameters C:*

1. $\varphi(x;b)$ divides over C.
2. For every Morley sequence $(b_i)_{i<\omega}$ over C with $b_0 = b$, $\{\varphi(x;b_i) : i < \omega\}$ is inconsistent.
3. There is a Morley sequence $(b_i)_{i<\omega}$ over C with $b_0 = b$ such that $\{\varphi(x;b_i) : i < \omega\}$ is inconsistent.

Kim's Lemma allows one to show that forking and dividing are the same, that forking is symmetric, and has other desirable properties [19,22]:

Theorem 1. *Assume T is a simple theory. Then non-forking independence satisfies the following properties:*

1. (Extension) If $a \downarrow_C^f b$ then, for all d, there is some $a' \equiv_{Cb} a$ such that $a' \downarrow_C^f bd$.
2. (Symmetry) $a \downarrow_C^f b$ if and only if $b \downarrow_C^f a$.
3. (Transitivity) If $B \subseteq C \subseteq D$, then $a \downarrow_B^f C$ and $a \downarrow_C^f D$ imply $a \downarrow_B^f D$.
4. (Independence Theorem over models) If $M \models T$, $a \equiv_M a'$, $a \downarrow_M^f B$, $a' \downarrow_M^f C$, and $B \downarrow_M^f C$, then there is some a_* with $a_* \equiv_{MB} a$, $a_* \equiv_{MC} a'$, and $a_* \downarrow_M^f BC$.

Later it was observed that Kim's Lemma, as well as symmetry and transitivity of non-forking individually characterize the simple theories [20]. In other words, simplicity is equivalent to the good behavior of non-forking independence. This theme was elaborated by a theorem known as the Kim-Pillay theorem [22]:

Theorem 2. *Assume there is an $\mathrm{Aut}(\mathbb{M})$-invariant ternary relation $\downarrow$ on small subsets of $\mathbb{M}$ satisfying the following properties:*

1. (Finite Character) $A \downarrow_C B$ if and only if $a \downarrow_C b$ for all finite tuples a and b from A and B, respectively.
2. (Monotonicity) $aa' \downarrow_C bb' \implies a \downarrow_C b$.
3. (Full existence) For all a, b and C, there is $a' \equiv_C a$ such that $a' \downarrow_C b$.
4. (Base Monotonicity) $a \downarrow_D bc \implies a \downarrow_{Dc} b$.
5. (Symmetry) $a \downarrow_C b \iff b \downarrow_C a$.
6. (Transitivity) If $a \downarrow_D c$ and $a \downarrow_{Dc} b$ then $a \downarrow_D cb$.
7. (Local character) For all A, there exist some cardinal κ such that, for any C, there is some $B \subseteq C$ with $|B| < \kappa$ and $A \downarrow_B C$.
8. (Independence Theorem over Models) If $M \models T$, A_0 and A_1 have the same type over M, $A_0 \downarrow_M B$, $A_1 \downarrow_M C$, and $B \downarrow_M C$, then there is $A_* \models \mathrm{tp}(A_0/MB) \cup \mathrm{tp}(A_1/MC)$ such that $A_* \downarrow_M BC$.

Then T is simple and $\downarrow = \downarrow^f$.

The Kim-Pillay theorem has two important consequences, one practical, the other theoretical. The practical utility of the theorem is that it gives a very convenient method for establishing that a concrete theory is simple, namely by checking that there is a notion of independence satisfying the required properties. Since, in many algebraic or combinatorial settings, there are already natural notions of independence around, this is usually the most straightforward and tractable method, rather than explicitly ruling out the tree property. Secondly, the theoretical significance of the theorem is that the non-forking independence is canonical: any independence relation satisfying the conditions of the theorem must agree with non-forking. Since the non-forking relation is determined by these properties in a simple theory, in most cases, one can avoid using the explicit definition of dividing and instead work axiomatically with these properties, which together are referred to as the 'forking calculus.' This allows for a great deal of compression in arguments in simplicity theory.

The good behavior of non-forking independence is the basis for the structure theory for simple theories. In the other direction, Shelah used the saturation spectrum to give an outside characterization of the simple theories, proving a nonstructure theorem for non-simple theories which showed that, consistently, it is harder to produce saturated models for them. Recall that a model M is called κ-saturated if, whenever p is a type over some set $A \subseteq M$ with $|A| < \kappa$, then p is realized in M. We say M is saturated if it is $|M|$-saturated. Saturated models are important for an array of model-theoretic arguments and they provide models in which everything that can happen does. The ability to construct saturated models is often a reflection of how complicated a theory can be. But it is also quite sensitive to set theory: saturated models of any theory at all can be obtained from sufficiently large instances of the generalized continuum hypothesis. In order to give a set-theoretic characterization of the simple theories, Shelah defined the saturation spectrum of a theory T to be the class of pairs of cardinals $\lambda \geq \kappa$ where $\lambda = \lambda^{|T|}$ and every model of T of size λ has a κ-saturated elementary extension of size κ. If T is stable and $\lambda = \lambda^{|T|}$, then T has a saturated model of size λ, so (λ, κ) is in the saturation spectrum of a stable theory T for all $\kappa \leq \lambda$. Similarly, regardless of whether T is stable or not, if additionally we assume $\lambda = \lambda^{<\kappa}$, then every model of T of size λ can be shown to have a κ-saturated elementary extension of size λ, by an easy union-of-chains argument. Shelah [29] showed via a forcing argument that it is consistent with ZFC that a theory T is simple if and only if there is a pair (λ, κ) in its saturation spectrum which does *not* satisfy $\lambda = \lambda^{<\kappa}$. In this model of set theory, then, non-simple theories will only have (λ, κ) in their saturation spectrum when they satisfy the obvious cardinal arithmetic constraint, and not otherwise. This gives an 'outside' characterization of the simple theories, purely in terms of the models.

1.3 A Dichotomy Theorem

We will now explain how to see the tree property, which characterizes the non-simple theories, as one instance of a family of model-theoretic tree properties.

Definition 4. We define two properties of a formula $\varphi(x;y)$.

1. We say $\varphi(x;y)$ has the tree property of the first kind (TP_1) if there is a collection of tuples $(a_\eta)_{\eta \in \omega^{<\omega}}$ satisfying the following conditions:
   (a) *(Paths are consistent)* For all $\eta \in \omega^\omega$, $\{\varphi(x; a_{\eta|i}) : i < \omega\}$ is consistent.
   (b) *(Incomparables are inconsistent)* For all $\eta \perp \nu$ in $\omega^{<\omega}$, $\{\varphi(x; a_\eta), \varphi(x; a_\nu)\}$ is inconsistent.
2. We say $\varphi(x;y)$ has the tree property of the second kind (TP_2) if there is an array of tuples $(a_{i,j})_{i,j<\omega}$ and $k < \omega$ satisfying the following conditions:
   (a) *(Paths are consistent)* For all functions $f : \omega \to \omega$, $\{\varphi(x; a_{i,f(i)}) : i < \omega\}$ is consistent.
   (b) *(Rows are k-inconsistent)* For all $i < \omega$, $\{\varphi(x; a_{i,j}) : j < \omega\}$ is k-inconsistent.

We say a theory is NTP_1 or NTP_2 if no formula has TP_1 or TP_2 modulo T, respectively.

The above properties were introduced by Shelah in [28][1] One may regard TP_1 and TP_2 as two extremal forms that the tree property might take: in TP_1, everything which is not forced to be consistent in the definition of the tree property is required to be inconsistent, in TP_2, everything which is not forced to be inconsistent in the definition of the tree property is required to be consistent. Remarkably, Shelah showed that in any theory with the tree property, one of these two extremal forms must be present [28]:

Theorem 3. *A theory T has the tree property if and only if it has TP_1 or TP_2.*

This theorem naturally suggests two possible ways to generalize from simplicity theory. One could weaken the hypothesis that T is simple to either T is NTP_1 or NTP_2 and try to hang on to as much of the structure theory while treating new examples. While this brief survey will focus on $NSOP_1$ and NTP_1, we should mention that a major pillar of model-theoretic tree properties is the analysis of NTP_2 theories, carried out in [4,5,33].

Example 5. The theory T_{sf} is the theory of a generic family of selector functions. It is a theory in the language L with sorts O and F for 'objects' and 'functions'. On O, there is a binary relation E and there is a binary function eval $: F \times O \to O$. The universal theory T_{sf}^- asserts that E is an equivalence relation on O and each $f \in F$ codes a selector function for E in the following sense:

- For all $f \in F$ and $a \in O$, $E(\text{eval}(f,a),a)$, that is, $\text{eval}(f,a)$ is an element in the same E-equivalence class as a.
- For all $f \in F$ and $a,b \in O$, if $E(a,b)$, then $\text{eval}(f,a) = \text{eval}(f,b)$, that is, $\text{eval}(f,-)$ takes on the same value on each E-class.

---

[1] The *properties* were introduced by Shelah but these *names* for them were coined by Kim in [20]. TP_1 and TP_2 were referred to by the rather unwieldy $\kappa_{\text{sct}}(T) = \infty$ and $\kappa_{\text{inp}}(T) = \infty$ in their first appearance.

The theory T_{sf} is the theory of the model companion of T_{sf}^{-}, which is also the theory of the Fraïssé limit of the (uniformly locally finite) Fraïssé class of finite models of T_{sf}^{-}, and is therefore $\aleph_0$-categorical with elimination of quantifiers. In a model of T_{sf}, we can pick an array $(a_{i,j})_{i,j<\omega}$ of distinct elements in the O sort such that $E(a_{i,j}, a_{i',j'})$ if and only if $i = i'$. If $f : \omega \to \omega$ is a function, then, since T_{sf} is the model companion of T_{sf}^{-} and the elements $(a_{i,f(i)})_{i<\omega}$ are pairwise E-inequivalent, we have that $\{\text{eval}(x, a_{i,f(i)}) = a_{i,f(i)} : i < \omega\}$ is consistent. However, for each i, we know that $\{\text{eval}(x, a_{i,j}) : j < \omega\}$ is 2-inconsistent, since the elements $(a_{i,j})_{j<\omega}$ are E-equivalent and distinct. This shows that T_{sf} has TP_2 (and it can be shown that it does not have TP_1).

Example 6. We have already shown that the theory DLO has TP_1 in Example 1(2). It can be shown that DLO does not have TP_2.

2 Kim-Independence and NSOP$_1$

2.1 The SOP$_n$ Hierarchy

Now we turn to NSOP$_1$ theories. We start with the definition.

Definition 5. *We say the formula $\varphi(x;y)$ has SOP$_1$ (the strong order property 1) if there is a tree $(a_\eta)_{\eta \in 2^{<\omega}}$ satisfying the following conditions:*

1. *(Paths are consistent) For all $\eta \in 2^\omega$, $\{\varphi(x; a_{\eta|i}) : i < \omega\}$ is consistent.*
2. *(The SOP$_1$ inconsistency condition) For all $\eta, \nu \in 2^{<\omega}$, if $\nu^\frown\langle 0\rangle \trianglelefteq \eta$, then $\{\varphi(x; a_\eta), \varphi(x; a_{\nu^\frown\langle 1\rangle})\}$ is inconsistent.*

We say a theory T is NSOP$_1$ if no formula has SOP$_1$ modulo T.

SOP$_1$ is an undeniably strange-looking property, one that is, moreover, very challenging to motivate. We can situate SOP$_1$ by seeing it as the bottom level of an infinite hierarchy of model-theoretic properties. The property SOP$_2$ is another name for the property TP$_1$[2]. The other properties are defined as follows:

Definition 6. *Let $n \geq 3$. We say a theory T has SOP$_n$ (the strong order property n) if there is a formula $\varphi(x, y)$ and a sequence $I = \langle a_i : i < \omega \rangle$ satisfying the following conditions:*

1. *$\varphi(x, y)$ witnesses the order property with respect to I — that is, $\mathbb{M} \models \varphi(a_i, a_j)$ if and only if $i < j$.*
2. *$\varphi(x, y)$ omits loops of size n — that is, the set of formulas $\{\varphi(x_i, x_{i+1}) : i < n-1\} \cup \{\varphi(x_{n-1}, x_0)\}$ is inconsistent.*

---

[2] Technically, the definition of TP$_1$ refers to a tree indexed by $\omega^{<\omega}$ and the definition of SOP$_2$ refers to a tree indexed by $2^{<\omega}$, but these are equivalent, even formula-by-formula, so we will ignore this difference.

These properties give increasingly better approximations to strict order as n increases (notice that in an infinite linear order $<$, no loops of any size are possible because of transitivity). The SOP_n hierarchy was introduced by Shelah in [30] for $n \geq 3$ and then, in subsequent work [11], SOP_1 and SOP_2 were added later by Džamonja and Shelah. They are defined very differently but they sit in the correct place in the chain of implications. It is clear from the definitions that SOP_2 implies SOP_1. In the definition of SOP_1, the index tree $2^{<\omega}$ could instead be replaced by $\omega^{<\omega}$. In this case, the consistency condition on paths would be identical, but the inconsistency condition would state that if $\eta, \nu \in \omega^{<\omega}$ and $\nu^\frown \langle i \rangle \trianglelefteq \eta$ for some $i < \omega$, then $\{\varphi(x; a_\eta), \varphi(x; a_{\nu^\frown \langle j \rangle})\}$ is inconsistent for all $j > i$. In this formulation, it is clear that children of a common node are 2-inconsistent, and thus SOP_1 implies 2-TP. This entails that the class of $NSOP_1$ theories contains the simple theories and is contained in the class of $NSOP_2 = NTP_1$ theories. It is a nice exercise to show that SOP_3 implies SOP_2. Thus we have the implications

$$\ldots \implies SOP_{n+1} \implies SOP_n \implies \ldots \implies SOP_1 \implies TP$$

Moreover, we know that the implications $SOP_1 \implies TP$ and, for $n \geq 3$, $SOP_{n+1} \implies SOP_n$ are strict, but it was left open by Džamonja and Shelah whether or not SOP_1, SOP_2, and SOP_3 might all be the same.

2.2 Independence at a Generic Scale

A structure theory for $NSOP_1$ theories was assembled out of ingredients remarkably similar to those that led to simplicity theory. Prior to any positive work on $NSOP_1$ theories, there were examples in the literature— the ω-free PAC fields studied by Chatzidakis [2], and generic vector spaces with an alternating bilinear form analyzed in Granger's dissertation [12]— of non-simple theories in which came equipped with a good notion of independence. These structures are closely related to the bounded PAC fields and smoothly approximable structures that motivated early work in simple theories. It turns out that these structures and the *ad hoc* notions of independence attached to them can be unified and explained by a notion of independence suited for the broader context of $NSOP_1$ theories, called *Kim-independence*.

Kim-independence corresponds to independence at a 'generic scale.' The idea behind this slogan comes from Kim's Lemma in simple theories, which says that dividing is always witnessed by Morley sequences. Since Kim's Lemma fails in every non-simple theory, we know that not every instance of dividing will be witnessed by Morley sequences. We think of the dividing that *is* seen by Morley sequences as living at the 'generic scale'. By zooming in on only what happens generically, we can recover a great deal of the good behavior of non-forking even for non-simple $NSOP_1$ theories.

Definition 7. *Suppose C is a set of parameters.*

1. We say a formula $\varphi(x;b)$ Kim-divides over C if there is a Morley sequence[3] $\langle b_i : i < \omega \rangle$ over C such that $\{\varphi(x;b_i) : i < \omega\}$ is inconsistent.
2. We say a formula $\varphi(x;b)$ Kim-forks over C if there are finitely many formulas $(\psi_i(x;d_i))_{i<k}$, each of which Kim-divides over C, such that $\varphi(x;b) \vdash \bigvee_{i<k} \psi_i(x;d_i)$.
3. We stipulate that the tuple a is Kim-independent from the tuple b over C, written $a \downarrow_C^K b$ if $\operatorname{tp}(a/Cb)$ contains no formula that Kim-forks over C.

As mentioned above, it follows immediately from Kim's Lemma that if T is simple, then $\downarrow^K = \downarrow^f$. However, outside of simple theories, Kim's Lemma fails so they necessarily differ. To see how this can happen in an example, we return to the theory of a generic family of selector functions from Example 5.

Example 7. In a model $M \models T_{\mathrm{sf}}$ pick an element $b \in O(M)$ and consider the formula $\operatorname{eval}(x,b) = b$. Taking $J = \langle c_i : i < \omega \rangle$ to be a non-constant indiscernible sequence in M with $c_0 = b$ and $M \models E(c_i, c_j)$ for all i, j, we see that $\{\operatorname{eval}(x, c_i) : i < \omega\}$ is 2-inconsistent. Thus J witnesses that $\operatorname{eval}(x,b) = b$ divides.

As the structure on O is just that of an equivalence relation with infinitely many classes, all of which are infinite, we know from Example 4 that if $I = \langle b_i : i < \omega \rangle$ is a Morley sequence then $M \models \neg E(b_i, b_j)$ for $i \neq j$. But this entails that $\{\operatorname{eval}(x, b_i) = b_i : i < \omega\}$ is consistent, so this sequence does not witness that $\operatorname{eval}(x, b) = b$ divides and, consequently, shows that $\operatorname{eval}(x, b) = b$ does not Kim-divide.

The following is the main theorem concerning Kim-independence in NSOP_1 theories:

Theorem 4. *The following are equivalent:*

1. T is NSOP_1.
2. $\downarrow^K$ satisfies Kim's Lemma for Kim-dividing over models: if $M \models T$, then $\varphi(x;b)$ Kim-divides over M if and only if $\{\varphi(x;b_i) : i < \omega\}$ is inconsistent for all Morley sequences $I = \langle b_i; i < \omega \rangle$ over M with $b_0 = b$.
3. $\downarrow^K$ is symmetric over models: for all $M \models T$, $a \downarrow_M^K b \iff b \downarrow_M^K a$.
4. $\downarrow^K$ is transitive over models: if $M \preceq N \models T$, then $a \downarrow_M^K N$ and $a \downarrow_N^K b$ imply $a \downarrow_M^K Nb$.
5. $\downarrow^K$ satisfies local character on a club: if $M \models T$, then given any finite tuple a, the set $\{N \prec M : a \downarrow_N^K M\}$ is a closed and unbounded subset of $[M]^{|T|}$.[4]

---

[3] For technical reasons, this differs from the definition given in [15], but it is proved there that, over models, it is always equivalent to it.

[4] Here, $[M]^{|T|}$ denotes the set of subsets of M of cardinality $|T|$. To say that $X \subseteq [M]^{|T|}$ is closed and unbounded means that it is closed under increasing union of sequences of length $|T|$ and that every subset of M of cardinality $\leq |T|$ is a subset of some element of X.

6. $\downarrow^K$ satisfies the Independence Theorem over Models: If $M \models T$, $a \equiv_M a'$, $a \downarrow^K_M B$, $a' \downarrow^K_M C$, and $B \downarrow^K_M C$, then there is some a_* with $a_* \equiv_{MB} a$, $a_* \equiv_{MC} a'$, and $a_* \downarrow^K_M BC$.

This theorem is really several theorems, drawn from a series of papers [7, 15–17]. This collection of results shows that the theory of Kim-independence provides a robust generalization of non-forking independence in simple theories, and also provides many conceptually distinct characterizations of the class of NSOP_1 theories in terms of the independence relation $\downarrow^K$. This establishes that the rather bizarre-looking property NSOP_1 may be characterized in more familar terms. The reason why we need to restrict to independence 'over models' in the statement of the above theorem is the fact that, in general, we only know that there are Morley sequences in types over models. Assuming that there are Morley sequences in types over every set (an assumption known as the 'existence axiom') suffices to extend the theory of Kim-independence to all sets [6,10], but it is also known that the existence axiom need not always hold in NSOP_1 theories [27].

As in simple theories, there is also an analogue of the Kim-Pillay theorem:

Theorem 5. *Assume there is an* $\mathrm{Aut}(\mathbb{M})$-*invariant ternary relation* $\downarrow$ *on small subsets of* $\mathbb{M}$ *satisfying the following properties when the base is a model:*

1. *(Finite Character)* $A \downarrow_M B$ *if and only if* $a \downarrow_M b$ *for all finite tuples* a *and* b *from* A *and* B, *respectively.*
2. *(Monotonicity)* $aa' \downarrow_M bb' \implies a \downarrow_M b$.
3. *(Full existence) For all* a, b *and* $M \models T$, *there is* $a' \equiv_M a$ *such that* $a' \downarrow_M b$.
4. *(Symmetry)* $a \downarrow_M b \iff b \downarrow_M a$.
5. *(Transitivity) If* $M \preceq N$, $a \downarrow_M N$ *and* $a \downarrow_N b$ *then* $a \downarrow_M Nb$.
6. *(Local character on a club) For all* $M \models T$ *and finite tuples* a, *there is a club* $\mathcal{C}$ *of elementary submodels of* M *of cardinality* $|T|$ *such that* $N \in \mathcal{C}$ *implies* $a \downarrow_N M$.
7. *(Independence Theorem over Models) If* $M \models T$, A_0 *and* A_1 *have the same type over* M, $A_0 \downarrow_M B$, $A_1 \downarrow_M C$, *and* $B \downarrow_M C$, *then there is* $A_* \models \mathrm{tp}(A_0/MB) \cup \mathrm{tp}(A_1/MC)$ *such that* $A_* \downarrow_M BC$.

Then T *is* NSOP_1 *and* $\downarrow = \downarrow^K$.

This theorem also is really several theorems. A subset of these axioms suffice to establish that a theory is NSOP_1 and a theorem stating this implication, prior to the definition of $\downarrow^K$, appeared in [7]. Subsequent versions, with the characterization of $\downarrow^K$, were included in [15,16] with a slightly different axiom set, and some simplifications were pointed out in [9]. This theorem provides the most common tool for showing that a given theory is NSOP_1 and likewise establishes the canonicity of Kim-independence.

3 TP$_1$ and $\trianglelefteq^*$-Maximality

Now we will discuss the nonstructure theory for theories with TP$_1$. We will view the result in this section as an analogue of Shelah's characterization of simple theories in terms of the saturation spectrum. Here too, the TP$_1$ theories will be characterized by the difficulty of producing saturated models, but instead of the saturation spectrum, this will be cashed out in terms of a relative measure of saturation difficulty, called the *interpretability order* $\trianglelefteq^*$.

We begin by recalling the notion of *interpretation* of one theory in another. Suppose T and $\tilde{T}$ are theories in the languages L and $\tilde{L}$, respectively. For simplicity, assume that the signature of L consists entirely of relation symbols. Denote this set of relation symbols by $\mathcal{R}_L$.

Suppose $\overline{\varphi} = \langle \varphi_R(x_R) : R \in \mathcal{R}_L \cup \{=\}\rangle$ is a sequence of $\tilde{L}$-formulas, where for some k, if $R \in \mathcal{R}_L \cup \{=\}$ is a n-ary, then x_R is a tuple of nk many variables. Assume that, in every model $\tilde{M} \models \tilde{T}$, $\varphi_=$ defines an equivalence relation on the set $D = \{a \in \tilde{M}^k : \tilde{M} \models (\exists x)\varphi_=(x,a)\}$ and that each formula $\varphi_R(x_R)$ is invariant under this relation, i.e. given a tuple of k tuples in $\tilde{M}$, replacing any one of these k-tuples with a coordinate-wise equivalent one does not change the truth value of φ_R evaluated at these tuples.

We define the model $\tilde{M}^{[\overline{\varphi}]}$ to be the L-structure whose underlying set is the set of equivalence classes of k-tuples of $\tilde{M}$ under the equivalence relation $\varphi_=$. Given an n-ary relation $R \in \mathcal{R}_L$, we interpret R on $\tilde{M}^{[\overline{\varphi}]}$ to be the set of n-tuples of equivalence classes $([a_0],\ldots,[a_{n-1}])$ such that $\tilde{M} \models \varphi_R(a_0,\ldots,a_{n-1})$ (which is well-defined by our assumption on φ_R). If, whenever $\tilde{M} \models \tilde{T}$, $\tilde{M}^{[\overline{\varphi}]} \models T$, then we say that $\overline{\varphi}$ is an *interpretation* of T in $\tilde{T}$. If there is some $\overline{\varphi}$ which is an interpretation of T in $\tilde{T}$, we say that T is *interpretable* in $\tilde{T}$.

The *interpretability order*[5] uses interpretations to compare the difficulty of producing saturated models of a theory. It is a kind of reduction notion for saturation problems. Basically, it says that a theory is more complicated than another theory if, using the resources of a first-order theory that interprets both, saturation from the complicated theory can be transferred to the less complicated one.

Definition 8. *Suppose T_0 and T_1 are complete first-order theories. We declare $T_0 \trianglelefteq^* T_1$ if and only if, for all sufficiently large cardinals λ, there is some theory $\tilde{T}$, with $|\tilde{T}| \leq |T_0| + |T_1|$, that interprets T_0 and T_1 via respective interpretations $\overline{\varphi}_0$ and $\overline{\varphi}_1$ with the following property: whenever $M \models \tilde{T}$ and $M^{[\overline{\varphi}_1]} \models T_1$ is λ-saturated, then $M^{[\overline{\varphi}_0]} \models T_0$ is λ-saturated.*

We have already seen via the saturation spectrum that the difficulty of producing saturated models of a theory can be a meaningful yardstick by which to measure its complexity. This order is closely related to (and indeed refines)

---

[5] It seems that the order received this name only recently in [25]. Before, model theorists just called it the 'triangle star order' due to the notation being $\trianglelefteq^*$. The new name is an obvious improvement.

Keisler's order, which compares the complexity of theories in terms of the difficulty of producing saturated ultrapowers of them and there is a great deal of overlap in the methods used to study both Keisler's order and the interpretability order. The interpretability order was first defined by Shelah in [30] but only later considered in detail by Džamonja and Shelah in [11]. Both orders have been the object of intensive study by Malliaris and Shelah, among others, in more recent work (see, e.g., [23,25]). Surveying this work would take us considerably outside the scope of the present article, but we will highlight one particular result of interest:

Theorem 6. *Assume GCH. A first-order theory T is maximal in the interpretability order $\trianglelefteq^*$ if and only if T has TP_1.*

When we say T is maximal in $\trianglelefteq^*$, we mean that $T' \trianglelefteq^* T$ for every other complete theory T'. This is a very strong nonstructure theorem for theories with TP_1 (or, equivalently, SOP_2), showing both that they are maximally complicated and that they may be given an outside characterization in terms of saturation. This theorem is really three theorems:

1. Džamonja and Shelah [11] prove, assuming GCH, that if T is maximal in $\trianglelefteq^*$, then it has a variant of TP_1 called SOP_2". This follows via a direct and technical argument, finding SOP_2" from the assumption that T is greater than or equal to the theory of trees in $\trianglelefteq^*$.
2. Shelah and Usvyatsov [31] show that a theory has TP_1 if and only if it has SOP_2", thus, at the level of theories, these properties are equivalent. Shelah and Usvyatsov manage to collapse these two properties by manipulating indiscernible trees.
3. Malliaris and Shelah [24] prove, without any additional set-theoretic assumptions, that any theory with TP_1 is maximal in $\trianglelefteq^*$. This was deduced as a consequence of their celebrated 'cofinality spectrum problem' technology [23], which also allowed them to prove that TP_1 implies maximality in Keisler's order and to show the equality $\mathfrak{p} = \mathfrak{t}$ of cardinal invariants of the continuum.

4 Mutchnik's Theorem

The results mentioned in Sects. 3 and 2 summarize our state of knowledge up to about 2022. There is a successful structure theory for $NSOP_1$ theories, built upon a theory of independence. There is also a successful non-structure theory for theories with SOP_2 (or, equivalently, TP_1), showing that they are maximal in Keisler's order and the interpretability order. When they defined the property SOP_1 (in the late 90s), Džamonja and Shelah observed that the $NSOP_1$ theories are contained in the $NSOP_2$ theories, but left open whether or not that containment is proper. Taken together, these facts led to an awkward situation. For either of the properties SOP_1 or SOP_2 to be a dividing line, there should also be a non-structure theory for theories with SOP_1 or a structure theory for $NSOP_2$ ($=NTP_1$) theories. If there exist $NSOP_2$ theories with SOP_1, then these theories would inhabit a gray zone in which neither the existing theories for $NSOP_1$ nor SOP_2 apply. This situation lent urgency to Džamonja and Shelah's question. Fortunately, this issue was definitively resolved by a breakthrough result due to Mutchnik [26]:

Theorem 7. *A theory has SOP_1 if and only if it has SOP_2.*

In other words, the $SOP_1/NSOP_1$ divide really marks a dividing line; it splits the universe of first-order theories into tame and wild, with interesting theorems on both sides of the division. Although describing Mutchnik's proof in any detail is well beyond the scope of this article, from a high-level vantage point, the argument proceeds via the following three steps:

1. **Prove a generalization of the Broom Lemma**— The Broom Lemma was a key technical ingredient in Chernikov and Kaplan's proof that forking and dividing coincide over models in NTP_2 theories [5]. In that setting, the Broom Lemma entailed the existence of a strengthened notion of Morley sequence, called a strict invariant Morley sequence, and Chernikov and Kaplan proved a variant of Kim's Lemma, where the hypothesis of simplicity was relaxed to T being an NTP_2 theory, at the cost of replacing Morley sequences with strict invariant Morley sequences, in its statement. Likewise, Mutchnik found an analogue of the Broom Lemma that allowed him to construct special kinds of Morley sequences— Morley sequences in 'canonical coheirs'— and used these to prove a variant of Kim's Lemma in the setting of $NSOP_2$ theories.
2. **Develop a theory of Conant-independence**— Recall that Kim-dividing is the kind of dividing that happens along *some* Morley sequence. Conant-independence, defined by Mutchnik, replaces the existential quantifier with a universal one, defining the Conant-dividing of a formula to mean that the formula divides along *every* Morley sequence[6]. In the context of $NSOP_1$ theories, Kim's Lemma for Kim-dividing says precisely that Kim-dividing and Conant-dividing are the same. In any theory with SOP_1, however, the notions necessarily disagree. Under the *a priori* weaker hypothesis of $NSOP_2$, Mutchnik develops a theory of Conant-independence, denoted $\downarrow^{K^*}$ that replicates many of the nice properties of Kim-independence in $NSOP_1$ theories, proving, for example, that $\downarrow^{K^*}$ is symmetric and satisfies a version of the independence theorem over models.
3. **Adapt free amalgamation arguments to upgrade from SOP_1 to SOP_3**— With the theory of Conant-independence in hand, the final step is derive a final contradiction by adapting an argument due to Conant, concerning free amalgamation. In [8], Conant introduced axioms, similar to those of the Kim-Pillay theorem, that axiomatized what it means for an independence relation $A \downarrow_C B$ to mean that A and B are 'freely amalgamated' over C. Theories with such a notion of independence are called *free amalgamation theories*. This notion is often quite different from forking-independence;

---

[6] This should really say *every Morley sequence in a coheir*, but we will gloss over this technical detail for simplicity's sake.

a familiar example comes from the random graph, with $A \downarrow_C B$ interpreted to mean that A and B are disjoint over C and there are no edges between $A \backslash C$ and $B \backslash C$. Conant proved that every modular free amalgamation theory is either simple or has SOP_3. Mutchnik abstracted the combinatorial essence of this argument and was able to replace free amalgamation with Conant-independence to prove that an $NSOP_2$ theory is either $NSOP_1$ or has SOP_3. This immediately implies that SOP_1 and SOP_2 coincide, since SOP_3 implies SOP_2 and thus any $NSOP_2$ theory with SOP_1 also has SOP_2, hence cannot actually be $NSOP_2$.

The proof of this theorem upended the general expectation within model theory that a proof that $SOP_1 = SOP_2$, if true, would look like a clever manipulation of indiscernible trees, rather than the outcome of a deep theory of independence relations. The argument, instead, suggests that logically defined notions of independence do more than systematize and unify phenomena observed in concrete examples, but they also can be brought to bear on hard questions of theoretical significance. Moreover, the proof shows the remarkable cross-pollination between the study of different model-theoretic tree properties, as in the connection between NTP_2 theories and $NSOP_2$ theories via the broom lemma, or with the abstract study of independence relations, via the connection to Conant's axioms for free amalgamation. This suggests that the area of model-theoretic tree properties has a certain unity and coherence.

Acknowledgments. The author was supported by NSF grant DMS-2246992. He also would like to thank Maryanthe Malliaris for helpful conversations regarding a closely related project.

References

1. Casanovas, E.: Simple theories and hyperimaginaries. Number 39. Cambridge University Press (2011)
2. Chatzidakis, Z.: Properties of forking in ω-free pseudo-algebraically closed fields. J. Symbolic Logic **67**(3), 957–996 (2002)
3. Cherlin, G.L., Hrushovski, E.: Finite Structures with Few Types. Princeton University Press (2003)
4. Chernikov, A.: Theories without the tree property of the second kind. Ann. Pure Appl. Logic **165**(2), 695–723 (2014)
5. Chernikov, A., Kaplan, I.: Forking and dividing in ntp2 theories. J. Symbolic Logic **77**(1), 1–20 (2012)
6. Chernikov, A., Kim, B., Ramsey, N.: Transitivity, lowness, and ranks in nsop1 theories. The Journal of Symbolic Logic, pp. 1–27 (2020)
7. Chernikov, A., Ramsey, N.: On model-theoretic tree properties. J. Math. Log. **16**(02), 1650009 (2016)
8. Conant, G.: An axiomatic approach to free amalgamation. J. Symbolic Logic **82**(2), 648–671 (2017)
9. Dobrowolski, J., Kamsma, M.: Kim-independence in positive logic. arXiv:2105.07788 (2021)

10. Dobrowolski, J., Kim, B., Ramsey, N.: Independence over arbitrary sets in nsop1 theories. Ann. Pure Appl. Logic **173**(2), 103058 (2022)
11. Džamonja, M., Shelah, S.: On $< |^*$-maximality. Ann. Pure Appl. Logic **125**(1–3), 119–158 (2004)
12. Granger, N.: Stability, Simplicity and the Model Theory of Bilinear Forms. The University of Manchester (United Kingdom) (1999)
13. Grossberg, R., Iovino, J., Lessmann, O.: A primer of simple theories. Arch. Math. Logic **41**(6), 541–580 (2002)
14. Hrushovski, E., et al.: Pseudo-finite fields and related structures. Model Theor. Appl. **11**, 151–212 (1991)
15. Kaplan, I., Ramsey, N.: On kim-independence. J. Eur. Math. Soc. **22**(5), 1423–1474 (2020)
16. Kaplan, I., Ramsey, N.: Transitivity of kim-independence. Adv. Math. **379**, 107573 (2021)
17. Kaplan, I., Ramsey, N., Shelah, S.: Local character of kim-independence. Proc. Amer. Math. Soc. **147**, 1719–1732 (2019)
18. Kim, B.: Simple First Order Theories. University of Notre Dame (1996)
19. Kim, B.: Forking in simple unstable theories. J. Lond. Math. Soc. **57**(2), 257–267 (1998)
20. Kim, B.: Simplicity, and stability in there. J. Symbolic Logic **66**(2), 822–836 (2001)
21. Kim, B.: Simplicity Theory, vol. 53. OUP Oxford (2013)
22. Kim, B., Pillay, A.: Simple theories. vol. 88, pp. 149–164 (1997). Joint AILA-KGS Model Theory Meeting (Florence, 1995)
23. Malliaris, M., Shelah, S.: Cofinality spectrum theorems in model theory, set theory, and general topology. J. Am. Math. Soc. **29**(1), 237–297 (2016)
24. Malliaris, M., Shelah, S.: Model-theoretic applications of cofinality spectrum problems. Israel J. Math. **220**(2), 947–1014 (2017). https://doi.org/10.1007/s11856-017-1526-7
25. Malliaris, M., Shelah, S.: A new look at interpretability and saturation. Ann. Pure Appl. Logic **170**(5), 642–671 (2019)
26. Mutchnik, S.: On nsop$_2$ theories. *arXiv preprint*arXiv:2206.08512 (2022)
27. Mutchnik, S.: An nsop$_1$ theory without the existence axiom. *arXiv preprint*arXiv:2407.13082 (2024)
28. Shelah, S.: Classification theory and the number of nonisomorphic models. Studies in Logic and the Foundations of Mathematics, vol. 92, 2nd edn. North-Holland Publishing Co., Amsterdam (1990)
29. Shelah, S.: Simple unstable theories. Ann. Math. Logic **19**(3) (1980)
30. Shelah, S.: Toward classifying unstable theories. *arXiv preprint*arXiv:math/9508205, (1995)
31. Shelah, S., Usvyatsov, A.: More on sop1 and sop2. Ann. Pure Appl. Logic **155**(1), 16–31 (2008)
32. Wagner, F.O.: Simple theories, vol. 260. Springer (2000)
33. Yaacov, I.B., Chernikov, A.: An independence theorem for ntp2 theories. J. Symbolic Logic **79**(1), 135–153 (2014)

Contributed Papers

Asynchronous Transition System Games for Two Processes and Their Analysis

Bharat Adsul[⬤] and Nehul Jain[✉][⬤]

Indian Institute of Technology Bombay, Mumbai, India
{adsul,nehul}@cse.iitb.ac.in

Abstract. We propose and investigate a new model of a distributed game played on a non-deterministic asynchronous transition system over two processes. This game is played between an environment and a distributed team of the two processes where each process has only partial information of the ongoing play – namely complete information up to the *last* synchronization and only its own *local* evolution since this last synchronization. The key algorithmic decision problem, for a given winning objective, is the existence of a distributed co-operative winning strategy for the team to meet that objective. We address this question for global safety, local reachability and global/simultaneous reachability objectives. We carry out a thorough analysis of these games and present natural fixpoint based algorithms for solving them. This allows us to construct distributed winning strategies with an explicit distributed finite-memory in the form of key past information, and also yields near optimal decision procedures. Specifically, our analysis shows that the decision problems for global safety and local reachability objectives are NP-complete. We also establish that the decision problem for global reachability objective is PSPACE-hard and provide an NEXPTIME algorithm for the same.

Keywords: Concurrency · Synthesis · Game-theoretic models

1 Introduction

The design and analysis of concurrent programs/protocols has always been a challenging task. Given the importance of concurrent/distributed applications and their pervasiveness, it is only natural to ask if a class of such protocols can be synthesized automatically starting from its specification. The initial work [16] proposed a setting for distributed synthesis and showed that the problem of realizing a given specification is, in general, undecidable. In this setting, the processes have access to only *local* information and are allowed to communicate only a limited amount of information. In contrast, recent works [1,1,5–11,13,15] have allowed the processes to access the entire *causal past*.

The work [8] introduced the distributed controller synthesis problem for series-parallel systems and showed that controlled reachability is decidable. In [13], the authors consider the case of connectedly communicating processes and show that the related distributed synthesis problems are decidable.

Now we briefly describe two recent lines of work, namely, Petri games and asynchronous games. Petri games [7] are played on Petri nets where tokens are considered as players. Places of the Petri net are divided into system places and environment places. In a distributed strategy, a token residing in a system place decides, based on its causal history, which of the transitions to allow next. Several interesting classes of Petri games [6,7] are shown to be decidable. However it has been recently shown [5] that Petri games with global winning conditions are undecidable, even with two system places and one environment place.

Asynchronous/control games from [9] are played on *deterministic* asynchronous automata due to Zielonka [18]. This setting involves a fixed number of processes. The set of actions in which these processes participate is partitioned into controllable and uncontrollable actions. In a distributed/control strategy each process decides, based on its causal past, the next set of control actions that it wants to allow. In [9,15], it has been shown asynchronous games over acyclic architectures are decidable. The work [10] introduces decomposable games which include games from [8,9,13], and proves their decidability. A recent surprising result from [11] shows that asynchronous games with local reachability objectives is undecidable even with six processes.

In this work, we propose a new model of distributed games, called ATS games, which are played on *non-deterministic* asynchronous transition systems. Such a game is played between an environment and a distributed team/system comprising of a fixed number of processes. The focus of this work is on a detailed analysis of the case of *two* processes. More precisely, we undertake the study of two process ATS games for a variety of *distributed* winning objectives such as global safety, local reachability and global reachability. We provide tight bounds on the computational complexity of the problem of deciding the existence of a distributed winning strategy for these objectives. Another important contribution of this work is the identification of *distributed memory* requirements to meet these objectives. It turns out that one process ATS games are essentially the well studied standard full-information two-player graph games [3,12] which play a key role in solutions of sequential synthesis problems. From this viewpoint, our results for two process ATS games may be seen as non-trivial generalizations of the corresponding results for one process ATS games.

The rest of the paper is organized as follows. After developing preliminaries in Sect. 2, we introduce our model (specialized to two processes) and formulate the key notions related to strategies in Sect. 3. The next two sections are devoted to the analysis of global safety, local reachability and global reachability winning conditions. In Sect. 6, we conclude by pointing out some connections with asynchronous control games.

2 Traces and Asynchronous Transition Systems

There is a rich and well-established theory [3,12,17] of standard full-information turn-based two-player games played on a game arena with a specified start-position and an appropriate winning condition. In this work, we introduce a new model of a distributed game, which is played between an environment and a distributed team consisting of two processes. Our game is played between these two entities on a non-deterministic **a**synchronous **t**ransition **s**ystem (ATS) over the two processes. We have two types of transitions in the ATS. On a purely local action of a process, only the local states of that process are updated. On the other hand, on a shared/joint/synchronizing action, the entire global state (a pair of local-states for each process) is jointly updated.

A play of our game consists of an ongoing *interaction* between the environment and the two processes. The key departure from the standard setting is that our game is a *partial* information game. During the play, each process has only partial knowledge of the ongoing interaction – namely its *causal past*.

We now develop some notation. Throughout this paper, we have a distributed team consisting of two processes. Let $\{1,2\}$ be this fixed set of two processes and we let i range over $\{1,2\}$. A distributed alphabet over this set is a pair $\widetilde{\Sigma} = (\Sigma_1, \Sigma_2)$ of finite sets of *actions*. The set Σ_i contains actions in which process i *participates*. Let $\Sigma = \Sigma_1 \cup \Sigma_2$ denote the total set of actions and $\Sigma_{12} = \Sigma_1 \cap \Sigma_2$ denote the set of all *shared/joint/synchronizing* actions. Thus, both processes participate/synchronize on actions in Σ_{12}. We write $\Sigma_i^\ell = \Sigma_i \setminus \Sigma_{12}$ for the set of *purely local* actions of process i. Clearly, *only* process i participates in actions from Σ_i^ℓ.

The interaction between the environment and the two processes during a play of our game is faithfully modelled by a Mazurkiewicz trace [2] which is a labelled partial order on the interaction events. This partial order represents the causality between the underlying events; events which are not ordered (that is, incomparable) are "concurrent".

Definition 1. *A trace over $\widetilde{\Sigma}$ is a labelled partial order $t = (E, \leq, \lambda)$ where,*

- *E is a possibly infinite set of* events.
- *$\lambda : E \to \Sigma$ labels each event $e \in E$ by an action $\lambda(e) \in \Sigma$.*
- *$\leq$ is a partial (causal) order on E such that*
  - *for each $e \in E$, $\downarrow e$[1] is finite; each event has a finite causal past.*
  - *the restriction $\leq_i$ of $\leq$ on $E_i = \{e \in E \mid \lambda(e) \in \Sigma_i\}$ is a total order.*
  - *the relation $\leq$ is the transitive closure of $\leq_1 \cup \leq_2$.*

Let $\mathrm{Tr}(\Sigma)$ denote the set of traces over $\widetilde{\Sigma}$ and $\mathrm{Tr}^*(\Sigma)$ denote the set of finite traces. Henceforth a trace means a trace over $\widetilde{\Sigma}$. Let $t = (E, \leq, \lambda)$ be a trace. A subset $c \subseteq E$ is a configuration of t iff c is finite and $\downarrow c = c$. We let $\mathcal{C}_t$ denote the set of configurations of trace t. Basic examples of configurations are $\emptyset$, $\downarrow e$, $\Downarrow e = (\downarrow e) \setminus \{e\}$ where $e \in E$. By slight abuse of notation, for a configuration c of t, we use c to also denote the trace obtained by restricting t to this configuration.

As mentioned before, we use an asynchronous transition system as the underlying game arena for our game. We now develop more notation to describe this fundamental notion due to Zielonka [14,18] specialized to our two-process setting. We equip each process i with a finite non-empty set of local i-states denoted

---
[1] For $X \subseteq E$, we let $\downarrow X = \bigcup_{e \in X} \downarrow e$ where $\downarrow e = \{f \in E \mid f \leq e\}$.

S_i. A global state is a tuple $s = (s_1, s_2)$ where $s_i \in S_i$ is an i-state. For $a \in \Sigma_i^\ell$, an a-state is an i-state and for $a \in \Sigma_{12}$, an a-state is a joint-state of the two processes, that is, a global state. By S_a we denote the set of all a-states. So, if $a \in \Sigma_i^\ell$, then $S_a = S_i$ and if $a \in \Sigma_{12}$, then $S_a = S_1 \times S_2$.

We use the notation $\{Y_a\}$ to denote the Σ-indexed family $\{Y_a\}_{a \in \Sigma}$.

Definition 2. *An (uninitialized two process) asynchronous transition system (ATS) over $\widetilde{\Sigma}$ is a tuple $\mathcal{A} = (\{S_i\}, \{\xrightarrow{a}\})$ where,*

- S_1, S_2 *are finite non-empty sets of local 1-states and 2-states.*
- *For $a \in \Sigma$, $\xrightarrow{a} \subseteq S_a \times S_a$ is a non-deterministic transition relation on a-states.*

We fix an ATS $\mathcal{A} = (\{S_i\}, \{\xrightarrow{a}\})$ for further discussion. It is important to note that, if $a \in \Sigma_i^\ell$, then $\xrightarrow{a} \subseteq S_i \times S_i$. In other words, purely local actions update only local-states of the sole participating process. In contrast, if $a \in \Sigma_{12}$, then $\xrightarrow{a}$ is a joint-update of global states on this synchronization. One naturally extends the "local" transition relations $\xrightarrow{a} \subseteq S_a \times S_a$ to global transition relations $\xrightarrow{a} \subseteq (S_1 \times S_2) \times (S_1 \times S_2)$, as follows. If $a \in \Sigma_1^\ell$, $(s_1, s_2) \xrightarrow{a} (s_1', s_2')$ if $s_1 \xrightarrow{a} s_1'$ and $s_2 = s_2'$; the definition for $a \in \Sigma_2^\ell$ is similar. For $a \in \Sigma_{12}$, $\xrightarrow{a} = \xrightarrow{a}$.

A run of the ATS $\mathcal{A}$ on a trace t is a truly concurrent notion. We formalize this notion with the help of underlying configurations.

Definition 3. *A (trace) run ρ of $\mathcal{A}$ on a trace $t = (E, \leq, \lambda) \in \mathrm{Tr}(\Sigma)$ from state $s_0 \in (S_1 \times S_2)$ is a map $\rho : \mathcal{C}_t \to (S_1 \times S_2)$ such that $\rho(\emptyset) = s_0$ and for configurations $c, c' \subseteq E$ with $c' = c \sqcup \{e\}$ and $\lambda(e) = a$, $\rho(c) \xrightarrow{a} \rho(c')$.*

Note that there can be multiple runs of $\mathcal{A}$ on a single trace. For ease of presentation, henceforth, we make the *simplifying assumption* that $\mathcal{A}$ is complete, that is, for every $a \in \Sigma$ and $s_a \in S_a$, there exists an $s_a' \in S_a$ such that $s_a \xrightarrow{a} s_a'$. Under this assumption, for every trace t, $\mathcal{A}$ has at least one run on t.

3 Two Process Asynchronous Transition System Games

We now formally introduce our model of a distributed game. It is played between an environment and a distributed team on an underlying ATS and works for any fixed set of processes. We call these games ATS games. In keeping with the focus of this paper, we only introduce two process ATS games here. We continue using the notations from the previous section. In particular, as before, $\widetilde{\Sigma} = (\Sigma_1, \Sigma_2)$ is a fixed distributed alphabet over two processes.

Definition 4. *We define a (two process) ATS game $\mathcal{G} = (\mathcal{A}, s_0, \mathrm{Win})$ as follows:*

- $\mathcal{A} = (\{S_i\}, \{\xrightarrow{a}\})$ *is an asynchronous transition system over $\widetilde{\Sigma}$.*
- $s_0 \in S_1 \times S_2$ *is an initial global state, where a play of the game begins.*
- Win *is a specification of the winning condition.*

Let $\mathcal{G}$ be an ATS game as in the above definition. An interaction/play of $\mathcal{G}$ begins at the initial state s_0 of $\mathcal{A}$. The environment chooses an action a at the current state. The distributed team responds to the choice a of the environment by choosing a *matching transition* of $\mathcal{A}$ on a (which is consistent with the current state) and advances the current state according to this transition. As a result of this interaction, the partial play advances to a new (global) state. Note that, if a is a purely local action (say $a \in \Sigma_1^\ell$), then only the local state of process 1 may change during this interaction. In fact, process 2 is not even "aware" of this interaction between the environment and process 1. This single "round" of interaction between the environment and the distributed team may be viewed as a single event trace labelled a, and a run of $\mathcal{A}$ on this trace resulting from the matching transition. The ongoing partial play proceeds by another round of interactions and this goes on and on.

So, a partial play naturally gives rise to a trace $t = (E, \leq, \lambda)$ where an event e with label $\lambda(e) = a$ of t corresponds to the environment choosing the action a during a round. The fact that actions in different rounds are *distributed* across the two processes gives a natural partial order on the associated events/rounds which is represented by $\leq$ in t. As a result, it is possible to view two $\leq$-incomparable events as concurrent choices of the environment. From this viewpoint, when the environment chooses a purely local action (say $a \in \Sigma_i^\ell$) at event e, process i is aware of only those events in t which are *causally* before e. This is precisely the causal past $\downarrow e$ of event e. In particular, process i is oblivious to events which are occurring concurrently with e. Note that there is no bound on the number of events which are concurrent with event e.

The responses of the distributed team during rounds in t are represented by a run ρ of $\mathcal{A}$ on t where, for an event e of t with label $\lambda(e) = a$ associated with a round, the matching transition on e is (s_a, s'_a) where $\rho(\Downarrow e) = s$ and $\rho(\downarrow e) = s'$. Recall that $\Downarrow e = (\downarrow e) \setminus \{e\}$. Thanks to locality of transitions of $\mathcal{A}$, the response of the distributed team to a purely local action *does not* change the local state of the other process, and is viewed as the response of the action owner process.

Definition 5. *A maximal (distributed) play of $\mathcal{G}$ is a pair (t, ρ) where*

- $t = (E, \leq, \lambda)$ is a maximal trace over $\widetilde{\Sigma}$, that is, for each i, $E_i = \{e \in E \mid \lambda(e) \in \Sigma_i\}$ is infinite. In short, both processes participate in infinite events.
- $\rho : \mathcal{C}_t \to (S_1 \times S_2)$ is a run of $\mathcal{A}$ on t starting at s_0, that is, for configurations c, c' of t with $c' = c \sqcup \{e\}$ and $\lambda(e) = a$, $\rho(c) \xrightarrow{a} \rho(c')$.

The winning condition Win specifies which maximal plays are won by the distributed team. As an example, consider a *global safety* winning condition described using a set $F \subset (S_1 \times S_2)$ of *global safe states*. A maximal play (t, ρ) is won by the distributed team (or satisfies the global safety condition) if for all $c \in \mathcal{C}_t$, $\rho(c) \in F$. In other words, to win a maximal play the distributed team needs to ensure that a global unsafe state *never* occurs.

Example 1. Consider the global safety game illustrated in Fig. 1. Here, we have $\Sigma_1 = \{a, b, c\}$ and $\Sigma_2 = \{a, b, d\}$. Thus c and d are purely local actions.

Figure 1(a) shows the underlying ATS. The local state sets are $S_1 = \{q_0, \ldots\}$ and $S_2 = \{r_0, \ldots\}$. The transitions on shared actions are $(q_0, r_0) \xrightarrow{a} (q_1, r_1)$ and $(q_0, r_0) \xrightarrow{b} (q'_1, r_1)$. The purely local transitions are clear from the figure. For instance, we have $q_3 \xrightarrow{c} q_4$, $q_3 \xrightarrow{c} q'_4$ and $r_2 \xrightarrow{d} r_3$. The initial global state is (q_0, r_0). The unsafe set is $\{q_2, q_4\} \times \{r'_2, r'_4\} \cup \{q'_2, q'_4\} \times \{r_2, r_4\}$. In other words, the set F of global safe states is the complement of the unsafe set.

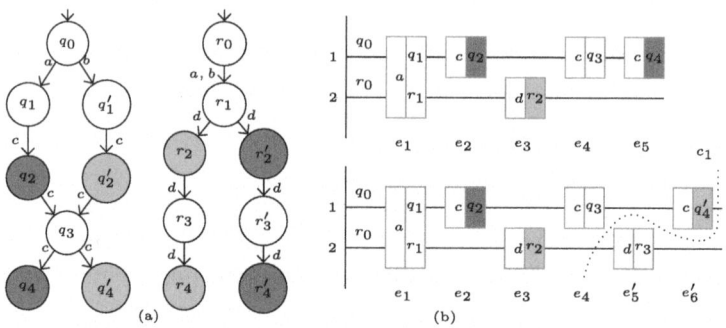

Fig. 1. Example of a game: the ATS with initial state (q_0, r_0) and some partial plays

Figure 1(b) shows some partial plays of the game. The top play begins with the environment choosing the action a at (q_0, r_0) and the distributed team advancing the global state to (q_1, r_1) by choosing the matching transition. This round is depicted in the figure by event e_1. The left-label on e_1 depict the trace label $\lambda(e_1) = a$ and the right-label records the response in terms of the target local/global state. In the second round, the environment selects c and process 1 advances its (local) state to q_2 from its previous state q_1. The locality of the c-transition ensures that, during this round, state of process 2 does not change. This round is depicted by event e_2 in the above figure. In the third round (shown by the event e_3), the environment selects d and process 2 advances its local state to r_2 from its previous state r_1. Note that events e_2 and e_3 are concurrent ($\leq$-incomparable) in the underlying play (t, ρ). Further, $c_1 = \{e_1, e_3\}$ is a valid configuration of t and $\rho(c_1) = (q_1, r_2)$. The set $c_2 = \{e_1, e_2, e_3\}$ is also a configuration and $\rho(c_2) = (q_2, r_2)$. Note that $\{e_1, e_2, e_3, e_5\}$ is *not* a configuration.

The bottom play proceeds as depicted. A configuration $c = \{e_1, e_2, e_3, e_4, e'_6\}$ of it is marked in the figure by a wavy dotted line. The global state occurring at c is (q'_4, r_2). As this is an unsafe state, this partial play is *already* lost by the distributed team and no maximal play extending it will be winning.

3.1 Distributed Strategies

We develop more notation to formulate the key notion of a distributed strategy. For a finite trace $t = (E, \leq, \lambda) \in \mathrm{Tr}^*(\Sigma)$ and $a \in \Sigma$, $t.a$ denotes the trace obtained by trace concatenation of t and a (see [2]). To define $t.a$, we introduce a fresh event f with label $\lambda(f) = a$ and set $t.a = (E \sqcup \{f\}, \leq', \lambda)$. The partial order $\leq'$ extends $\leq$ as follows. If $a \in \Sigma_{12}$, for all $e \in E$, $e \leq' f$. So, in this case, f becomes the unique maximal event of $t.a$. If $a \in \Sigma_i^\ell$, then for all $e \in \downarrow E_i$, $e \leq' f$. Thus, f simply becomes the last i-event of $t.a$. We refer [2] for the definition of the concatenation $t_1.t_2$ of arbitrary finite traces t_1 and t_2. We primarily use $t_1.w$ where w is a sequence of actions, that is, a word over Σ. In this case, we can simply define $t_1.\epsilon = t_1$ and $t.(w.a) = (t_1.w)a$.

A distributed strategy is an advice function for the distributed team to respond to choices of the environment. We clearly require that this response depends only on the collective causal history of participating processes and must be consistent with "current" state.

Definition 6. *A distributed strategy in $\mathcal{G}$ is a function $\sigma : \mathrm{Tr}^*(\Sigma) \to S_1 \times S_2$ such that $\sigma(\epsilon) = s_0$ and, for every $t \in \mathrm{Tr}^*(\Sigma)$ and $a \in \Sigma$, $\sigma(t) \xrightarrow{a} \sigma(t.a)$.*

Definition 7. *Let $\sigma : \mathrm{Tr}^*(\Sigma) \to S_1 \times S_2$ be a distributed strategy in $\mathcal{G}$. A maximal play (t, ρ) of $\mathcal{G}$ is said to conform σ if, for all $c \in \mathcal{C}_t$, $\rho(c) = \sigma(c)$. Recall that for a configuration c of t, c also denotes the finite trace $t_c = (c, \leq, \lambda)$. A distributed strategy σ is winning if all maximal plays conforming it are won (as per Win) by the distributed team.*

Fix a distributed strategy σ in $\mathcal{G}$. Let t be a finite trace with a unique maximal event whose label is in Σ_{12}. So, the *last* event in t is a synchronizing event in which both processes participate. Note that at t both processes have identical causal past. The strategy σ induces two natural "local" strategies "rooted" at t. Intuitively, these local strategies capture the σ-response of individual processes to sequences of purely local actions which extend the trace t.

Definition 8. *Continuing above notation, we define $\sigma_i[t] : (\Sigma_i^\ell)^* \to S_i$ as follows: for $w \in (\Sigma_i^\ell)^*$, $\sigma_i[t](w)$ is the i-local component of the global state $\sigma(t.w)$.*

Remark 1. Let $\sigma(t) = (s_1, s_2)$, $w_1 \in (\Sigma_1^\ell)^*$ and $w_2 \in (\Sigma_2^\ell)^*$. Thanks to the locality of $\mathcal{A}$, we have $\sigma(t.w_1) = (\sigma_1[t](w_1), s_2)$ and $\sigma(t.w_2) = (s_1, \sigma_2[t](w_2))$. Furthermore, $\sigma_1[t](w_1) = s_1'$ and $\sigma_2[t](w_2) = s_2'$ iff $\sigma(t.(w_1 \| w_2)) = (s_1', s_2')$ where $t.(w_1 \| w_2) = t.w_1.w_2 = t.w_2.w_1$ is the extension of t by the *parallel* composition $w_1 \| w_2$. More precisely, $w_1 \| w_2$ is the trace $w_1.w_2 = w_2.w_1$ and this trace equality is a consequence of the fact that only process i participates in w_i.

It will be useful to think of a distributed strategy σ as made up of a family $\{(\sigma_1[t], \sigma_2[t])\}_{t \in \mathrm{Tr}^*(\Sigma).\Sigma_{12}}$ of pairs of local strategies parameterized by finite traces with a unique maximal synchronizing event. A play conforming σ is initially played according to the pair $(\sigma_1[\epsilon], \sigma_2[\epsilon])$ of *local* strategies until the *first* synchronizing event e_1 occurs which is responded by the matching transition $(\sigma(\Downarrow e_1), \sigma(\downarrow e_1))$. After this point, the new pair $(\sigma_1[\downarrow e_1], \sigma_2[\downarrow e_1])$ of local strategies is employed until the second synchronizing event e_2 occurs and so on.

In this work, we also use distributed strategies with *local memory*.

Definition 9. *A distributed memory-based strategy in $\mathcal{G}$ is an initialized deterministic ATS $\mathcal{M} = (\{S_i \times M_i\}, \{\delta_a\}, (s_0, m_0)^2)$ over $\widetilde{\Sigma}$ where $m_0 \in M_1 \times M_2$ is*

---
[2] $(s_0 = (s_0^1, s_0^2), m_0 = (m_0^1, m_0^2))$ is identified with global state $((s_0^1, m_0^1), (s_0^2, m_0^2))$ of $\mathcal{M}$.

an initial memory *state and* s_0 *is the initial state of* $\mathcal{G}$ *such that, for every* $a \in \Sigma$ *and an* a-*state* $(s_a, m_a) \in S_a \times M_a$, $\delta_a(s_a, m_a) = (s'_a, m'_a)$ *implies that* $s_a \xrightarrow{a} s'_a$.

A maximal play (t, ρ) *is said to conform* $\mathcal{M}$ *if for all* $c \in \mathcal{C}_t$, $\rho(c) = \Pi_{\mathcal{G}}(\rho_{\mathcal{M}}(c))$ *where* $\rho_{\mathcal{M}} : \mathcal{C}_t \to (S_1 \times M_1) \times (S_2 \times M_2)$ *is the* unique *run of* $\mathcal{M}$ *on* t *from the initial state* (s_0, m_0) *of* $\mathcal{M}$ *and* $\Pi_{\mathcal{G}} : (S_1 \times M_1) \times (S_2 \times M_2) \to (S_1 \times S_2)$ *is the natural projection* $\Pi_{\mathcal{G}}(((s_1, m_1), (s_2, m_2))) = (s_1, s_2)$ *from global states of* $\mathcal{M}$ *to that of* $\mathcal{A}$. *It is easy to see that the* function $\rho_{\mathcal{G}} : \mathcal{C}_t \to S_1 \times S_2$; $\rho_{\mathcal{G}}(c) = \Pi_{\mathcal{G}}(\rho_{\mathcal{M}}(c))$, *induced by the run* $\rho_{\mathcal{M}}$, *is in fact a run of* $\mathcal{A}$ *on* t *from* s_0. *We call* $\mathcal{M}$ winning *if all maximal plays conforming it are won by the distributed team.*

Example 2. We revisit the global safety game from Fig. 1 and explained in Example 1. Each process stores in its memory the state of process 1 after the *last synchronization in the past*. If the environment chooses a (resp. b) then it is q_1 (resp. q'_1). With this bit of information, they can ensure the safety objective as follows. If the memory state is q_1, then both processes move *left* at their respective decision points. That is, process 1 at q_3 moves to q_4 on c, and process 2 at r_1 moves to r_2 on d. In case, the memory state is q'_1, then both processes move *right*. It is easy to see that this is indeed a winning distributed strategy which can be realized by a distributed memory-based strategy.

The key algorithmic question is to efficiently decide for a class $\mathcal{C}$ of ATS games whether there exists a distributed winning strategy for games in $\mathcal{C}$.

4 Global Safety Objective

In this section, we present a fixpoint based algorithm for solving two process global safety ATS games. Fix such a game $\mathcal{G}$ with ATS $\mathcal{A} = (\{S_i\}, \{\xrightarrow{a}\})$ over $\widetilde{\Sigma}$, the initial state $s_0 \in S_1 \times S_2$ and the safe set $F \subseteq S_1 \times S_2$. A maximal play (t, ρ) is won by the distributed team/system if for all $c \in \mathcal{C}_t$, $\rho(c) \in F$.

A key element in our solution is the notion of a trap. The construction of traps (see [3,12]) is central to the solution of standard two-player safety games. We define a subset $X_i \subseteq S_i$ to be an i-*trap* if, for every $s \in X_i$ and for each $a \in \Sigma_i^\ell$, there exists $s' \in X$ such that $s \xrightarrow{a} s'$. The crucial property of an i-trap is, within X_i, that process i has a "uniform local strategy" to ensure that plays on purely local actions stay within X_i. To do so, for an i-state $s \in X_i$ and for each $a \in \Sigma_i^\ell$, process i simply selects an i-state s' such that $s \xrightarrow{a} s'$. Below we assume a fixed choice of such a uniform local strategy f_{X_i} for an i-trap X_i. A *rectangle* R is a subset of global states of the form $X_1 \times X_2 \subseteq S_1 \times S_2$ where X_i is an i-trap.

We now describe a fixpoint argument to construct the *winning region* for our *uninitialized* global safety game $\mathcal{G}$. The key idea is to consider a restricted environment which is allowed to play at most j *synchronizing* actions and build a winning region for this restriction by induction on j.

We inductively define, as illustrated in Fig. 2(a), for $j \geq 0$, two interleaved decreasing families of sets of global states V_j and U_j as follows: $V_0 = F$,

$$U_j = \{s \in V_j | \exists \text{ a rectangle } R = X_1 \times X_2 \text{ such that } s \in R \text{ and } R \subseteq V_j\}$$
$$V_{j+1} = \{s \in U_j \mid \forall a \in \Sigma_{12} \, \exists s' \in U_j \text{ such that } s \xrightarrow{a} s'\}$$

Observe that $F = V_0 \supseteq U_0 \supseteq V_1 \supseteq U_1 \supseteq V_2 \supseteq U_2 \ldots$. It is important to note that, if $s \in U_j$ with a "witness" rectangle R as in the above definition, then $R \subseteq U_j$. So, if $s \in V_j \setminus U_j$ then there is no rectangle containing s which is completely inside V_j. Also note that, if $s \in U_j \setminus V_{j+1}$ then there is an action $a \in \Sigma_{12}$, such that for all s' with $s \xrightarrow{a} s'$, $s' \notin U_j$.

Lemma 1. *Let $j \geq 0$. If $s \notin U_j$, then for every distributed strategy from s there exists a loosing (partial) play conforming it in which the environment chooses at most j synchronizing actions.*

Proof Sketch: We prove the statement by induction on j. The base case is skipped. So assume $j > 0$, $s \notin U_j$ and let σ^s be a distributed strategy from s. Let $(f_1 = \sigma_1^s[\epsilon], f_2 = \sigma_2^s[\epsilon])$ be the initial pair of local strategies (see Definition 8) of σ^s. It is easy to see that $X_i = f_i(\Sigma_i^{\ell*})$ is an i-trap and s belongs to the rectangle $R = X_1 \times X_2$. The fact that $s \notin U_j$ implies that $R \not\subseteq V_j$. Therefore there exist words $w_i \in (\Sigma_i)^*$ and $s'_i = f_i(w_i)$ such that $(s'_1, s'_2) \notin V_j$. So, with $t = w_1 \| w_2$, $\sigma(t) = s' = (s'_1, s'_2) \notin V_j$. Now we consider the following two exhaustive cases.
Case $[s' \in U_{j-1} \setminus V_j]$: In this case, there is an $a \in \Sigma_{12}$ such that for all s'' with $s' \xrightarrow{a} s''$, $s'' \notin U_{j-1}$. The environment can extend the partial play t by such an a. So, with $t' = t.a$, we have $s'' = \sigma^s(t.a) \notin U_{j-1}$. We now view σ^s rooted at $t.a$ as a strategy from s'' and use induction to find a loosing play t' of it with at most $j - 1$ synchronizing actions. This implies that the play $t.a.t'$ is a loosing play of σ^s with at most j synchronizations and we are done.
Case $[s' \notin U_{j-1}]$: Note that σ^s rooted at t may also be viewed as strategy from s' and we can use induction to finish the proof in this case. □

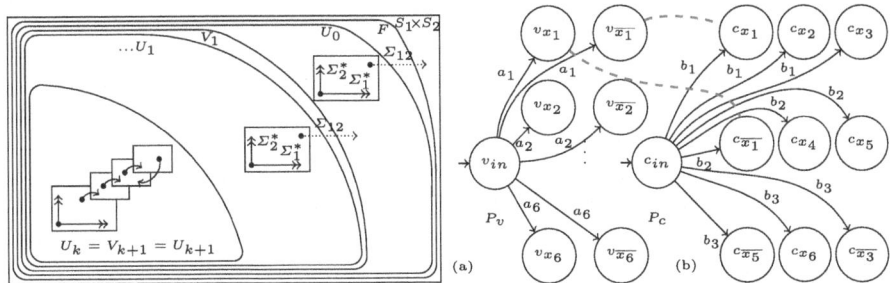

Fig. 2. (a) Fixpoint construction: $F = V_0 \supseteq U_0 \supseteq V_1 \supseteq U_1 \ldots$ and $U_k = V_{k+1} = U_{k+1}$ (b) Hardness reduction: Two of the unsafe states are connected by red dashed lines. (Color figure online)

Let $k \geq 0$ be the least index so that $U_k = U_{k+1}$. Clearly, $U_k = V_{k+1} = U_{k+1}$. For every $s \in U_k$, we fix a rectangle $R_s = X_1^s \times X_2^s \subseteq U_k$. Let $f_i^s : X_i^s \times \Sigma_i^\ell \to X_i^s$ be a fixed choice of uniform local strategy for the i-trap X_i^s. Further, for every $s \in U_k$ and $a \in \Sigma_{12}$, we fix $s' \in U_k$ such that $s \xrightarrow{a} s'$ and denote this a-successor of s in U_k by $\mathrm{succ}(s,a)$.

We now construct a uniform distributed memory-based strategy from U_k. Towards this, we define the deterministic ATS $\mathcal{M} = (\{S_i \times M_i\}, \{\delta_a\})$ where $M_i = S_1 \times S_2$. The transition function δ_a of $\mathcal{M}$ is defined as follows: if $a \in \Sigma_i^\ell$, then $\delta_a((m, s_i)) = f_i^m(s_i)$. If $a \in \Sigma_{12}$, $\delta_a(((m_1, s_1), (m_2, s_2))) = ((s', s_1'), (s', s_2'))$ where $s' = (s_1', s_2') = \mathrm{succ}((s_1, s_2), a))$. An important observation is that the local memory states in $\mathcal{M}$ simply store the global state (m) of $\mathcal{G}$ at the *last synchronization*. Both processes agree on a rectangle at this synchronization and continue using the local strategies (f_i^m) until the next synchronization.

Theorem 1. *A state $s \in U_k$ iff there exists a distributed winning strategy from s. Moreover, if $s \in U_k$, then $\mathcal{M}$ with initial memory state $(s, s) \in M_1 \times M_2$ is a distributed memory-based winning strategy in $\mathcal{G}$ from the initial state s.*

Proof Sketch: Lemma 1 shows that if $s \notin U_k$ then there does not exist a distributed winning strategy from s. If $s = (s_1, s_2) \in U_k$, then it is not difficult to show that all reachable global states of $\mathcal{M}$ from the initial state $((s, s_1), (s, s_2))$ are of the form $((m, s_1'), (m, s_2'))$ and, $(s_1', s_2') \in U_k$. As $U_k \subseteq F$, global states of $\mathcal{A}$ encountered during plays conforming $\mathcal{M}$ are safe. So $\mathcal{M}$ is winning. □

Theorem 2. *Two process global safety ATS games are NP-complete.*

Proof Sketch: By Theorem 1, if there is a distributed winning strategy, there is a distributed memory-based strategy whose local memory state stores the global state at the *last* synchronization. This serves as a polynomial size certificate for yes instances of the decision problem of the existence of a distributed winning strategy. It is easy to see that the resulting verification task can be achieved by a poly-time algorithm. This shows membership in NP.

To show NP-hardness, we provide a reduction from 3-SAT. Consider a 3-SAT instance φ with n variables and m clauses. We construct a two process safety ATS game $\mathcal{G}_\varphi$ with $|S_1| = 2n + 1$, $|S_2| = 3m + 1$ over the alphabet $|\Sigma_1| = n$, $|\Sigma_2| = m$ and $\Sigma_{12} = \emptyset$. Figure 2(b) illustrates the reduction for the formula

$$\varphi = (x_1 \lor x_2 \lor x_3) \land (\overline{x_1} \lor x_4 \lor x_5) \land (\overline{x_5} \lor x_6 \lor \overline{x_3})$$

As seen from the figure, the initial state of $\mathcal{G}_\varphi$ is (v_{in}, c_{in}). At this initial state, environment can play a purely local action $a_i \in \Sigma_1$ of process 1 "asking" it to decide the truth value of variable x_i. Process 1 can respond by moving to either v_{x_i} or $v_{\bar{x}_i}$. This response naturally corresponds to setting the truth value of x_i. Concurrently, environment can play a purely local action $b_j \in \Sigma_2$ of process 2 "asking" it to pick the term in the clause C_j which makes the clause true. As shown in the Fig. 2(b), process 2 can respond by choosing one of the three terms from clause C_j. The unsafe global states correspond to "conflicting" choices by the two processes (see Fig. 2(b)). It is not difficult to see that the resulting game $\mathcal{G}_\varphi$ has a distributed winning strategy iff φ is satisfiable. □

5 Local and Global Reachability Objectives

Let $\mathcal{G}$ be an uninitialized game with ATS $\mathcal{A} = (\{S_i\}, \{\stackrel{a}{\to}\})$. A local reachability objective is given by two subsets $F_i \subseteq S_i$. A maximal play (t, ρ) is won by the distributed team if there exists configurations c and c' such that, with $\rho(c) = s = (s_1, s_2)$, $\rho(c') = s' = (s_1', s_2')$, we have $s_1 \in F_1$ and $s_2' \in F_2$. In short, to win a maximal play, process i must ensure a visit to some local state in F_i.

In contrast, a global/simultaneous reachability objective is given by a single subset $F \subseteq S_1 \times S_2$ and maximal play (t, ρ) is won by the team if there exists a configuration c such that $\rho(c) \in F$. So, to meet the global reachability objective, the team needs to ensure a visit to a global state in F.

5.1 Local Reachability

We continue with the above notation where local reachability objective is specified by a pair of subsets $F_i \subseteq S_i$. We first perform an *asynchronous game simulation* wherein each process simply maintains additional information in its local state to record if it has *already* visited a state in F_i. So, we set $Q_i = S_i \times \{0, 1\}$ and *extend* naturally $\stackrel{a}{\to}$ to Q_a as follows: for $a \in \Sigma_i^\ell$, for $s \stackrel{a}{\to} s'$, we add a-transitions $((s, 1), (s', 1))$, $((s, 0), (s', 1))$ if $s' \in F_i$, $((s, 0), (s', 0))$ if $s' \notin F_i$. For $a \in \Sigma_{12}$, and $s \stackrel{a}{\to} s'$ with $s = (s_1, s_2)$ and $s' = (s_1', s_2')$, we add a-transitions $(((s_1, b_1), (s_2, b_2)), ((s_1', b_1'), (s_2', b_2')))$ where $b_i' = b_i \vee (s_i' \stackrel{?}{\in} F_i)$. Let $\mathcal{A}' = (\{Q_i\}, \{\stackrel{a}{\rightarrowtail}\})$ be the resulting ATS. Observe that, we have used $\stackrel{a}{\rightarrowtail}$ to denote the extension of $\stackrel{a}{\to}$ from S_a to Q_a.

We also define a map $\theta : S_1 \times S_2 \to Q_1 \times Q_2$ as follows: for $s = (s_1, s_2)$, we set $\theta(s) = ((s_1, b_1), (s_2, b_2))$ where $b_i = 1$ iff $s_i \in F_i$. Further, we set new local target set to be $F_i' = S_i \times \{1\} \subseteq Q_i$. Let $\mathcal{G}'$ be a new uninitialized game with underlying ATS $\mathcal{A}'$ and local reachability condition $\{F_i'\}$. It is easy to see that $\mathcal{G}$ has a distributed winning strategy from s iff $\mathcal{G}'$ has a distributed winning strategy from $\theta(s)$. It is important to note that, in order to win $\mathcal{G}'$, process i needs to ensure a visit to a local state with bit-component 1, that is, a state of the form $(s, 1)$ where $s \in S_i$.

Our solution relies on the notion of an i-attractor. See [3,12] for this notion which plays a key role in the solution of standard two-player reachability games. A subset $X_i \subseteq Q_i$ is an i-attractor (wrt F_i') if it is an i-trap and from *every* state in X_i, process i has a (sequential/local) strategy to *force* a visit to $F_i' = S_i \times \{1\}$. It follows from standard results about reachability games [3,12] that process i, in fact, has a zero-memory strategy to force a visit to F_i'.

We now provide a fixpoint construction to compute the winning region in $\mathcal{G}'$. This construction is similar in spirit and *dual* to that of global safety games. The key idea is to consider a *stronger* winning requirement to meet the local reachability objective of $\mathcal{G}'$ within at most j synchronizations. Starting with $j = 0$, we progressively increase j to relax the winning requirement. Finally, we reach a fixpoint where the stronger winning requirement matches the local reachability objective.

We inductively define, as illustrated in Fig. 3, for $j \geq 0$, an increasing chain of sets $U_j \subseteq Q_1 \times Q_2$ as follows: $U_0 = F_1' \times F_2' = (S_1 \times \{1\}) \times (S_2 \times \{1\})$.

$$U_j = \left\{ q \in Q_1 \times Q_2 \;\middle|\; \begin{array}{l} \exists\; i\text{-attractor pair}(X_i) \text{ such that } q \in X_1 \times X_2 \text{ and} \\ \forall q' \in (X_1 \times X_2)\; \forall a \in \Sigma_{12}\; \exists q'' : (q' \xrightarrow{a} q'' \;\&\; q'' \in U_{j-1}) \end{array} \right\}$$

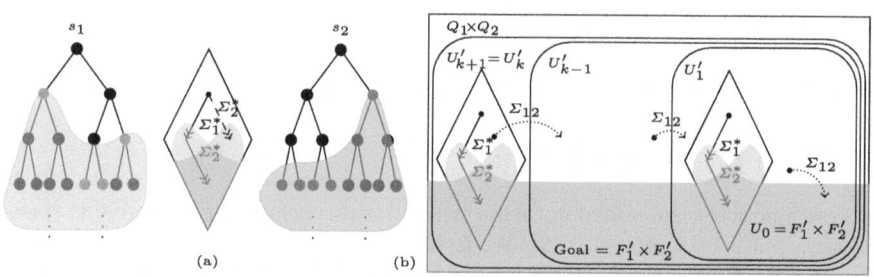

Fig. 3. Reachability: (a) A pair of i-attractors. (b) Fixpoint sets $U_j = U_j' \cup U_0$.

It is easy to see that $U_0 \subseteq U_1 \subseteq U_2 \subseteq U_3 \ldots$. Let U_k be the fixpoint of this increasing chain. Our main results about local reachability games are captured in the next two theorems. We skip their proofs which are adaptations of the proofs of the corresponding results about global safety games.

Theorem 3. *There exists a distributed winning strategy in $\mathcal{G}$ from s iff the state $q = \theta(s) \in U_k$. Moreover, if $\theta(s) \in U_k$, there is a distributed memory-based strategy in $\mathcal{G}$ from s, whose local memory state of process i stores the global state of $\mathcal{A}$ at the last synchronization and a bit-record of whether or not process i has already visited a state in F_i.*

Theorem 4. *Two process local reachability ATS games are NP-complete.*

5.2 Global Reachability

We now focus on global reachability objective. As usual, we fix an uninitialized game $\mathcal{G}$ with ATS $\mathcal{A} = (\{S_i\}, \{\xrightarrow{a}\})$ and consider the global reachability objective specified by a set $F \subseteq S_1 \times S_2$. Recall that, the distributed team meets the global reachability objective along a maximal play (t, ρ) if there exists a configuration c such that $\rho(c) \in F$.

We now relate the global reachability game $\mathcal{G}$ to a local reachability game $\mathcal{G}'$ on ATS $\mathcal{A}'$. In $\mathcal{A}'$, we set local state set to be $Q_i = S_i \times 2^{S_i} \times \{0, 1\}$ and *extend* a-transitions $\xrightarrow{a}$ on S_a to a-transitions $\xrightarrow{a}$ on Q_a.

Let $a \in \Sigma_i^\ell$ and $s \xrightarrow{a} s'$. Further, let $Y \subseteq S_i$ and $b \in \{0, 1\}$. With $q = (s, Y, b)$, we set $q \xrightarrow{a} q'$ where $q' = (s', Y \cup \{s'\}, b)$. On local action a, process i simply updates the set Y of local states that has been visited *since the last synchronization.* Now let $a \in \Sigma_{12}$ and $s \xrightarrow{a} s'$ where $s = (s_1, s_2)$ and $s' = (s'_1, s'_2)$. For $Y_i \subseteq S_i$ and $b_i \in \{0,1\}$, we set

$$((s_1, Y_1, b_1), (s_2, Y_2, b_2)) \xrightarrow{a} ((s'_1, Y'_1, b'_1), (s'_2, Y'_2, b'_2))$$

where $Y'_1 = \{s'_1\}, Y'_2 = \{s'_2\}$ and the bits b'_1 and b'_2 are set according to the following rule. If $((Y_1 \times Y_2) \cup \{(s'_1, s'_2)\}) \cap F \neq \emptyset$, then $b'_1 = 1, b'_2 = 1$. Otherwise $b'_1 = b_1$ and $b'_2 = b_2$. It is important to note that, the condition $((Y_1 \times Y_2) \cup \{(s'_1, s'_2)\}) \cap F \neq \emptyset$ asserts that *since the previous synchronization, a target global state in F has been visited.* We include the current global state (s'_1, s'_2) as part of this check. As Y_i represents the set of local i-states since the last synchronization, $Y_1 \times Y_2$ represents the set of global states that have occurred since then and it may not include the global state (s'_1, s'_2) at the current synchronization. Further, note that, we *reset* $Y'_i = \{s'_i\}$ in order to correctly keep track of only local states of process i since the *last* synchronization.

As expected, we define $\mathcal{A}'$ to be the ATS $(\{Q_i\}, \{\xrightarrow{a}\})$. We also define (an initial state simulation) map $\theta : S_1 \times S_2 \to Q_1 \times Q_2$ as: for $s = (s_1, s_2)$, if $(s_1, s_2) \notin F$, $\theta(s) = ((s_1, \{s_1\}, 0), (s_2, \{s_2\}, 0))$; otherwise $\theta(s) = ((s_1, \{s_1\}, 1), (s_2, \{s_2\}, 1))$.

The above setup captures plays of $\mathcal{G}$ where a global state from the target set F is *visited* before a synchronization where both processes flag it by setting their control bits to 1 in $\mathcal{G}'$. We also need to worry about maximal plays where a visit to F occurs *after* the last synchronization when individual processes are involved in purely local actions. Along such plays, a visit to F is *not observed* due to lack of synchronizations. It is indeed possible to handle these "decoupled" purely local scenarios by first computing the corresponding winning region. We can then use it to bootstrap a fixpoint construction over the global states of $\mathcal{G}'$ to compute the winning region of $\mathcal{G}$. The details are completely skipped in this paper. A key consequence of this analysis is the next theorem.

Theorem 5. *If there is a distributed winning strategy in $\mathcal{G}$ from s then there is also a distributed memory-based strategy whose local memory state of process i stores the global state of $\mathcal{A}$ at the last synchronization, the set of local states of $\mathcal{A}$ visited since the last synchronization and a bit-record indicating whether or not a visit to a state in F has occurred in its causal past.*

The fixpoint construction implies that global reachability ATS games are in NEXPTIME. There is a somewhat surprising connection between two process ATS games and *generalized* two-player reachability games studied in [4]. This allows us to provide lower bounds on computational and memory complexity of global reachability games. Due to lack of space, these proofs are omitted.

Theorem 6. *We have the following lower bounds for global reachability games.*

1. *Two process global reachability ATS games are PSPACE-hard.*
2. *One can construct two process global reachability ATS games with each process having $O(n)$ local states in which a distributed winning strategy exists. However, any distributed memory-based winning strategy needs at least $2^n - 1$ local memory states.*

6 Conclusion

We have proposed ATS games on two processes and presented their detailed analysis. We have considered winning objectives such as global safety, local reachability and global reachability, and established near optimal algorithms for solving them. We have also constructed distributed memory-based winning strategies with an *explicit form of the key past information* that needs to be maintained.

We plan to analyze two process ATS games for infinitary objectives such as local parity and pin down the precise decision complexity and memory complexity. Another promising future direction is to extend the precise results of this paper to three/four process ATS games and identify the precise nature of the key past information needed to win them.

This paper has focused on two process ATS games but it is not difficult to see that the model generalizes to the setting of multiple processes. It turns out that asynchronous control games from [8,9] are *expressively equivalent* to ATS games. In other words, there are *efficient* back-and-forth translations between control games and ATS games and these translations *preserve* the number of processes. Therefore, it is possible to inter-transfer the results between the two models. In particular, it appears that the results of two process ATS games in this work lead to *similar complexity and memory bounds* for solving corresponding two process control games. An interesting future direction is to precisely work out these inter-relations between ATS games and control games, and use them to transfer positive as well as negative results.

A very important consequence of the above mentioned equivalence and the recent breakthrough result [11] for control games, is that, six process ATS games with local reachability as well as global safety objectives are undecidable. As mentioned in the introduction, there are already several works which have identified interesting decidable classes of distributed games. It is only natural to investigate more restricted and natural classes of distributed games which render the corresponding decision problems tractable.

References

1. Beutner, R., Finkbeiner, B., Hecking-Harbusch, J.: Translating asynchronous games for distributed synthesis. In: 30th International Conference on Concurrency Theory (CONCUR 2019). Leibniz International Proceedings in Informatics (LIPIcs), vol. 140, pp. 26:1–26:16. Schloss Dagstuhl – Leibniz-Zentrum für Informatik, Dagstuhl, Germany (2019)
2. Diekert, V., Rozenberg, G. (eds.): The Book of Traces. World Scientific (1995)
3. Fijalkow, N., et al.: Games on Graphs (2023)
4. Fijalkow, N., Horn, F.: The surprizing complexity of generalized reachability games (2012). https://arxiv.org/abs/1010.2420
5. Finkbeiner, B., Gieseking, M., Hecking-Harbusch, J., Olderog, E.R.: Global winning conditions in synthesis of distributed systems with causal memory. In: 30th

EACSL Annual Conference on Computer Science Logic (CSL 2022). Leibniz International Proceedings in Informatics (LIPIcs), vol. 216, pp. 20:1–20:19. Schloss Dagstuhl – Leibniz-Zentrum für Informatik, Dagstuhl, Germany (2022)
6. Finkbeiner, B., Gölz, P.: Synthesis in distributed environments. In: 37th IARCS Annual Conference on Foundations of Software Technology and Theoretical Computer Science (FSTTCS 2017). Leibniz International Proceedings in Informatics (LIPIcs), vol. 93, pp. 28:1–28:14. Schloss Dagstuhl – Leibniz-Zentrum für Informatik, Dagstuhl, Germany (2018)
7. Finkbeiner, B., Olderog, E.R.: Petri games: synthesis of distributed systems with causal memory. Inf. Comput. **253**, 181–203 (2017), gandALF 2014
8. Gastin, P., Lerman, B., Zeitoun, M.: Distributed games with causal memory are decidable for series-parallel systems. In: Lodaya, K., Mahajan, M. (eds.) FSTTCS 2004. LNCS, vol. 3328, pp. 275–286. Springer, Heidelberg (2004). https://doi.org/10.1007/978-3-540-30538-5_23
9. Genest, B., Gimbert, H., Muscholl, A., Walukiewicz, I.: Asynchronous games over tree architectures. In: Automata. Languages, and Programming, pp. 275–286. Springer, Berlin, Heidelberg (2013)
10. Gimbert, H.: On the control of asynchronous automata. In: 37th IARCS Annual Conference on Foundations of Software Technology and Theoretical Computer Science (FSTTCS 2017). Leibniz International Proceedings in Informatics (LIPIcs), vol. 93, pp. 30:1–30:15. Schloss Dagstuhl – Leibniz-Zentrum für Informatik, Dagstuhl, Germany (2018)
11. Gimbert, H.: Distributed asynchronous games with causal memory are undecidable. Log. Methods Comput. Sci. **18**(3) (2022)
12. Grädel, E., Thomas, W., Wilke, T. (eds.): Automata logics, and infinite games: a guide to current research. Springer-Verlag, Berlin, Heidelberg (2002)
13. Madhusudan, P., Thiagarajan, P.S., Yang, S.: The MSO theory of connectedly communicating processes. In: Sarukkai, S., Sen, S. (eds.) FSTTCS 2005. LNCS, vol. 3821, pp. 201–212. Springer, Heidelberg (2005). https://doi.org/10.1007/11590156_16
14. Mukund, M.: Automata on distributed alphabets. Modern Appl. Automata Theor. **2**, 257–288 (2012)
15. Muscholl, A., Walukiewicz, I.: Distributed synthesis for acyclic architectures. In: 34th International Conference on Foundation of Software Technology and Theoretical Computer Science (FSTTCS 2014). Leibniz International Proceedings in Informatics (LIPIcs), vol. 29, pp. 639–651. Schloss Dagstuhl - Leibniz-Zentrum für Informatik, Dagstuhl, Germany (2014)
16. Pneuli, A., Rosner, R.: Distributed reactive systems are hard to synthesize. In: Proceedings [1990] 31st Annual Symposium on Foundations of Computer Science, vol. 2, pp. 746–757 (1990)
17. Thomas, W.: Solution of church's problem: a tutorial. New Perspectives on Games and Interaction **4** (01 2008)
18. Zielonka, W.: Notes on finite asynchronous automata. RAIRO-Theor. Inf. Appl. **21**(2), 99–135 (1987)

Relational Companions of Logics

Sankha S. Basu[✉][iD] and Sayantan Roy[iD]

Department of Mathematics, Indraprastha Institute of Information Technology-Delhi, New Delhi, India
sankha@iiitd.ac.in

Abstract. The variable inclusion companions of logics have lately been thoroughly studied by multiple authors. There are broadly two types of these companions: the left and the right variable inclusion companions. Another type of companions of logics induced by Hilbert-style presentations (Hilbert-style logics) were introduced in [1]. A sufficient condition for the restricted rules companion of a Hilbert-style logic to coincide with its left variable inclusion companion was proved there, while a necessary condition remained elusive. The present article has two parts. In the first part, we give a necessary and sufficient condition for the left variable inclusion and the restricted rules companions of a Hilbert-style logic to coincide. In the rest of the paper, we recognize that the variable inclusion restrictions used to define variable inclusion companions of a logic $\langle \mathcal{L}, \vdash \rangle$ are relations from $\mathcal{P}(\mathcal{L})$ to $\mathcal{L}$. This leads to a more general idea of a relational companion of a logical structure, a framework that we borrow from the field of universal logic. We end by showing that even Hilbert-style logics and the restricted rules companions of these can be brought under the umbrella of the general notions of logical structures and their relational companions that are discussed here.

Keywords: Companion logics · Universal logic · Logics of variable inclusion

1 Introduction

The *logics of variable inclusion* have recently been rigorously studied, e.g., in [7,8,12]. These companion logics come in four flavors, viz., the *left*, the *right*, the *pure left*, and the *pure right variable inclusion companion* logics. The definitions and various examples of each of these classes can be found in the above references. It is well-known that the left variable inclusion companion of classical propositional logic (CPC) is the paraconsistent weak Kleene logic (PWK). A simple Hilbert-style presentation of PWK, consisting of the same set of axioms as CPC and a restricted version of the classical rule of modus ponens, was presented in [6]. This led to the following natural question. Can we always obtain a Hilbert-style presentation of the left variable inclusion companion logic from that of the original logic (if it has one, of course) by just restricting the rules of inference? The answer to this question was shown to be negative in [1]. In the course of the argument, the *restricted rules companion* of a Hilbert-style logic, i.e., a logic induced syntactically by a Hilbert-style presentation, was introduced. A sufficient condition for the restricted rules companion to coincide with the left variable inclusion companion of a Hilbert-style logic was proved as well. However, a necessary condition for the same remained unattained.

The present article has two main parts. In the first, a necessary and sufficient condition for the left variable inclusion and the restricted rules companions of a Hilbert-style logic to coincide, is presented.

In the second part, we generalize the notions of variable inclusion companion and restricted rules companion logics. This is done using the framework of *logical structures* from universal logic [4,5].

A logical structure is a pair $\mathcal{S} = \langle \mathcal{L}, \vdash \rangle$, where $\mathcal{L}$ is a set and $\vdash \subseteq \mathcal{P}(\mathcal{L}) \times \mathcal{L}$. A *logic* is a logical structure $\mathcal{S} = \langle \mathcal{L}, \vdash \rangle$, where $\mathcal{L}$ is the set of formulas defined inductively, in the usual way, over a set of variables V, using a finite set of connectives/operators called the *signature/type*. In other words, $\mathcal{L}$ is the formula algebra over V of some type. The formula algebra has the universal mapping property for the class of all algebras of the same type as $\mathcal{L}$ over V, i.e., any function $f : V \to A$, where A is the universe of an algebra $\mathbf{A}$ of the same type as $\mathcal{L}$, can be uniquely extended to a homomorphism from $\mathcal{L}$ to $\mathbf{A}$ (see [10,11] for more details). We, however, do not assume any condition on the $\vdash$-relation. For any $\alpha \in \mathcal{L}$, $\mathrm{var}(\alpha)$ denotes the set of all the variables occurring in α, and for any $\Delta \subseteq \mathcal{L}$, $\mathrm{var}(\Delta) = \bigcup_{\alpha \in \Delta} \mathrm{var}(\alpha)$.

2 Left Variable Inclusion and Restricted Rules Companions

In this section, we will deal exclusively with logics, in the usual sense, as described in the previous section. As mentioned earlier, the *logics of variable inclusion* have recently been rigorously studied, e.g., in [7,8,12]. The following definition of a left variable inclusion companion can be found in these.

Definition 1. *Suppose $\mathcal{S} = \langle \mathcal{L}, \vdash \rangle$ is a logic. The* left variable inclusion companion *of $\mathcal{S}$, denoted by $\mathcal{S}^l = \langle \mathcal{L}, \vdash^l \rangle$, is defined as follows. For any $\Gamma \cup \{\alpha\} \subseteq \mathcal{L}$,*

$\Gamma \vdash^l \alpha$ *iff there is a $\Delta \subseteq \Gamma$ such that* $\mathrm{var}(\Delta) \subseteq \mathrm{var}(\alpha)$ *and $\Delta \vdash \alpha$.*

The restricted rules companion of a Hilbert-style logic, i.e., a logic induced syntactically by a Hilbert-style presentation, was introduced in [1] as follows.

Definition 2. *Suppose $\mathcal{S} = \langle \mathcal{L}, \vdash \rangle$ is a Hilbert-style logic with $A \subseteq \mathcal{L}$ as the set of axioms and $R_\mathcal{S} \subseteq \mathcal{P}(\mathcal{L}) \times \mathcal{L}$ as the set of rules of inference. The* restricted rules companion *of $\mathcal{S}$, denoted by $\mathcal{S}^{re} = \langle \mathcal{L}, \vdash^{re} \rangle$, is then defined as the Hilbert-style logic with the following sets of axioms and rules.*

$$\text{Set of axioms} \quad = A, \text{and}$$
$$\text{set of rules of inference} = R_{\mathcal{S}^{re}} = \left\{ \frac{\Gamma}{\alpha} \in R_\mathcal{S} \mid \mathrm{var}(\Gamma) \subseteq \mathrm{var}(\alpha) \right\}.$$

Suppose $\mathcal{S} = \langle \mathcal{L}, \vdash \rangle$ is a Hilbert-style logic and $\mathcal{S}^l = \langle \mathcal{L}, \vdash^l \rangle$, $\mathcal{S}^{re} = \langle \mathcal{L}, \vdash^{re} \rangle$ be its left variable inclusion and restricted rules companion logics, respectively. It was established that, $\vdash^{re} \subseteq \vdash^l$ ([1, Theorem 3.6]), but the reverse inclusion, i.e., $\vdash^l \subseteq \vdash^{re}$ is not guaranteed ([1, Remark 3.7]). The following sufficient condition for the latter inclusion was also given in this paper.

Theorem 1. *([1, Theorem 4.3]) Suppose $\mathcal{S} = \langle \mathcal{L}, \vdash \rangle$ is a finitary[1] Hilbert-style logic such that $\dfrac{\alpha, \alpha \longrightarrow \beta}{\beta}$ (modus ponens [MP]) is a rule of inference in $\mathcal{S}$. Suppose further that the Deduction theorem holds in $\mathcal{S}$. Then the restricted rules companion of $\mathcal{S}$ coincides with the left variable inclusion companion of $\mathcal{S}$, i.e., $\vdash^{re} = \vdash^l$.*

However, a necessary condition for $\vdash^l = \vdash^{re}$ remained elusive. In the rest of this section, we investigate this further and provide a necessary and sufficient condition for the two companion logics to coincide. The following lemmas list some straightforward inferences that can be drawn from the definitions of left variable inclusion and restricted rules companions.

Lemma 1. *Suppose $\mathcal{S} = \langle \mathcal{L}, \vdash \rangle$ is a logic.*

(i) $\mathcal{S}^l = \langle \mathcal{L}, \vdash^l \rangle$ is monotonic[2]. Moreover, if $\mathcal{S}$ is monotonic, then $\vdash^l \subseteq \vdash$.
(ii) $(\mathcal{S}^l)^l = \mathcal{S}^l$, i.e., $(\vdash^l)^l = \vdash^l$.
(iii) Suppose $\mathcal{S}$ is a Hilbert-style logic. Then, $\mathcal{S}^{re}$ is monotonic and moreover, $\vdash^{re} \subseteq \vdash$.
(iv) Suppose $\mathcal{S}$ is a Hilbert-style logic. Then, $(\mathcal{S}^{re})^{re} = \mathcal{S}^{re}$, i.e., $(\vdash^{re})^{re} = \vdash^{re}$.

Proof. Parts (i) and (ii) follow straightforwardly from the definition of left variable inclusion companions. For part (iii), we recall that every Hilbert-style logic is monotonic. Thus, in fact, $\mathcal{S}$ and $\mathcal{S}^{re}$, being Hilbert-style logics, are both monotonic. That $\vdash^{re} \subseteq \vdash$, follows from the definition of $\mathcal{S}^{re}$.

For part (iv), let A and $R_\mathcal{S}$ as its sets of axioms and rules of inference, respectively, of $\mathcal{S}$. Then $\mathcal{S}^{re}$ has A and $R_{\mathcal{S}^{re}}$, as its sets of axioms and rules, where $R_{\mathcal{S}^{re}}$ is as described in the definition of a restricted rules companion. Now, since any rule in $R_{\mathcal{S}^{re}}$ is already restricted, A and $R_{\mathcal{S}^{re}}$ also comprise a Hilbert-style presentation for $(\mathcal{S}^{re})^{re}$. Thus, $(\mathcal{S}^{re})^{re} = \mathcal{S}^{re}$, i.e., $(\vdash^{re})^{re} = \vdash^{re}$.

Theorem 2. *Suppose $\mathcal{S}_1 = \langle \mathcal{L}, \vdash_1 \rangle$ and $\mathcal{S}_2 = \langle \mathcal{L}, \vdash_2 \rangle$ are two logics such that $\vdash_1 \subseteq \vdash_2$. Then, $\vdash_1^l \subseteq \vdash_2^l$. Moreover, if $\mathcal{S}_1$ and $\mathcal{S}_2$ are Hilbert-style logics, such that for any $\dfrac{\Gamma}{\alpha} \in R_{\mathcal{S}_1^{re}}$, $\Gamma \vdash_2^{re} \alpha$, then $\vdash_1^{re} \subseteq \vdash_2^{re}$.*

---

[1] A logic $\langle \mathcal{L}, \vdash \rangle$ is said to be *finitary* if the following property holds. For all $\Gamma \cup \{\alpha\} \subseteq \mathcal{L}$, if $\Gamma \vdash \alpha$ then there exists a finite $\Gamma' \subseteq \Gamma$ such that $\Gamma' \vdash \alpha$.
[2] A logic $\langle \mathcal{L}, \vdash \rangle$, and also the consequence relation $\vdash$, is said to be *monotonic* if the following property holds. For all $\Gamma \cup \Sigma \cup \{\alpha\} \subseteq \mathcal{L}$, if $\Gamma \vdash \alpha$ and $\Gamma \subseteq \Sigma$ then $\Sigma \vdash \alpha$.

Proof. Suppose $\Sigma \cup \{\alpha\} \subseteq \mathcal{L}$ such that $\Sigma \vdash_1^l \alpha$. Then there exists $\Delta \subseteq \Sigma$ with $\text{var}(\Delta) \subseteq \text{var}(\alpha)$ such that $\Delta \vdash_1 \alpha$. Since $\vdash_1 \subseteq \vdash_2$, $\Delta \vdash_2 \alpha$. Thus, $\Sigma \vdash_2^l \alpha$. Hence, $\vdash_1^l \subseteq \vdash_2^l$.

Next, suppose $\mathcal{S}_1, \mathcal{S}_2$ are Hilbert-style logics such that for any $\dfrac{\Gamma}{\alpha} \in R_{\mathcal{S}_1^{re}}$, $\Gamma \vdash_2^{re} \alpha$. Now, let $\Sigma \cup \{\alpha\} \subseteq \mathcal{L}$ such that $\Sigma \vdash_1^{re} \alpha$. Then, there exists a derivation $D = \langle \alpha_1, \ldots, \alpha_n = \alpha \rangle$ of α from Σ, where for each $1 \leq i \leq n$, α_i is either an axiom of $\mathcal{S}_1$, or an element of Σ, or is obtained by applying a rule of inference in $R_{\mathcal{S}_1^{re}}$. If α_i is an axiom, then $\vdash_1 \alpha_i$, and since $\vdash_1 \subseteq \vdash_2$, $\vdash_2 \alpha_i$. Then, by [1, Theorem 3.4], $\vdash_2^{re} \alpha_i$. So, there exists a derivation of α_i in $\mathcal{S}_2^{re}$. Now, suppose α_i is obtained by applying a rule of inference $\dfrac{\Delta}{\alpha_i} \in R_{\mathcal{S}_1^{re}}$, where $\Delta \subseteq \{\alpha_1, \ldots, \alpha_{i-1}\}$. Then, by assumption, $\Delta \vdash_2^{re} \alpha_i$, and hence, there exists a derivation of α_i from Δ in $\mathcal{S}_2^{re}$. So, we can translate the derivation D of α from Σ as follows. For each $1 \leq i \leq n$, if α_i is an axiom, then we replace α_i by the elements of a derivation of α_i in $\mathcal{S}_2^{re}$. If α_i is obtained from $\Delta \subseteq \{\alpha_1, \ldots, \alpha_{i-1}\}$, by using a rule of inference in $\mathcal{S}_1^{re}$, then we replace α_i by the elements of a derivation of α_i from Δ in $\mathcal{S}_2^{re}$. Finally, if $\alpha_i \in \Sigma$, then we keep it unchanged. Let D' be the resulting sequence of formulas. Clearly, D' is a derivation of α from Σ in $\mathcal{S}_2^{re}$. Thus, $\Sigma \vdash_2^{re} \alpha$. Hence, $\vdash_1^{re} \subseteq \vdash_2^{re}$.

Remark 1. It is not true, in general, that if $\mathcal{S}_1 = \langle \mathcal{L}, \vdash_1 \rangle$ and $\mathcal{S}_2 = \langle \mathcal{L}, \vdash_2 \rangle$ are two Hilbert-style logics with $\vdash_1 \subseteq \vdash_2$, then $\vdash_1^{re} \subseteq \vdash_2^{re}$. This can be seen from the following example.

Suppose $\mathcal{L}$ is the formula algebra over a countable set of variables V of type $\{\wedge, \vee\}$. Let $\mathcal{S}_1 = \langle \mathcal{L}, \vdash_1 \rangle$ be the Hilbert-style logic with an empty set of axioms and the following two rules of inference.

$$R_1 : \frac{\alpha \wedge \beta}{\alpha} \quad \text{and} \quad R_2 : \frac{\alpha}{\alpha \vee \beta}, \quad \text{where } \alpha, \beta \in \mathcal{L}.$$

$\mathcal{S}_1$ is the same logic that was used in [1, Remark 3.7] to show that the left variable inclusion companion of a Hilbert-style logic can differ from its restricted rules companion.

Let $\mathcal{S}_2 = \langle \mathcal{L}, \vdash_2 \rangle$ be the Hilbert-style logic with an empty set of axioms and the following rule of inference in addition to R_1, R_2 above.

$$R_3 : \frac{\alpha \wedge \beta}{\alpha \vee \beta}, \quad \text{where } \alpha, \beta \in \mathcal{L}.$$

Clearly, $\vdash_2 \subseteq \vdash_1$, since R_3 can be derived in $\mathcal{S}_1$ as follows. For any $\alpha, \beta \in \mathcal{L}$,

$$\alpha \wedge \beta \vdash_1 \; 1.\, \alpha \wedge \beta$$
$$ 2.\, \alpha \qquad [R_1 \text{ on } (1)]$$
$$ 3.\, \alpha \vee \beta \qquad [R_2 \text{ on } (2)]$$

We note that $\mathcal{S}_1^{re} = \langle \mathcal{L}, \vdash_1^{re} \rangle$, is the logic induced by the same set of axioms and the following two rules of inference.

$$R_1' : \frac{\alpha \wedge \beta}{\alpha} \text{ such that } \mathrm{var}(\alpha \wedge \beta) \subseteq \mathrm{var}(\alpha), \text{ i.e., } \mathrm{var}(\beta) \subseteq \mathrm{var}(\alpha), \text{ and}$$

$$R_2 : \frac{\alpha}{\alpha \vee \beta},$$

while $\mathcal{S}_2^{re} = \langle \mathcal{L}, \vdash_1^{re} \rangle$, is the logic induced by the same set of axioms and the rules R_1', R_2, and R_3. (R_2 and R_3 do not need to be restricted as $\mathrm{var}(\alpha) \subseteq \mathrm{var}(\alpha \vee \beta)$ and $\mathrm{var}(\alpha \wedge \beta) = \mathrm{var}(\alpha \vee \beta)$ for any $\alpha, \beta \in \mathcal{L}$.)

Now, suppose p, q are distinct variables. Then, while $p \wedge q \vdash_2^{re} p \vee q$, $p \wedge q \not\vdash_1^{re} p \vee q$, since we cannot apply R_1' and R_2 to derive $p \vee q$ from $p \wedge q$ in $\mathcal{S}_1$, as shown in [1, Remark 3.7].

Thus, $\vdash_2^{re} \not\subseteq \vdash_1^{re}$ although $\vdash_2 \subseteq \vdash_1$.

Theorem 3. *Suppose $\mathcal{S} = \langle \mathcal{L}, \vdash \rangle$ is a Hilbert-style logic. Then, $\mathcal{S}^l = \mathcal{S}^{re}$ iff $(\mathcal{S}^{re})^l = \mathcal{S}^l$. In other words, $\vdash^l = \vdash^{re}$ iff $(\vdash^{re})^l = \vdash^l$, i.e., for all $\Gamma \cup \{\alpha\} \subseteq \mathcal{L}$, $\Gamma \vdash^l \alpha$ iff there exists $\Delta \subseteq \Gamma$ such that $\mathrm{var}(\Delta) \subseteq \mathrm{var}(\alpha)$ and $\Delta \vdash^{re} \alpha$.*

Proof. Suppose $\mathcal{S}^l = \mathcal{S}^{re}$, i.e., $\vdash^l = \vdash^{re}$. Now, by Lemma 1 (iii), $\vdash^{re} \subseteq \vdash$. So, by Theorem 2, $(\vdash^{re})^l \subseteq \vdash^l$. Now, suppose $\Gamma \cup \{\alpha\} \subseteq \mathcal{L}$ such that $\Gamma \vdash^l \alpha$. Then, there exists $\Delta \subseteq \Gamma$ such that $\mathrm{var}(\Delta) \subseteq \mathrm{var}(\alpha)$ and $\Delta \vdash \alpha$. Clearly, $\Delta \vdash^l \alpha$ as well. This implies that $\Delta \vdash^{re} \alpha$, since $\vdash^l = \vdash^{re}$. Now, as $\Delta \subseteq \Gamma$, $\mathrm{var}(\Delta) \subseteq \mathrm{var}(\alpha)$, and $\Delta \vdash^{re} \alpha$, $\Gamma (\vdash^{re})^l \alpha$. Thus, $\vdash^l \subseteq (\vdash^{re})^l$. Hence, $(\vdash^{re})^l = \vdash^l$.

Conversely, suppose $(\vdash^{re})^l = \vdash^l$. We know that $\vdash^{re} \subseteq \vdash^l$. Thus, we only need to show that $\vdash^l \subseteq \vdash^{re}$. Suppose $\Gamma \cup \{\alpha\} \subseteq \mathcal{L}$ such that $\Gamma \vdash^l \alpha$. Then, by our assumption, $\Gamma (\vdash^{re})^l \alpha$. So, there exists $\Delta \subseteq \Gamma$ such that $\mathrm{var}(\Delta) \subseteq \mathrm{var}(\alpha)$ and $\Delta \vdash^{re} \alpha$. Now, as $\mathcal{S}^{re}$ is a Hilbert-style logic, it is monotonic. This implies that $\Gamma \vdash^{re} \alpha$. Thus, $\vdash^l \subseteq \vdash^{re}$. Hence $\vdash^l = \vdash^{re}$.

3 Relational Companions of a Logical Structure

We now let go of the formula algebras and land in the arena of logical structures that were described in the introduction. The attempt here would be to generalize the notion of a logic of variable inclusion. In doing so, we will be able to capture a lot more than just the logics of left variable inclusion.

Definition 3. *Suppose $\mathcal{S} = \langle \mathcal{L}, \vdash \rangle$ is a logical structure and $\varrho \subseteq \mathcal{P}(\mathcal{L}) \times \mathcal{L}$.*

(i) *The ϱ-companion of $\mathcal{S}$ is the logical structure $\mathcal{S}^\varrho = \langle \mathcal{L}, \vdash^\varrho \rangle$, where $\vdash^\varrho \subseteq \mathcal{P}(\mathcal{L}) \times \mathcal{L}$ is defined as follows. For any $\Gamma \cup \{\alpha\} \subseteq \mathcal{L}$,*

$$\Gamma \vdash^\varrho \alpha \text{ iff there is a } \Delta \subseteq \Gamma \text{ such that } (\Delta, \alpha) \in \varrho \text{ and } \Delta \vdash \alpha.$$

(ii) *The **pure** ϱ-companion of $\mathcal{S}$ is the logical structure $\mathcal{S}^{p\varrho} = \langle \mathcal{L}, \vdash^{p\varrho} \rangle$, where $\vdash^{p\varrho} \subseteq \mathcal{P}(\mathcal{L}) \times \mathcal{L}$ is defined as follows. For any $\Gamma \cup \{\alpha\} \subseteq \mathcal{L}$,*

$$\Gamma \vdash^{p\varrho} \alpha \text{ iff there is a } \Delta \subseteq \Gamma \text{ such that } \Delta \neq \emptyset, (\Delta, \alpha) \in \varrho \text{ and } \Delta \vdash \alpha.$$

Any such ϱ-companion ($p\varrho$-companion) $\mathcal{S}^{\varrho}$ ($\mathcal{S}^{p\varrho}$) is called a relational companion *(pure relational companion) of $\mathcal{S}$.*

Example 1. Suppose $\mathcal{S} = \langle \mathcal{L}, \vdash \rangle$ is a logic (in the usual sense).

(i) The left variable inclusion companion $\mathcal{S}^l$ is the relational companion $\mathcal{S}^L = \langle \mathcal{L}, \vdash^L \rangle$ of $\mathcal{S}$, where the relation $L \subseteq \mathcal{P}(\mathcal{L}) \times \mathcal{L}$ is defined by variable inclusion from left to right, i.e., $(\Delta, \alpha) \in L$ iff $\mathrm{var}(\Delta) \subseteq \mathrm{var}(\alpha)$. Thus, $\mathcal{S}^l = \mathcal{S}^L$, the L-companion of $\mathcal{S}$.

(ii) The *pure left variable inclusion companion* of $\mathcal{S}$, denoted by $\mathcal{S}^{pl} = \langle \mathcal{L}, \vdash^{pl} \rangle$, is defined (e.g., in [8,12]) as follows. For any $\Gamma \cup \{\alpha\} \subseteq \mathcal{L}$,

$\Gamma \vdash^{pl} \alpha$ iff there is a $\Delta \subseteq \Gamma$ such that $\Delta \neq \emptyset, \mathrm{var}(\Delta) \subseteq \mathrm{var}(\alpha)$, and $\Delta \vdash \alpha$.

Clearly, $\mathcal{S}^{pl}$ is the pure L-companion of $\mathcal{S}$, where the relation L is as defined in (i), i.e., $\mathcal{S}^{pl} = \mathcal{S}^{pL}$.

Example 2. Suppose $\mathcal{S} = \langle \mathcal{L}, \vdash \rangle$ is a monotonic logic (in the usual sense).

(i) The right variable inclusion companion of $\mathcal{S}$, denoted by $\mathcal{S}^r = \langle \mathcal{L}, \vdash^r \rangle$ is defined (e.g., in [12]) as follows. For any $\Gamma \cup \{\alpha\} \subseteq \mathcal{L}$,

$\Gamma \vdash^r \alpha$ iff Γ contains an $\mathcal{S}$-antitheorem, or $\Gamma \vdash \alpha$ and $\mathrm{var}(\alpha) \subseteq \mathrm{var}(\Gamma)$.

An $\mathcal{S}$-antitheorem is a set $\Sigma \subseteq \mathcal{L}$ such that for every substitution σ, $\sigma(\Sigma) \vdash \varphi$ for all $\varphi \in \mathcal{L}$, i.e., $\sigma(\Sigma)$ *explodes* in $\mathcal{S}$ for every σ[3].
$\mathcal{S}^r$ can be seen as the relational companion $\mathcal{S}^R = \langle \mathcal{L}, \vdash^R \rangle$ of $\mathcal{S}$, where the relation $R \subseteq \mathcal{P}(\mathcal{L}) \times \mathcal{L}$ is defined as follows. $(\Delta, \alpha) \in R$ iff either Δ is an $\mathcal{S}$-antitheorem or $\mathrm{var}(\alpha) \subseteq \mathrm{var}(\Delta)$.
The proof that $\mathcal{S}^r = \mathcal{S}^R$ can be given as follows. Let $\Gamma \cup \{\alpha\} \subseteq \mathcal{L}$. Suppose $\Gamma \vdash^R \alpha$. Then there exists a $\Delta \subseteq \Gamma$ such that $(\Delta, \alpha) \in R$ and $\Delta \vdash \alpha$. Now, since $(\Delta, \alpha) \in R$, either Δ is an $\mathcal{S}$-antitheorem, in which case Γ contains an $\mathcal{S}$-antitheorem, or $\mathrm{var}(\alpha) \subseteq \mathrm{var}(\Delta)$ and $\Delta \vdash \alpha$. In the latter case, as $\Delta \subseteq \Gamma$, $\mathrm{var}(\alpha) \subseteq \mathrm{var}(\Gamma)$ and since $\mathcal{S}$ is monotonic, $\Gamma \vdash \alpha$ as well. Thus, $\Gamma \vdash^r \alpha$.
Conversely, suppose $\Gamma \vdash^r \alpha$. Then, either Γ contains an $\mathcal{S}$-antitheorem or $\mathrm{var}(\alpha) \subseteq \mathrm{var}(\Gamma)$ and $\Gamma \vdash \alpha$. If Γ contains an $\mathcal{S}$-antitheorem Δ, then $\Delta \subseteq \Gamma$ such that $(\Delta, \alpha) \in R$ and $\Delta \vdash \alpha$ as Δ is an $\mathcal{S}$-antitheorem. On the other hand, if $\Gamma \vdash \alpha$ and $\mathrm{var}(\alpha) \subseteq \mathrm{var}(\Gamma)$, then Γ is its own subset such that $(\Gamma, \alpha) \in R$ and $\Gamma \vdash \alpha$. Thus, in either case, there exists $\Delta \subseteq \Gamma$ such that $(\Delta, \alpha) \in R$ and $\Delta \vdash \alpha$, and so, $\Gamma \vdash^R \alpha$. Hence $\vdash^r = \vdash^R$, i.e., $\mathcal{S}^r = \mathcal{S}^R$.

---

[3] Suppose $\mathcal{L}$ is a formula algebra over a set of variables V (which is the case when discussing variable inclusion logics). A *substitution* is any function $\sigma : V \to \mathcal{L}$ that extends to a unique endomorphism (also denoted by σ) from $\mathcal{L}$ to itself via the universal mapping property.

(ii) The *pure right variable inclusion companion* of $\mathcal{S}$, denoted by $\mathcal{S}^{pr} = \langle \mathcal{L}, \vdash^{pr} \rangle$, is defined (e.g., in [8,12]) as follows. For any $\Gamma \cup \{\alpha\} \subseteq \mathcal{L}$,

$$\Gamma \vdash^{pr} \alpha \text{ iff } \Gamma \vdash \alpha \text{ and } \text{var}(\alpha) \subseteq \text{var}(\Gamma).$$

$\mathcal{S}^{pr}$ can be seen as a relational companion $\mathcal{S}^{PR} = \langle \mathcal{L}, \vdash^{PR} \rangle$ of $\mathcal{S}$, where the relation $PR \subseteq \mathcal{P}(\mathcal{L}) \times \mathcal{L}$ is defined by variable inclusion from right to left, i.e., $(\Delta, \alpha) \in PR$ iff $\text{var}(\alpha) \subseteq \text{var}(\Delta)$. This can be shown with a similar argument as for the right variable inclusion companions above. The monotonicity of $\mathcal{S}$ is required for showing that $\vdash^{PR} \subseteq \vdash^{pr}$.

Although $\mathcal{S}^{pr}$ can be seen as a relational companion of $\mathcal{S}$, it cannot be described as a pure relational companion of $\mathcal{S}$, unless $\mathcal{S}$ is a logic without antitheorems. However, if $\mathcal{S}$ is a logic without antitheorems, then $\mathcal{S}^{pr} = \mathcal{S}^{pR}$, where R is the relation used in (i).

Example 3. Some more inclusion logics have been introduced in [9] as generalizations of the left and right variable inclusion logics. In these companion logics, the containment requirement is extended to classes of subformulas. We might call these the *left* and *right subformula inclusion companion logics* (the actual names require some additional machinery and hence we avoid these). While the left subformula inclusion companions of a logic can be seen as relational companions of it, the right subformula inclusion companions of a monotonic logic can be seen as relational companions of it. This is much like the case for left and right variable inclusion companions.

It is easy to see that the left, pure left, and pure right variable inclusion companions of a logic $\mathcal{S} = \langle \mathcal{L}, \vdash \rangle$, with a unary (negation) operator $\neg$ in the signature, are all paraconsistent with respect to $\neg$, i.e., there exists $\alpha, \beta \in \mathcal{L}$ such that $\{\alpha, \neg\alpha\} \not\vdash \beta$. In other words, the principle of explosion, ECQ, with respect to $\neg$ (we call this $\neg$-ECQ), fails in these companion logics (see [3] for more on the paraconsistency of these companion logics). The right variable inclusion companion of a logic is, however, not necessarily paraconsistent with respect to a $\neg$. Thus, not all relational companions of a logic, with such a unary operator $\neg$, are paraconsistent with respect to $\neg$. It can also be observed from the definition of a relational companion that the matter of paraconsistency depends on the relation. The following example from [14] can be seen as a relational companion created with the intention of obtaining a paraconsistent logical structure. The authors have called this process *paraconsistentization*.

Example 4. Suppose $\mathcal{S} = \langle \mathcal{L}, \vdash \rangle$ is a logical structure. Then $\mathcal{S}^{\mathbb{P}} = \langle \mathcal{L}, \vdash^{\mathbb{P}} \rangle$ is the logical structure, where $\vdash^{\mathbb{P}} \subseteq \mathcal{P}(\mathcal{L}) \times \mathcal{L}$ is defined as follows. For any $\Gamma \cup \{\alpha\} \subseteq \mathcal{L}$,

$$\Gamma \vdash^{\mathbb{P}} \alpha \text{ iff there is a } \Delta \subseteq \Gamma \text{ such that } \Delta \text{ is } \mathcal{S}\text{-nontrivial and } \Delta \vdash \alpha.$$

A set Δ is $\mathcal{S}$-nontrivial, if there exists $\varphi \in \mathcal{L}$ such that $\Delta \not\vdash \varphi$.

Theorem 4. *Suppose $\mathcal{S} = \langle \mathcal{L}, \vdash \rangle$ is a logical structure and $\varrho \subseteq \mathcal{P}(\mathcal{L}) \times \mathcal{L}$.*

(i) Then, $\mathcal{S}^{\varrho} = \langle \mathcal{L}, \vdash^{\varrho} \rangle$ and $\mathcal{S}^{p\varrho} = \langle \mathcal{L}, \vdash^{p\varrho} \rangle$ are monotonic.

(ii) If $\mathcal{S}$ is monotonic, then $\vdash^\varrho, \vdash^{p\varrho} \subseteq \vdash$.
(iii) $(\mathcal{S}^\varrho)^\varrho = \mathcal{S}^\varrho$, i.e., $(\vdash^\varrho)^\varrho = \vdash^\varrho$, and $(\mathcal{S}^{p\varrho})^{p\varrho} = \mathcal{S}^{p\varrho}$, i.e., $(\vdash^{p\varrho})^{p\varrho} = \vdash^{p\varrho}$.

Proof. (i) Suppose $\Gamma \cup \Sigma \cup \{\alpha\} \subseteq \mathcal{L}$ such that $\Gamma \subseteq \Sigma$ and $\Gamma \vdash^\varrho \alpha$. Then, there exists $\Delta \subseteq \Gamma$ such that $(\Delta, \alpha) \in \varrho$ and $\Delta \vdash \alpha$. So, $\Delta \subseteq \Sigma$ as well. Thus, $\Sigma \vdash^\varrho \alpha$. Hence, $\mathcal{S}^\varrho$ is monotonic.

An almost identical argument can be used to show that $\mathcal{S}^{p\varrho}$ is monotonic.

(ii) Suppose $\mathcal{S}$ is monotonic. Let $\Gamma \cup \{\alpha\} \subseteq \mathcal{L}$ and $\Gamma \vdash^\varrho \alpha$. Then, there exists $\Delta \subseteq \Gamma$ such that $(\Delta, \alpha) \in \varrho$ and $\Delta \vdash \alpha$. So, by monotonicity, $\Gamma \vdash \alpha$. Hence, $\vdash^\varrho \subseteq \vdash$. It is easy to see that $\vdash^{p\varrho} \subseteq \vdash^\varrho$. Thus, if $\vdash^\varrho \subseteq \vdash$, then $\vdash^{p\varrho} \subseteq \vdash$ as well.

(iii) By (i), $\mathcal{S}^\varrho$ is monotonic. So, by (ii), $(\vdash^\varrho)^\varrho \subseteq \vdash^\varrho$.
Suppose $\Gamma \cup \{\alpha\} \subseteq \mathcal{L}$ and $\Gamma \vdash^\varrho \alpha$. Then, there exists $\Delta \subseteq \Gamma$ such that $(\Delta, \alpha) \in \varrho$ and $\Delta \vdash \alpha$. Clearly, $\Delta \vdash^\varrho \alpha$ as well. Thus, $\Gamma (\vdash^\varrho)^\varrho \alpha$, and so, $\vdash^\varrho \subseteq (\vdash^\varrho)^\varrho$. Hence, $(\vdash^\varrho)^\varrho = \vdash^\varrho$.

The argument for $\mathcal{S}^{p\varrho}$ is identical, except for the non-emptiness requirement on the Δ in the converse part.

Remark 2. The above theorem generalizes the results concerning the left variable inclusion companion of a logic in Lemma 1.

Theorem 5. *Suppose $\mathcal{S}_1 = \langle \mathcal{L}, \vdash_1 \rangle, \mathcal{S}_2 = \langle \mathcal{L}, \vdash_2 \rangle$ are logical structures such that $\vdash_1 \subseteq \vdash_2$, and $\varrho, \sigma \subseteq \mathcal{P}(\mathcal{L}) \times \mathcal{L}$ such that $\varrho \subseteq \sigma$. Then, $\vdash_1^\varrho \subseteq \vdash_2^\sigma$ and $\vdash_1^{p\varrho} \subseteq \vdash_2^{p\sigma}$.*

Proof. Suppose $\Gamma \cup \{\alpha\} \subseteq \mathcal{L}$ such that $\Gamma \vdash_1^\varrho \alpha$. Then, there exists $\Delta \subseteq \Gamma$ such that $(\Delta, \alpha) \in \varrho$ and $\Delta \vdash_1 \alpha$. Since $\varrho \subseteq \sigma$, $(\Delta, \alpha) \in \sigma$ and as $\vdash_1 \subseteq \vdash_2$, $\Delta \vdash_2 \alpha$. Thus, $\Gamma \vdash_2^\sigma \alpha$. Hence, $\vdash_1^\varrho \subseteq \vdash_2^\sigma$.

It can be proved that $\vdash_1^{p\varrho} \subseteq \vdash_2^{p\sigma}$ with similar arguments.

Corollary 1. *The following are some immediate observations from the above theorem.*

(i) *Suppose $\mathcal{S}_1 = \langle \mathcal{L}, \vdash_1 \rangle, \mathcal{S}_2 = \langle \mathcal{L}, \vdash_2 \rangle$ are logical structures such that $\vdash_1 \subseteq \vdash_2$ and $\varrho \subseteq \mathcal{P}(\mathcal{L}) \times \mathcal{L}$. Then, $\vdash_1^\varrho \subseteq \vdash_2^\varrho$ and $\vdash_1^{p\varrho} \subseteq \vdash_2^{p\varrho}$.*
(ii) *Suppose $\mathcal{S} = \langle \mathcal{L}, \vdash \rangle$ is a logical structure and $\varrho, \sigma \subseteq \mathcal{P}(\mathcal{L}) \times \mathcal{L}$. Then $\vdash^\varrho, \vdash^\sigma \subseteq \vdash^{\varrho \cup \sigma}$ and $\vdash^{\varrho \cap \sigma} \subseteq \vdash^\varrho, \vdash^\sigma$.*

Theorem 6. *Suppose $\mathcal{S} = \langle \mathcal{L}, \vdash \rangle$ is a logical structure and $\varrho, \sigma \subseteq \mathcal{P}(\mathcal{L}) \times \mathcal{L}$.*

(i) $(\vdash^\varrho)^\sigma \subseteq \vdash^\varrho$. *The equality holds if $\varrho \subseteq \sigma$.*
(ii) *If $\mathcal{S}$ is monotonic, then $(\vdash^\varrho)^\sigma \subseteq \vdash^\sigma$.*
(iii) *If $\varrho \subseteq \sigma$, then $(\vdash^\varrho)^\sigma \subseteq \vdash^\sigma$.*
(iv) *If $\vdash^\sigma \subseteq \vdash^\varrho$, then $\vdash^\sigma \subseteq (\vdash^\varrho)^\sigma$.*
(v) *If $\varrho \subseteq \sigma$, then $\vdash^\varrho = \vdash^\sigma$ iff $(\vdash^\varrho)^\sigma = \vdash^\sigma$.*

Analogous statements hold for pure relational companions of $\mathcal{S}$.

Proof. (i) The ϱ-companion of $\mathcal{S}$, $\mathcal{S}^\varrho$ is monotonic, by Theorem 4(i). So, by Theorem 4(ii), $(\vdash^\varrho)^\sigma \subseteq \vdash^\varrho$.

Now, suppose $\varrho \subseteq \sigma$ and $\Gamma \cup \{\alpha\} \subseteq \mathcal{L}$ such that $\Gamma \vdash^\varrho \alpha$. Then, there exists $\Delta \subseteq \Gamma$ such that $(\Delta, \alpha) \in \varrho$ and $\Delta \vdash \alpha$. Clearly, $\Delta \vdash^\varrho \alpha$. Since $\varrho \subseteq \sigma$, $(\Delta, \alpha) \in \sigma$. Thus, $\Gamma (\vdash^\varrho)^\sigma \alpha$. So, $\vdash^\sigma \subseteq (\vdash^\varrho)^\sigma$. Hence, $(\vdash^\varrho)^\sigma = \vdash^\varrho$.

(ii) Suppose $\mathcal{S}$ is monotonic. Then, by Theorem 4(ii), $\vdash^\varrho \subseteq \vdash$. Hence, by Corollary 1(i), $(\vdash^\varrho)^\sigma \subseteq \vdash^\sigma$.

(iii) Suppose $\varrho \subseteq \sigma$ and $\Gamma \cup \{\alpha\} \subseteq \mathcal{L}$ such that $\Gamma(\vdash^\varrho)^\sigma \alpha$. Then, there exists $\Delta \subseteq \Gamma$ such that $(\Delta, \alpha) \in \sigma$ and $\Delta \vdash^\varrho \alpha$. So, there exists $\Delta' \subseteq \Delta$ such that $(\Delta', \alpha) \in \varrho$ and $\Delta' \vdash \alpha$. Now, $\Delta' \subseteq \Gamma$, and as $\varrho \subseteq \sigma$, $(\Delta', \alpha) \in \sigma$. So, $\Gamma \vdash^\sigma \alpha$. Thus, $(\vdash^\varrho)^\sigma \subseteq \vdash^\sigma$.

(iv) Suppose $\vdash^\sigma \subseteq \vdash^\varrho$ and $\Gamma \cup \{\alpha\} \subseteq \mathcal{L}$ such that $\Gamma \vdash^\sigma \alpha$. Then, there exists $\Delta \subseteq \Gamma$ such that $(\Delta, \alpha) \in \sigma$ and $\Delta \vdash \alpha$. Clearly, $\Delta \vdash^\sigma \alpha$. Now, as $\vdash^\sigma \subseteq \vdash^\varrho$, $\Delta \vdash^\varrho \alpha$ as well. Thus, $\Gamma(\vdash^\varrho)^\sigma \alpha$. Hence, $\vdash^\sigma \subseteq (\vdash^\varrho)^\sigma$.

(v) Suppose $\varrho \subseteq \sigma$. Moreover, suppose $\vdash^\varrho = \vdash^\sigma$. Then, using parts (iii) and (iv) above, we have $(\vdash^\varrho)^\sigma = \vdash^\sigma$.

Conversely, suppose $(\vdash^\varrho)^\sigma = \vdash^\sigma$. Let $\Gamma \cup \{\alpha\} \subseteq \mathcal{L}$ such that $\Gamma \vdash^\sigma \alpha$. So, $\Gamma(\vdash^\varrho)^\sigma \alpha$. Then, there exists $\Delta \subseteq \Gamma$ such that $(\Delta, \alpha) \in \sigma$ and $\Delta \vdash^\varrho \alpha$. This implies that there exists $\Delta' \subseteq \Delta$ such that $(\Delta', \alpha) \in \varrho$ and $\Delta' \vdash \alpha$. Since $\Delta' \subseteq \Gamma$, $\Gamma \vdash^\varrho \alpha$. Thus. $\vdash^\sigma \subseteq \vdash^\varrho$. Now, as $\varrho \subseteq \sigma$, $\vdash^\varrho \subseteq \vdash^\sigma$, by Theorem 5. Hence, $\vdash^\varrho = \vdash^\sigma$.

The proofs for the pure relational companions of $\mathcal{S}$ can be constructed with similar arguments.

Definition 4. *Suppose A is a set and $\varrho \subseteq \mathcal{P}(A) \times A$. Then, ϱ is said to be downward directed if for any $\Delta \cup \{\alpha\} \subseteq \mathcal{L}$, $(\Delta, \alpha) \in \varrho$ implies that $(\Delta', \alpha) \in \varrho$ for all $\Delta' \subseteq \Delta$.*

Theorem 7. *Suppose $\mathcal{S} = \langle \mathcal{L}, \vdash \rangle$ is a logical structure and $\varrho, \sigma \subseteq \mathcal{P}(\mathcal{L}) \times \mathcal{L}$. If σ is downward directed, then $(\vdash^\varrho)^\sigma \subseteq (\vdash^\sigma)^\varrho$. Hence, if ϱ, σ are both downward directed, then $(\vdash^\varrho)^\sigma = (\vdash^\sigma)^\varrho$.*

Proof. Suppose σ is downward directed. Let $\Gamma \cup \{\alpha\} \subseteq \mathcal{L}$ such that $\Gamma(\vdash^\varrho)^\sigma \alpha$. Then, there exists $\Delta \subseteq \Gamma$ such that $(\Delta, \alpha) \in \sigma$ and $\Delta \vdash^\varrho \alpha$. This implies that there exists $\Delta' \subseteq \Delta$ such that $(\Delta', \alpha) \in \varrho$ and $\Delta' \vdash \alpha$. Now, as σ is downward directed, $(\Delta', \alpha) \in \sigma$. Then, $\Delta' \vdash^\sigma \alpha$. Now, since $(\Delta', \alpha) \in \varrho$ and $\Delta' \subseteq \Gamma$, this implies that $\Gamma(\vdash^\sigma)^\varrho \alpha$. Hence, $(\vdash^\varrho)^\sigma \subseteq (\vdash^\sigma)^\varrho$.

Thus, if ϱ is also downward directed, then $(\vdash^\sigma)^\varrho \subseteq (\vdash^\varrho)^\sigma$, and hence, in that case, $(\vdash^\varrho)^\sigma = (\vdash^\sigma)^\varrho$.

It is not hard to see that, for any logic $\mathcal{S} = \langle \mathcal{L}, \vdash \rangle$, $\vdash \alpha$ iff $\vdash^l \alpha$ for all $\alpha \in \mathcal{L}$. The following theorem generalizes this to relational companions of logical structures.

Theorem 8. *Suppose $\mathcal{S} = \langle \mathcal{L}, \vdash \rangle$ is a logical structure and $\varrho \subseteq \mathcal{P}(\mathcal{L}) \times \mathcal{L}$ such that $(\emptyset, \alpha) \in \varrho$ for all $\alpha \in \mathcal{L}$. Then, for any $\alpha \in \mathcal{L}$, $\emptyset \vdash \alpha$ (written as $\vdash \alpha$) iff $\emptyset \vdash^\varrho \alpha$ (written as $\vdash^\varrho \alpha$).*

Proof. Suppose $\alpha \in \mathcal{L}$ such that $\vdash \alpha$. Then, as $(\emptyset, \alpha) \in \varrho$, $\vdash^\varrho \alpha$. Conversely, suppose $\vdash^\varrho \alpha$. Then, there exists $\Delta \subseteq \emptyset$ such that $(\Delta, \alpha) \in \varrho$ and $\Delta \vdash \alpha$. Clearly, $\Delta = \emptyset$. Thus, $\vdash \alpha$.

As discussed above, the left, pure left, and pure right variable inclusion companions of a logic are paraconsistent (see [3] for a detailed study on this). The following theorem generalizes this to certain relational companions of logical structures.

Definition 5. *Suppose A, B are sets and $\varrho \subseteq A \times B$. Then, ϱ is said to have finite reach if for every $a \in A$, the set $\varrho_a = \{b \in B \mid (a,b) \in \varrho\}$ is finite.*

Theorem 9. *Suppose $\mathcal{S} = \langle \mathcal{L}, \vdash \rangle$ is a logical structure such that $\mathcal{L}$ is infinite and $\varrho \subseteq \mathcal{P}(\mathcal{L}) \times \mathcal{L}$ has finite reach. Then, for every finite $\Gamma \subseteq \mathcal{L}$, there exists $\beta \in \mathcal{L}$ such that $\Gamma \nvdash^\varrho \beta$, i.e., every finite $\Gamma \subseteq \mathcal{L}$ is nontrivial in $\mathcal{S}^\varrho$. The same is true for the pure ϱ-companion of $\mathcal{S}$.*

Proof. Suppose ϱ has finite reach. Let Γ be a finite subset of $\mathcal{L}$. So, $\mathcal{P}(\Gamma)$ is finite. Now, as ϱ has finite reach, $\varrho_\Delta = \{\alpha \in \mathcal{L} \mid (\Delta, \alpha) \in \varrho\}$ is finite for every $\Delta \in \mathcal{P}(\Gamma)$. Thus, $\bigcup_{\Delta \in \mathcal{P}(\Gamma)} \varrho_\Delta$, being a finite union of finite sets, is finite. Let $\beta \in \mathcal{L} \setminus \bigcup_{\Delta \in \mathcal{P}(\Gamma)} \varrho_\Delta$ (such a β exists since $\mathcal{L}$ is infinite). Then, $(\Delta, \beta) \notin \varrho$ for every $\Delta \subseteq \Gamma$. Thus, $\Gamma \nvdash^\varrho \beta$. Hence, every finite subset of $\mathcal{L}$ is non-trivial.

An almost identical argument proves the same result for $\mathcal{S}^{p\varrho} = \langle \mathcal{L}, \vdash^{p\varrho} \rangle$.

Corollary 2. *Suppose $\mathcal{S} = \langle \mathcal{L}, \vdash \rangle$ is a logic (in the usual sense) such that there is a unary (negation) operator $\neg$ in the signature. Moreover, let $\varrho \subseteq \mathcal{P}(\mathcal{L}) \times \mathcal{L}$ be a relation with finite reach. Then, $\neg$-ECQ fails in $\mathcal{S}^\varrho$ and $\mathcal{S}^{p\varrho}$. Thus, $\mathcal{S}^\varrho$ and $\mathcal{S}^{p\varrho}$ are paraconsistent with respect to $\neg$.*

Definition 6. *A logical structure $\mathcal{S} = \langle \mathcal{L}, \vdash \rangle$ is said to be finitely trivializable if there exists a finite $\Gamma \subseteq \mathcal{L}$ such that $\Gamma \vdash \alpha$ for all $\alpha \in \mathcal{L}$, i.e., there is a finite trivial subset of $\mathcal{L}$.*

Remark 3. Suppose $\mathcal{S} = \langle \mathcal{L}, \vdash \rangle$ is a logical structure such that $\mathcal{L}$ is infinite and $\varrho \subseteq \mathcal{P}(\mathcal{L}) \times \mathcal{L}$ is a relation with finite reach. Then, by Theorem 9, $\mathcal{S}^\varrho$ and $\mathcal{S}^{p\varrho}$ are not finitely trivializable.

Suppose $\mathcal{S} = \langle \mathcal{L}, \vdash \rangle$ is a logical structure. The generalized principle of explosion (gECQ) was introduced in [2] as follows. For every $\alpha \in \mathcal{L}$, there exists $\beta \in \mathcal{L}$ such that $\{\alpha, \beta\} \vdash \gamma$ for all $\gamma \in \mathcal{L}$. A logic or logical structure in which gECQ fails is called NF-paraconsistent (NF stands for Negation-Free).

spECQ (where sp stands for set-point), a principle of explosion introduced in [3], can be described as follows. For each $\Gamma \subsetneq \mathcal{L}$, there exists $\alpha \in \mathcal{L}$ such that $\Gamma \cup \{\alpha\} \subsetneq \mathcal{L}$ and $\Gamma \cup \{\alpha\} \vdash \beta$ for all $\beta \in \mathcal{L}$.

It is easy to see that, if $\mathcal{S} = \langle \mathcal{L}, \vdash \rangle$ is a logical structure that is not finitely trivializable, then gECQ and spECQ fail in it. This is also discussed in [3].

Corollary 3. *Suppose $\mathcal{S} = \langle \mathcal{L}, \vdash \rangle$ is a logical structure such that $\mathcal{L}$ is infinite and $\varrho \subseteq \mathcal{P}(\mathcal{L}) \times \mathcal{L}$ is a relation with finite reach. Then, by Remark 3, gECQ and spECQ fail in $\mathcal{S}^\varrho$ and $\mathcal{S}^{p\varrho}$.*

3.1 Hilbert-Type Logical Structures and Their Restrictions

In this subsection, we first generalize Hilbert-style logics to logical structures and then discuss 'restricted' companions of these as generalizations of the restricted rules companions of Hilbert-style logics discussed in Sect. 2. We first note that any axiom in a Hilbert-style logic can also be regarded as a rule with an empty set of hypotheses. Thus, it is sufficient to deal with Hilbert-style logics induced by only a set of rules. Secondly, for the restricted rules companion of a logic, we used the variable inclusion restriction only to restrict all the rules of inference. This can be generalized to a situation where different rules are restricted in different ways.

Definition 7. *Suppose $\mathcal{L}$ is a set and $\mathcal{R} \subseteq \mathcal{P}(\mathcal{L}) \times \mathcal{L}$. Let $\vdash \, \subseteq \mathcal{P}(\mathcal{L}) \times \mathcal{L}$ be defined as follows. For any $\Gamma \cup \{\alpha\} \subseteq \mathcal{L}$, $\Gamma \vdash \alpha$ if there exists a finite sequence $(\beta_0, \ldots, \beta_n)$ of elements of $\mathcal{L}$ with $\beta_n = \alpha$, and for each $0 \leq i \leq n$, either $\beta_i \in \Gamma$, or there exists $\Gamma' \subseteq \{\beta_0 \ldots, \beta_{i-1}\}$ such that $(\Gamma', \beta_i) \in \mathcal{R}$. Then $\mathcal{S} = \langle \mathcal{L}, \vdash \rangle$ is called the* Hilbert-type logical structure induced by $\mathcal{R}$.

A Hilbert-type logical structure $\langle \mathcal{L}, \vdash \rangle$ *is any logical structure that is induced by some $\mathcal{R} \subseteq \mathcal{P}(\mathcal{L}) \times \mathcal{L}$.*

Remark 4. It is clear from the above definition that a Hilbert-type logical structure $\mathcal{S} = \langle \mathcal{L}, \vdash \rangle$ is finitary and Tarski-type, i.e., reflexive, monotonic, and transitive (see [13] for the definitions).

Definition 8. *Suppose $\mathcal{S} = \langle \mathcal{L}, \vdash \rangle$ is a Hilbert-type logical structure induced by $\mathcal{R} \subseteq \mathcal{P}(\mathcal{L}) \times \mathcal{L}$ and Π is a collection of relations from $\mathcal{P}(\mathcal{L})$ to $\mathcal{L}$, i.e., for any $\sigma \in \Pi$, $\sigma \subseteq \mathcal{P}(\mathcal{L}) \times \mathcal{L}$. Let $\mathcal{R}^\Pi = \{(\Gamma, \alpha) \in \mathcal{R} \mid \text{there exists } \sigma \in \Pi \text{ such that } (\Gamma, \alpha) \in \sigma\}$. Then, the logical structure induced by $\mathcal{R}^\Pi$, denoted by $\mathcal{S}^\Pi = \langle \mathcal{L}, \vdash^\Pi \rangle$, is called the Π-restricted companion of $\mathcal{S}$.*

Remark 5. Suppose $\mathcal{S} = \langle \mathcal{L}, \vdash \rangle$ is a Hilbert-style logic with $\mathcal{R}$ as the set of rules of inference. Let $\Pi = \{\sigma\}$, where $\sigma \subseteq \mathcal{P}(\mathcal{L}) \times \mathcal{L}$ is defined by $(\Gamma, \alpha) \in \sigma$ iff $\mathrm{var}(\Gamma) \subseteq \mathrm{var}(\alpha)$. Then, $\mathcal{R}^\Pi = \{(\Gamma, \alpha) \in \mathcal{R} \mid \mathrm{var}(\Gamma) \subseteq \mathrm{var}(\alpha)\}$, and hence $\mathcal{S}^\Pi = \langle \mathcal{L}, \vdash^\Pi \rangle$, the Π-restricted companion of $\mathcal{S}$, is the restricted rules companion of $\mathcal{S}$, $\mathcal{S}^{re} = \langle \mathcal{L}, \vdash^{re} \rangle$.

We can also see a Π-restricted companion of a Hilbert-type logical structure as a relational companion of a logical structure as follows.

Suppose $\mathcal{S} = \langle \mathcal{L}, \vdash \rangle$ is a logical structure induced by $\mathcal{R} \subseteq \mathcal{P}(\mathcal{L}) \times \mathcal{L}$ and Π is a collection of relations from $\mathcal{P}(\mathcal{L})$ to $\mathcal{L}$. Then, $\mathcal{S}^\Pi = \langle \mathcal{L}, \vdash^\Pi \rangle$ is the Hilbert-type logical structure induced by $\mathcal{R}^\Pi$, where $\mathcal{R}^\Pi$ is as described in the above definition. Let $\varrho \subseteq \mathcal{P}(\mathcal{L}) \times \mathcal{L}$ be defined as follows. $(\Delta, \alpha) \in \varrho$ iff $\Delta \vdash^\Pi \alpha$. Then, $\mathcal{S}^\varrho = \langle \mathcal{L}, \vdash^\varrho \rangle$ is the ϱ-companion of $\mathcal{S}$.

Now, suppose $\Gamma \cup \{\alpha\} \subseteq \mathcal{L}$ and $\Gamma \vdash^\Pi \alpha$. So, there exists a finite sequence $(\beta_0, \ldots, \beta_n)$ of elements of $\mathcal{L}$ with $\beta_n = \alpha$, and for each $0 \leq i \leq n$, either $\beta_i \in \Gamma$, or there exists $\Gamma' \subseteq \{\beta_0 \ldots, \beta_{i-1}\}$ such that $(\Gamma', \beta_i) \in \mathcal{R}^\Pi$. Let $\Delta = \Gamma \cap \{\beta_0, \ldots, \beta_n\}$. Clearly, $\Delta \vdash^\Pi \alpha$, i.e., $(\Delta, \alpha) \in \varrho$. So, $\Gamma \vdash^\varrho \alpha$. Conversely, suppose $\Gamma \vdash^\varrho \alpha$. Then, there exists a $\Delta \subseteq \Gamma$ such that $(\Delta, \alpha) \in \varrho$, i.e., $\Delta \vdash^\Pi \alpha$ and

$\Delta \vdash \alpha$. Now, since $\mathcal{S}^\Pi$ is a Hilbert-type logical structure, and hence, monotonic by Remark 4, $\Delta \vdash^\Pi \alpha$ implies $\Gamma \vdash^\Pi \alpha$. Hence, $\vdash^\Pi = \vdash^\varrho$.

Thus, the restricted rules companions of a logic can also be seen as relational companions of the logic.

4 Conclusions and Future Directions

In this article, we have proposed the relational companions of logical structures as generalizations of the variable inclusion companions of logics and the restricted rules companions of Hilbert-style logics. More properties of these companion logics, especially those of the Π-restricted companions of Hilbert-type logics, could be investigated further.

Another, perhaps interesting, observation is the possibility of using a logical structure $\mathcal{S}_1 = \langle \mathcal{L}, \vdash_1 \rangle$ to define a companion to another logical structure $\mathcal{S}_2 = \langle \mathcal{L}, \vdash_2 \rangle$ in the following way. Since $\vdash_1 \subseteq \mathcal{P}(\mathcal{L}) \times \mathcal{L}$ is also a relation from $\mathcal{P}(\mathcal{L})$ to $\mathcal{L}$, we can define the $\vdash_1$-companion of $\mathcal{S}_2$. This is already the case in the way we have described a Π-restricted companion of a Hilbert-type logical structure as a relational companion at the end of the previous section. This line of investigation can be interesting from the perspective of combining logical structures as well, since this is, in a way, merging the two relations $\vdash_1$ and $\vdash_2$.

References

1. Basu, S.S., Chakraborty, M.K.: Restricted rules of inference and paraconsistency. Logic J. IGPL **30**(3), 534–560 (2021). https://doi.org/10.1093/jigpal/jzab019
2. Basu, S.S., Roy, S.: Negation-free definitions of paraconsistency. In: Indrzejczak, A., Zawidzki, M. (eds.) Proceedings of the 10th International Conference on Non-Classical Logics. Theory and Applications, Łódź, Poland, 14–18 March 2022. Electronic Proceedings in Theoretical Computer Science, vol. 358, pp. 150–159. Open Publishing Association (2022). https://doi.org/10.4204/EPTCS.358.11
3. Basu, S.S., Roy, S.: Generalized explosion principles. Studia Logica, p. 36 (2024). https://doi.org/10.1007/s11225-024-10153-x
4. Béziau, J.Y.: Universal logic. In: Childers, T., Majer, O. (eds.) Logica–94 - Proceedings of the 8th International Symposium, pp. 73–93. Prague (1994)
5. Béziau, J.Y.: 13 questions about universal logic. University of Łódź. Department of Logic. Bull. Section Logic **35**(2-3), 133–150 (2006), questions by Linda Eastwood
6. Bonzio, S., Gil-Férez, J., Paoli, F., Peruzzi, L.: On paraconsistent weak kleene logic: axiomatisation and algebraic analysis. Stud. Logica. **105**(2), 253–297 (2016). https://doi.org/10.1007/s11225-016-9689-5
7. Bonzio, S., Moraschini, T., Pra Baldi, M.: Logics of left variable inclusion and Płonka sums of matrices. Arch. Math. Logic (2), 49–76 (2020). https://doi.org/10.1007/s00153-020-00727-6
8. Bonzio, S., Paoli, F., Pra Baldi, M.: Logics of Variable Inclusion, Trends in Logic—Studia Logica Library, vol. 59. Springer, Cham (2022)
9. Caleiro, C., Marcelino, S., Filipe, P.: Infectious semantics and analytic calculi for even more inclusion logics. In: 2020 IEEE 50th International Symposium on Multiple-Valued Logic—ISMVL 2020, pp. 224–229. IEEE Computer Society, Los Alamitos (2020)

10. Font, J.M.: Abstract Algebraic Logic: An Introductory Textbook. Studies in Logic and the Foundations of Mathematics, College Publications (2016)
11. Font, J.M., Jansana, R., Pigozzi, D.: A survey of abstract algebraic logic. Stud. Logica. **74**(1), 13–97 (2003)
12. Paoli, F., Pra Baldi, M., Szmuc, D.: Pure variable inclusion logics. Logic Logical Philosop. **30**(4), 631–652 (2021). https://doi.org/10.12775/llp.2021.015
13. Roy, S., Basu, S.S., Chakraborty, M.K.: Lindenbaum-type logical structures. Log. Univers. **17**(1), 69–102 (2023)
14. de Souza, E.G., Costa-Leite, A., Dias, D.: Paraconsistent orbits of logics. Log. Univers. **15**(3), 271–289 (2021)

Bounded Henkin Quantifiers and the Exponential Time Hierarchy

Abhishek De[✉]

School of Computer Science, University of Birmingham, Birmingham, UK
a.de@bham.ac.uk

Abstract. In logic, quantifiers typically have an implicit linear order of dependencies. Henkin introduced a more general framework for quantifier prefixes, allowing for partial orderings among quantifiers. These quantifiers, now known as *Henkin quantifiers*, have been extensively studied in logic and have found significant applications in linguistics, arithmetic, and descriptive complexity.

Surprisingly, bounded versions of Henkin quantifiers, which restrict the range of these quantifiers, have not been explored. This paper defines bounded Henkin quantifiers and examines their properties from a complexity-theoretic perspective. In particular, we show that the set of predicates definable by quantifier-free formulas prefixed by a bounded Henkin quantifier is exactly NEXPTIME. The proof goes via machine models defined using bounded Henkin quantifiers. Finally, we show that formulas with Henkin quantifiers define a complexity class contained in Δ_2 in the exponential hierarchy and define a natural complete problem for this class.

Keywords: Henkin Quantifier · Bounded Quantifier · Time Complexity · DQBF · Exponential Hierarchy · Imperfect Information

1 Introduction

Consider the sentence, "Some relative of each villager and some relative of each townsman hate each other." due to Hintikka [12]. To express its meaning, we require *non-linear quantification*:

$$\varphi := \begin{pmatrix} \forall x_1 \exists y_1 \\ \forall x_2 \exists y_2 \end{pmatrix} (V(x_1) \wedge T(x_2)) \to (R(x_1, y_1) \wedge R(x_2, y_2) \wedge H(y_1, y_2))$$

where unary predicates V and T denote the set of villagers and the set of townsmen, respectively and the binary predicates R and H denote the symmetric relations 'relatives' and 'hate each other' respectively.

Henkin quantifiers (also known as branching quantifiers and partially ordered quantifiers), proposed by Henkin [11], are a generalisation of the notion of quantifiers where the order of dependencies between quantifiers need not be linear.

© The Author(s), under exclusive license to Springer Nature Switzerland AG 2025
C. Aiswarya et al. (Eds.): ICLA 2025, LNCS 15402, pp. 97–110, 2025.
https://doi.org/10.1007/978-3-031-89610-1_7

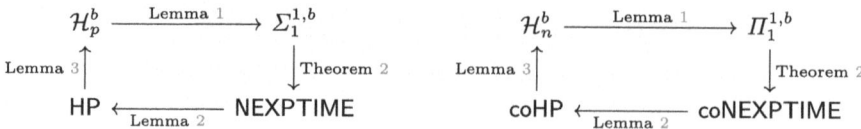

Fig. 1. Summary of Main Contributions

According to Henkin's semantics, φ is equivalent to the following existential second-order sentence:

$$\exists f \exists g \forall x_1 \forall x_2 \, (V(x_1) \wedge T(x_2)) \to (R(x_1, f(x_1)) \wedge R(x_2, g(x_2)) \wedge H(f(x_1), g(x_2))$$

Functions f and g (so-called *Skolem functions*) choose relatives for every villager and every townsman, respectively. Notice that the value of $f(\bullet)$ and $g(\bullet)$ is determined only by the choice of a certain villager and townsman respectively. In other words, to satisfy the formula, relatives have to be chosen independently.

Walkoe [13] studied the basic model-theoretical properties of an extension $L(H)$ of ordinary first-order languages in which every sentence is a first-order sentence prefixed with a Henkin quantifier. Ehrenfeucht observed that $L(H)$ is powerful enough to define the quantifier $Q_{\geq \mathbb{N}}$ (i.e. "there are infinitely many"): $(Q_{\geq \mathbb{N}} x)\varphi(x)$ can be written as

$$(\exists a) \begin{pmatrix} \forall x_1 \exists y_1 \\ \forall x_2 \exists y_2 \end{pmatrix} (\varphi(a) \wedge (x_1 = x_2 \leftrightarrow y_1 = y_2) \wedge (\varphi(x_1) \to (\varphi(y_1) \wedge y_1 \neq a)))$$

Several things follow from this, including the non-axiomatisability of $L(H)$, and its equivalence to existential second-order logic – the latter result published independently by Enderton [7] and Walkoe [13]. Blass and Gurevich [5] showed that over finite structure $L(H)$ formulas can express NP-complete properties (coupled with Fagin's result [8] that NP $= \Sigma_1^1$, this is a finitary version of the Enderton-Walkoe result above). For a survey on Henkin quantifiers, see [16].

Bounded quantifiers restrict the range of the quantified variable. They appear in various areas of mathematical logic such as recursion theory (the class of formulas with bounded quantifiers is the first level $\Sigma_0^0 = \Pi_0^0 = \Delta_0^0$ of the *arithmetic hierarchy*) and set theory (the class of formulas with bounded quantifiers is the first level $\Sigma_0^0 = \Pi_0^0 = \Delta_0^0$ of the *Lévy hierarchy*). In Peano Arithmetic, depending on whether $\sharp$ (read *smash*) is in the language, a relation on natural numbers is definable by a bounded formula if and only if it is computable in the linear-time hierarchy [17] or the polynomial-time hierarchy [14].

Contribution. In this paper, we define bounded Henkin quantifiers in the language of Bounded Arithmetic with smash. In the vein of Enderton-Walkoe [7,13] and Blass-Gurevich [5], we obtain that the class of first-order sentences prefixed with a bounded Henkin quantifier corresponds to existential second-order bounded formulas. Our proof technique is summarised in Fig. 1. It goes through bespoke machine models and a complete problem for NEXPTIME. Finally, we generalise this to obtain a hierarchy of formulas with bounded Henkin quantifiers that is contained in the Δ_2 level of second-order bounded formulas. This motivates complete problems for the class defined by bounded Henkin quantifiers.

Organisation of the Paper. In Sect. 2, we recall the language and formulas of (second-order) Bounded Arithmetic. In Sect. 3, we introduce bounded Henkin quantifiers and show that positive formulas are equivalent to existential second-order bounded formulas. In Sect. 4, we introduce machine models corresponding to positive formulas with bounded Henkin quantifiers and prove that every existential second-order bounded formula corresponds to a positive formula. We define normal forms for our formulas in Sect. 5. Finally, in Sect. 6, we generalise our results to arbitrary formulas with bounded Henkin quantifiers and define a complete problem for the corresponding class. We conclude and discuss some pertinent questions in Sect. 7.

Notation of Complexity Classes. The class of languages definable by deterministic Turing Machines is denoted P. The levels of the polynomial hierarchy are denoted Σ_i and Π_i for $i \in \mathbb{N}$ and levels of the exponential hierarchy are denoted Σ_i^{EXP} and Π_i^{EXP} for $i \in \mathbb{N}$. For basic definitions of complexity classes, see standard textbooks such as [1].

2 (Second-Order) Bounded Formulas

This section presents a bare-bones background on the formulas of (second-order) bounded arithmetic (BA and BA^2 respectively). Meatier expositions can be found in standard textbooks of proof complexity [6,15,27]. We start by adding some function symbols to the language of arithmetic $L = \{0, S, +, \cdot, \leq\}$[1].

Definition 1. *Fix a countable set of variables* Var. **Terms,** *denoted $s, t, \ldots$, are defined inductively as follows.*

$$t, t' ::= x \in \mathsf{Var} \mid 0 \mid S(t) \mid t + t' \mid t \cdot t' \mid |t| \mid t \sharp t' \mid \lfloor t/2 \rfloor$$

In the standard interpretation of terms in $\mathbb{N}$: $|t|^{\mathcal{I}}$ is the number of digits in the binary representation of $t^{\mathcal{I}}$ i.e. $\lceil \log(t^{\mathcal{I}}+1) \rceil$, $\lfloor \frac{t}{2} \rfloor^{\mathcal{I}} := \lfloor \frac{t^{\mathcal{I}}}{2} \rfloor$ and $(t \sharp t')^{\mathcal{I}} := 2^{|t|^{\mathcal{I}} |t'|^{\mathcal{I}}}$.

Definition 2. *The **first-order bounded formulas,** denoted $\varphi, \psi, \ldots$, are defined inductively as follows.*

$$\varphi, \psi ::= s \leq t \mid \neg \varphi \mid \varphi \vee \psi \mid \varphi \wedge \psi \mid \exists x \leq s\, \varphi \mid \forall x \leq s\, \varphi$$

*Quantifiers of the form $Qx \leq |s|\, \varphi$ are said to be **sharply bounded**. A formula is sharply bounded if all its quantifiers are sharply bounded.*

---

[1] In bounded arithmetic literature, more symbols might be used. However, since we do not discuss *theories*, we omit them from the language.

The hierarchy of Σ_i^b and Π_i^b formulas is defined inductively:

- Let $\Sigma_0^b = \Pi_0^b$ be the set of sharply bounded formulas.
- Σ_{i+1}^b is the closure of Π_i^b under existential bounded quantification and arbitrary sharply bounded quantification, modulo prenex operations.
- Π_{i+1}^b is the closure of Σ_i^b under universal bounded quantification and arbitrary sharply bounded quantification, modulo prenex operations.

First-order bounded formulas capture the polynomial-time hierarchy. Recall that a predicate $R \subseteq \mathbb{N}^k$ is said to be $\mathcal{C}$-definable for some complexity class $\mathcal{C}$ if $L_R \in \mathcal{C}$ for some effective encoding L_R of R in binary.

Theorem 1 ([14]). *Let $i \geq 1$ and $Q \in \{\Sigma, \Pi\}$. A predicate $R \subseteq \mathbb{N}^k$ is Q_i-definable if and only if there is a formula $\varphi(x_1, \ldots, x_k) \in Q_i^b$ such that for all $\overline{n} \in \mathbb{N}^k$, $R(\overline{n}) \iff \overline{n} \models \varphi(x_1, \ldots, x_k)$.*

In proof complexity, BA (and BA^2 respectively) is the collective name for the families of weak sub-theories of Peano arithmetic (and second-order arithmetic respectively). Second-order theories extend the first-order theories by adding second-order variables, $X, Y, Z, \ldots$ intended to range over sets.

Definition 3. *The **second-order bounded formulas**, denoted $\alpha, \beta, \ldots$, are defined inductively as follows (where φ is a first-order bounded formula).*

$$\alpha, \beta ::= s \leq t \mid X(t) \mid \neg \alpha \mid \alpha \vee \beta \mid \alpha \wedge \beta \mid \exists x \leq s\, \alpha \mid \forall x \leq s\, \alpha \mid \exists X \alpha \mid \forall X \alpha$$

The formula $X(t)$ denotes that t is in X. At first glance, it may be surprising that second-order variables are not explicitly bounded. However, they implicitly have polynomial bounds on their members (corresponding to polynomial upper bounds on the size of oracle queries). The hierarchy of $\Sigma_i^{1,b}$ and $\Pi_i^{1,b}$ formulas is defined inductively:

- Let $\Sigma_0^{1,b} = \Pi_0^{1,b}$ be the set of first-order bounded formulas.
- $\Sigma_{i+1}^{1,b}$ is the closure of $\Pi_i^{1,b}$ under existential second-order quantification and arbitrary first-order bounded quantification, modulo prenex operations.
- $\Pi_{i+1}^{1,b}$ is the closure of $\Sigma_i^{1,b}$ under universal second-order quantification and arbitrary first-order bounded quantification, modulo prenex operations.

Second-order bounded formulas capture the exponential hierarchy[2].

Theorem 2 (Folklore). *Let $i \geq 1$ and $Q \in \{\Sigma, \Pi\}$. A predicate $R \subseteq \mathbb{N}^k$ is Q_i^{EXP}-definable if and only if there is a formula $\alpha(x_1, \ldots, x_k) \in Q_i^{1,b}$ such that for all $\overline{n} \in \mathbb{N}^k$, $R(\overline{n}) \iff \overline{n} \models \alpha(x_1, \ldots, x_k)$.*

---

[2] There is some confusion in the literature about the exact definition of the hierarchy (*cf.* [10,20]). Here we mean what is sometimes known as the *weak exponential hierarchy*.

3 Bounded Henkin Quantifiers

This section will develop the theory of *bounded Henkin quantifiers*. We will first recall Henkin quantifiers.

3.1 Henkin Quantifiers

Definition 4. *A **Henkin quantifier** is a triple $Q = (A, E, d)$, where A and E are disjoint finite sets of variables called respectively universal and existential variables, and $d \subseteq (A \times E)$ is a relation called the **dependency relation**. We say that Q is **linear** if there is a linear ordering $<$ on $A \cup E$ such that $< \cap (A \times E) = d$.*

Following Walkoe [13], we call a Henkin quantifier **standard** if for all (x, y), $(x, y') \in d$, we have $y = y'$. Such a quantifier can be written as *jagged matrices* where each row corresponds to an existential variables and the set of universal variables it depends on:

$$\begin{pmatrix} \forall x_{11} & \forall x_{12} & \ldots & \forall x_{1n_1} \exists y_1 \\ \forall x_{21} & \forall x_{22} & \ldots & \forall x_{2n_2} \exists y_2 \\ \vdots & \vdots & \vdots & \vdots \\ \forall x_{m1} & \forall x_{m2} & \ldots & \forall x_{mn_m} \exists y_m \end{pmatrix}$$

Note that the order of the rows and the order of universal variables within a row is irrelevant. For each $m, n \in \mathbb{N}$, we denote the standard branching quantifier with $n_i = n$ for all $1 \leq i \leq m$ by H_m^n. The smallest non-trivial (*i.e.* non-linear) standard branching quantifier is

$$H_2^2 = \begin{pmatrix} \forall x_1 \exists y_1 \\ \forall x_2 \exists y_2 \end{pmatrix}$$

which will be used exemplarily throughout this paper to avoid tedious details.

We extend the syntax of first-order logic with formulas with Henkin quantifiers to obtain **H-formulas**. Define H-formulas by structural induction: if φ is an H-formula, then $Q\varphi(\overline{u})$ is an H-formula where $\overline{u}$ is the set of variables in Q. The semantics of H-formulas are defined by *Skolemisation*:

Definition 5. *The **Skolemisation** $sk(\varphi)$ of a formula φ with Henkin quantifiers is defined by structural induction:*

- *$sk(\varphi) = \varphi$ if φ is atomic;*
- *$sk(\neg \varphi) = \neg sk(\varphi)$;*
- *$sk(\varphi \circ \psi) = sk(\varphi) \circ sk(\psi)$ where $\circ \in \{\wedge, \vee\}$;*
- *$sk(\mathcal{E}\varphi) = \mathcal{E}sk(\varphi)$ where $\mathcal{E} \in \{\exists, \forall\}$;*
- *$sk(Q\varphi) = \exists f_1 \ldots \exists f_m \forall x_1 \ldots \forall x_n \varphi'$ where $Q = (A = \{x_1, \ldots, x_n\}, E = \{y_1, \ldots, y_m\}, d)$ and φ' is φ where every $y_i \in E$ is replaced by $f_i(x_{j_1}, \ldots, x_{j_k})$ such that $\{(x_{j_1}, y_i), \ldots, (x_{j_k}, y_i)\} = (A \times \{y_i\}) \cap d$.*

*The function symbols $f_1, \ldots, f_m$ are called **Skolem functions**.*

Example 1. The interpretation of $H_2^2\varphi(x_1, y_1, x_2, y_2)$ is defined as the interpretation of the second-order formula

$$\exists f_1.\exists f_2.\forall x_1.\forall x_2.\varphi(x_1, f_1(x_1), x_2, f_2(x_2))$$

Definition 6. *An H-formula is said to be **positive** (**negative** respectively) if Henkin quantifiers occur under an even (odd respectively) number of negations.*

Proposition 1. *Let φ be a positive (negative respectively) H-formula. There exists an H-formula $Q\psi$ ($\neg Q\psi$ respectively) such that (i) Q is a standard Henkin quantifier, (ii) ψ is a quantifier-free first-order formula, and (iii) φ and $Q\psi$ ($\neg Q\psi$ respectively) are equivalent.*

Proof (Sketch). Assume φ contains only positive occurrences of Henkin quantifiers (the negative case in symmetric). We first observe the following tautologies: $Q\varphi \circ \psi \leftrightarrow Q(\varphi \circ \psi)$, $\neg Q\varphi \wedge \psi \leftrightarrow \neg Q(\varphi \vee \neg \psi)$, and $\neg Q\varphi \vee \psi \leftrightarrow \neg Q(\varphi \wedge \neg \psi)$ where $\circ \in \{\vee, \wedge\}$. Then, observe that consecutive Henkin quantifiers can be replaced by a single Henkin quantifier: for example, $H_2^2 H_2^2$ can be replaced by the following quantifier

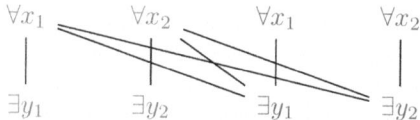

Finally, for any Henkin quantifier $Q = (A, E, d)$ with $(x, y), (x, y') \in D$ for distinct $y, y' \in E$, we have the equivalent quantifier $Q' = (A', E, d')$ where $A' = A \cup \{x'\}$ and $d' = (d \setminus \{(x, y')\}) \cup \{(x', y')\}$. Then, $Q\varphi$ is semantically equivalent to $Q'(x = x' \to \varphi)$. By applying this trick repeatedly, one can obtain a standard Henkin quantifier equivalent to an arbitrary Henkin quantifier.

This prenexing can be generalised to arbitrary H-formulas as follows.

Proposition 2. *Every H-formula is equivalent to an H-formula of the form $R\neg Q_0 \neg Q_1 \ldots \neg Q_n \varphi$ where R is either $\neg Q$ or Q, and $Q, Q_0, \ldots, Q_n$ are standard Henkin quantifiers.*

3.2 Bounded Henkin Quantifiers

We will now define bounded Henkin quantifiers. Fix the language L defined in Sect. 2 and recall terms as defined in Definition 1.

Definition 7. *A **bounded Henkin quantifier** is a pair (Q, Φ) where Q is a Henkin quantifier and Φ maps variables in Q to terms.*

Bounded formulas with Henkin quantifiers are defined by induction: if φ is a bounded formula, then $(Q, \Phi)\varphi(\overline{u})$ is a bounded formula where $\overline{u}$ is the set of variables in Q. As before, the semantics is defined by Skolemisation: for example, the interpretation of $(H_2^2, \Phi)\varphi(x_1, y_1, x_2, y_2)$ is defined as the interpretation of the following second-order bounded formula.

$$\exists f_1.\exists f_2.\forall x_1 \leq \Phi(x_1).\forall x_2 \leq \Phi(x_2).\varphi(x_1, f_1(x_1), x_2, f_2(x_2)) \tag{1}$$

Definition 8. *The set of predicates definable by sharply bounded, positive bounded, and negative bounded H-formulas is called $\mathcal{H}^{sb}$, $\mathcal{H}^b_p$, and $\mathcal{H}^b_n$ respectively.*

The main result of the paper is:

Theorem 3. $\mathcal{H}^b_p = $ NEXPTIME *and* $\mathcal{H}^b_n = $ coNEXPTIME.

We will first prove the easy direction of Theorem 3.

Lemma 1. $\mathcal{H}^b_p \subseteq $ NEXPTIME *and* $\mathcal{H}^b_n \subseteq $ coNEXPTIME.

Proof (Sketch). By Proposition 1, any H^b_p formula is equivalent to a formula φ of the form $(Q, \Phi)\psi$ where (Q, Φ) is a bounded standard Henkin quantifier and ψ is quantifier-free. We will show that the Skolemisation $sk(\varphi)$ is a $\Sigma^{1,b}_1$ formula. First note that the language of bounded arithmetic can be 'bootstrapped' with the *pairing function* β defined as follows:

$$\beta(i, \ulcorner \langle a_1, \ldots, a_k \rangle \urcorner) = \begin{cases} n & \text{if } i = 0; \\ a_i & \text{if } 1 \leq i \leq k \end{cases}$$

where $\ulcorner \bullet \urcorner$ denotes a Gödel numbering for sequences. It is a standard result in bounded arithmetic that there are polynomially bounded Gödel numbering for sequences that can be defined using basic functions (for instance, one can define $\ulcorner \langle m, n \rangle \urcorner := \lfloor \frac{m+n}{2} \rfloor m + n$). Consequently, the Skolem functions in $sk(\varphi)$ can replaced by predicates, such that first-order bound variables are still bounded by terms of polynomial growth rate. For example, Eq. (1), can be rewritten as[3],

$$\exists X_1. \exists X_2. \forall z_1 \leq t_1. \forall z \leq t_2. X_1(z_1) \wedge X_2(z_2) \leftrightarrow \varphi(\beta(1, z_1), \beta(2, z_1), \beta(1, z_2), \beta(2, z_2))$$

where $t_i = \left\lfloor \frac{\Phi(x_i) + \Phi(y_i)(\Phi(x_i))}{2} \right\rfloor \Phi(x_i) + \Phi(y_i)(\Phi(x_i))$ for $i = 1, 2$. Note that this formula is in $\Sigma^{1,b}_1$. Applying Theorem 2, we are done. The proof is symmetric for $\mathcal{H}^b_n$. □

A similar idea allows us to prove the following:

Proposition 3. $\mathcal{H}^{sb} \subseteq $ EXP

Proof (Sketch). Let $\varphi \in \mathcal{H}^{sb}$. We will prove by induction on φ. If φ contains no Henkin quantifier, it defines a polytime predicate. Suppose $\varphi(\overline{x}) = (Q, \Phi)\psi(\overline{x}, \overline{y})$ such that $\psi(\overline{x}, \overline{y})$ defines a exponential predicate. Note that the Skolem functions in $sk(\varphi)$ are bounded in size of input and output such that their Gödel encoding is polynomially bounded. For example, if $Q = H^2_2$, then the Skolem functions f_i in Eq. (1) can be encoded as sequences in size $O(\Phi(y_i)(\Phi(x_i))^{\Phi(x_i)})$ which is polynomial (for $i = 1, 2$) since Φ is a sharp bound. Finally, EXP is closed under polynomial oracles; hence, we are done. □

---

[3] Note that $\Phi(x_i)$ has no free variables and $\Phi(y_i)$ potentially has x_i free.

We are still left to prove the reverse inclusion of Theorem 3. One possibility is adapting the *Enderton-Walkoe translating procedure* [7,13] to our setting. However, this requires several finickity bootstrapping in BA^2; instead, we introduce new machine models and use a complete problem of NEXPTIME as a blackbox to obtain a much cleaner proof.

4 Machine Models

Just like complexity classes can be defined as the set of predicates definable by some class of logical formulas, they can also be defined[4] as subclasses of models of computations (such as Turing Machines (TMs)) with explicit bounds on computation. Consider the two equivalent definitions of NP:

1. It is the set of languages recognised by a non-deterministic polytime TM
2. It is the set of languages such that for every word in the language, there is a polynomial-sized certificate of membership of that word.

Observe that, the second definition has a bounded quantification: *there is a polynomial-sized...* and this observation can be formalised as Theorem 1. It is natural to ask if we can do the same for $\mathcal{H}_p^b$ and $\mathcal{H}_n^b$.

Definition 9. *Let Q be a Henkin quantifier and $\mathcal{C}$ be a time complexity class. A language $L \in \{0,1\}^*$ is in $Q\mathcal{C}$ if there exists a function Φ that maps variables in Q to polynomials and a $\mathcal{C}$-time TM M such that*

$$x \in L \iff (Q, \Phi)\, M(x, \overline{u}) = 1.$$

where $\overline{u} = (u_i)$ are the variables in Q and $u_i \in \{0,1\}^{\Phi(u_i)(|x|)}$. Analogously, a language $L \in \{0,1\}^*$ is in $\text{co}Q\mathcal{C}$ if there exists a function Φ that maps variables in Q to polynomials and a $\mathcal{C}$-time TM M such that

$$x \notin L \iff (Q, \Phi)\, M(x, \overline{u}) = 0.$$

where $\overline{u} = (u_i)$ are the variables in Q and $u_i \in \{0,1\}^{\Phi(u_i)(|x|)}$.

Clearly, $\text{co}Q\mathcal{C} = \{L \mid \{0,1\}^* \setminus L \in Q\mathcal{C}\}$. Define $\mathsf{HP} = \bigcup_Q Q\mathsf{P}$ and $\text{coHP} = \bigcup_Q \text{co}Q\mathsf{P}$.

Definition 10. *A **dependency quantified Boolean formula (DQBF)** is a formula φ of the form $Q\psi$ where Q is a Henkin quantifier and ψ is a quantifier-free Boolean formula called the **matrix** of φ.*

Theorem 4 ([24,25]). *The DQBF satisfiability problem is NEXPTIME-complete.*

This allows us to prove the following,

---

[4] In fact, machine-based definitions are more common.

Lemma 2. NEXPTIME $\subseteq$ HP *and* coNEXPTIME $\subseteq$ coHP.

Proof (Sketch). Let $Q\varphi$ be a DQBF. Clearly, satisfiability and validity of $Q\varphi$ are in QP and coQP respectively (because checking an assignment on the matrix is polytime checkable). Noting that HP and coHP are closed under polynomial many-one reductions, we are done. □

In the following lemma, we construe HP as the set of predicates definable in HP under the binary encoding of sequences of natural numbers.

Lemma 3. HP $\subseteq \mathcal{H}_p^b$ *and* coHP $\subseteq \mathcal{H}_n^b$.

Proof (Sketch). Let R be a predicate definable in HP. We need to show that $\{\overline{n} \mid R(\overline{n})\} \in$ HP if and only if R is definable in $\mathcal{H}_p^b$. By definition, we have that there is a deterministic polytime TM M, Henkin quantifier Q, and Φ mapping variables in Q to polynomials such that $\{\overline{n} \mid R(\overline{n})\} = \{\overline{n} \mid (Q, p(\overline{n}))M(\overline{n}, \overline{u}) = 1\}$. Since M is polytime, it is definable by a first-order bounded formula (say $\psi(\overline{x}, \overline{y})$) and we have that R is defined by $(Q, \Phi')\psi(\overline{x}, \overline{y})$ where Φ' is Φ with polynomials written as terms. The proof is symmetric for coHP. □

Stitching together Lemmas 1 to 3, we have the proof of Theorem 3.

Remark 1. Aposteriori, we already have more standard machine models for $\mathcal{H}_p^b$ (and $\mathcal{H}_n^b$ respectively) *viz.* the set of languages recognised by an exponential-time non-deterministic (and co-non-deterministic respectively) TMs. Another way to look at the result is that it provides alternative machine models for NEXPTIME and coNEXPTIME.

5 Taming Bounded Henkin Quantifiers

In this section, we will show that H_2^2 is enough to capture the power of all Henkin quantifiers. The idea is similar to the *normal forms* for Henkin quantifiers defined by Walkoe [13].

Proposition 4. *Any H_p^b formula is equivalent to a formula of the form $H_2^2\varphi$ where φ is quantifier-free.*

Proof (Sketch). By Proposition 1, we can assume we have a formula $Q\varphi$ where Q is a standard Henkin quantifier and φ is quantifier-free. Let $\forall x_1 \forall x_2 \ldots \forall x_n \exists y$ be a row of Q in the jagged matrix presentation of Q. Now asserting $x_i \leq t_i$ for all $i \in \{1, \ldots, n\}$ is the same as asserting $\ulcorner \langle x_1, x_2, \ldots, x_n \rangle \urcorner \leq O(\sum_{i=1}^n p_i(|x|))$. This is a precise bound that can be computed from the Gödel numbering of sequences and can be written as a term. Therefore $Q'\varphi$ is equivalent to $H_k^2 \varphi'$ where φ' is φ with each occurence of x_i replaced by $\beta(i, x)$. For obtaining 2 from k, the idea is that k functions from A to A can be defined using one function from A^k into A^k. In particular,

$$\begin{pmatrix} \forall x_1 \exists y_1 \\ \ldots \\ \forall x_k \exists y_k \end{pmatrix} \psi \leftrightarrow \begin{pmatrix} \forall x_1 \ldots \forall x_k \exists y_1 \ldots y_k \\ \forall z_1 \ldots \forall z_k \exists w_1 \ldots w_k \end{pmatrix} \left(\bigwedge_{i=1}^{k} x_i = z_i \rightarrow y_i = w_i \right) \wedge \psi$$

Finally, using the pairing function trick as before, we are done. □

Proposition 5. *Let Q be a non-linear Henkin quantifier Q and $\mathcal{C}$ be a time complexity class larger than linear time. Then, $Q\mathcal{C} = H_2^2 \mathcal{C}$.*

Proof (Sketch). The proof follows exactly like above however we need to check equality of strings at various points; thus need $\mathcal{C}$ to be at least linear time to do that for free. □

Remark 2. $H_2^2\mathsf{P}$ was defined by Voss [28]. As it turns out, $H_2^2\mathsf{P} = \mathsf{HP}$ as defined in Sect. 4. Furthermore, we have that $\mathsf{coHP} = \mathsf{co}H_2^2\mathsf{P}$.

6 Generalisations

Till now we only considered positive or negative bounded Henkin quantified formulas. In this section, we will generalise to arbitrary bounded H-formulas. Define a hierarchy of H-formulas in the spirit of Sect. 2 as follows.

- $H_0^{p,b} = H_0^{n,b} = H^{sb}$.
- $H_{i+1}^{p,b}$ is the closure of $H_i^{n,b}$ under bounded Henkin quantification, arbitrary first-order quantification modulo prenex operations.
- $H_{i+1}^{n,b}$ is the set of formulas such that their negations are in $H_{i+1}^{p,b}$.

Clearly, $\{H_i^{p,b}\}_{i \in \mathbb{N}} \cup \{H_i^{n,b}\}_{i \in \mathbb{N}}$ defines a partition of H-formulas by Proposition 2. Let $\mathcal{H}$ be the set of predicates definable by bounded H-formulas.

Proposition 6. $\mathcal{H} \subseteq \Delta_2^{1,b}$.

Proof (Sketch). Let φ be a bounded H-formula. By Proposition 2, we can write φ as $R \neg Q_0 \neg Q_1 \ldots \neg Q_n \psi$ where R is either Q or $\neg Q$. Assume $R = Q$ (the proof is symmetric for the other case.) We will induct on n. If $n = 0$, we are done by Lemma 1. By induction, suppose $\varphi' = \neg Q_0 \neg Q_1 \ldots \neg Q_n \psi$ can be expressed as a $\Delta_2^{1,b}$ formula. In particular, φ' is equivalent to a $\Sigma_2^{1,b}$ formula $\exists X \forall Y \varphi_0$. Consequently, $sk(Q\varphi')$ is a formula of the form $\exists f \forall x \exists X \forall Y \varphi_1$. Consider the second-order tautology $\forall x \exists X \varphi_2 \equiv \exists Z \forall x \varphi_2'$, where arity of Z is one more than that of X, and φ_2' is φ_2 where every occurrence of $X(t_1, \ldots, t_n)$ has been replaced by $Z(x, t_1, \ldots, t_n)$. Using this tautology that essentially encodes a form of choice, we have that φ is a $\Sigma_2^{1,b}$ formula. However, φ' is also equivalent to a $\Pi_2^{1,b}$ formula. A priori $sk(Q\varphi')$ is a $\Sigma_3^{1,b}$ formula but note that the second-order universal quantifier does not depend on the outermost second-order existential quantifier and can thus be transformed into an equivalent $\Pi_2^{1,b}$ formula using the following second-order tautology

$$\exists f \forall x \forall g \exists y \varphi(x,y,f(x),g(y)) \equiv \forall g \exists f \forall x \exists y \varphi(x,y,f(x),g(x,f(x),y)) \qquad \square$$

Remark 3. Some readers might be surprised by this result. However, it is consistent with the theory of Henkin quantifiers in other settings (*cf.* [5,7]). In fact, to obtain full second-order one needs to have quantifiers and their negations in parallel like $\begin{pmatrix} Q \\ \neg Q' \end{pmatrix}$ (*cf.* [21,23]). Such considerations are beyond the scope of this paper.

Analogously, we can generalise the machine models in Sect. 4. Fix an arbitrary Q and $\mathcal{C}$ and define the following hierarchy where arrows denote the $\subseteq$ relation.

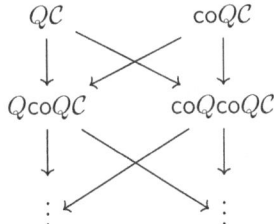

In Sect. 5, we showed that it is enough to consider H_2^2 instead of arbitrary Henkin quantifiers. For the rest of this section, we simply write H instead of H_2^2.

$$\mathsf{H}_i^p\mathsf{P} := \underbrace{\mathsf{H}\,\mathsf{coH}\ldots\mathsf{coH}}_{i-1 \text{ times}}\mathsf{P} \qquad \mathsf{H}_i^n\mathsf{P} := \underbrace{\mathsf{H}\,\mathsf{coH}\ldots\mathsf{coH}}_{i \text{ times}}\mathsf{P}$$

With essentially the same technique as Lemma 3, it is not difficult to show:

Proposition 7. *For each $i \in \mathbb{N}$, $\mathsf{H}_i^p\mathsf{P} = \mathcal{H}_i^{n,b}$ and $\mathsf{H}_i^n\mathsf{P} = \mathcal{H}_i^{p,b}$.*

Consequently, we have that the hierarchy of $\mathsf{H}_i^p\mathsf{P}$ and $\mathsf{H}_i^n\mathsf{P}$ also collapses at Δ_2^{EXP}. We conjecture that this is a strict inclusion. Mostowski showed this strictness of inclusion in the setting of Peano Arithmetic [22], and it is not clear how to adapt his example to our setting.

If we were to study $\mathcal{H}$ as a bona fide complexity class, a natural step is to define complete problems for it.

Definition 11. *A **generalised dependency quantified Boolean formula** (**GDQBF**) is a formula φ of the form $R\neg Q_0 \ldots Q_n \psi$ where $R = Q$ or $R = \neg Q$, $Q_0, \ldots, Q_n$ are Henkin quantifiers, and ψ is a quantifier-free Boolean formula.*

We can interpret a TQBF formula by a game of perfect information played by two players $\forall$ (Abelard) and $\exists$ (Eloise) who take turns choosing values of their respective bound variables. Eloise wins if the matrix is satisfied. In [24,25], this concept is generalised to DQBFs where formulas are interpreted as *imperfect information* games between Abelard and Eloise. For example given $\begin{pmatrix} \forall x_1 \exists y_1 \\ \forall x_2 \exists y_2 \end{pmatrix} \varphi$, when Eloise makes the choice for y_1, it is made without knowledge of what Abelard chose for x_2. It is straightforward to generalise the DQBF game to GDBFs: when players encounter a negation, they switch roles.

Proposition 8. *Let φ be a GDQBF. Eloise has a winning strategy in the GDQBF game on φ if and only if φ is satisfiable.*

The proof is routine and follows the soundness and completeness argument of the game semantics of predicate logic [19].

Theorem 5. *The satisfiability of GDQBFs (equivalently, the decision problem of whether a GDQBF game has an Eloise-winning strategy) is $\mathcal{H}$-complete.*

This proof is also routine and the coding is exactly the same as the *Cook-Levin* coding of Turing Machines for SAT [1]. In [9] DQBFs were generalised to obtain complete problems for each level of the exponential hierarchy. We conjecture that under suitable complexity assumptions, GDQBFs are strictly weaker than *alternating DQBFs* defined in [9].

7 Conclusion and Future Work

In this work, we initiated an investigation of bounded Henkin quantifiers. We took two different approaches to defining complexity classes using bounded Henkin quantifiers: firstly, we defined bounded Henkin quantified formulas over the language of Bounded Arithmetic and secondly, we defined machine models where oracles to deterministic Turing Machines are given by Henkin quantifiers. We showed that these notions coincide. In particular, we showed that positive formulas define exactly the NEXPTIME predicates and arbitrary formulas do not give much extra power and define Δ_2^{EXP} predicates. We conjecture that they do *not* define all Δ_2^{EXP} predicates, and one can prove that conditional to some standard complexity assumptions such as the exponential hierarchy does not collapse. We conclude with a discussion on avenues of future work.

Bounded Arithmetic. Extensions of Peano Arithmetic with the induction scheme for all formulas with Henkin quantifiers have been considered [22]. It would be natural to consider extensions of Bounded Arithmetic with induction on all formulas with *bounded* Henkin quantifiers. Recent advances in the structural proof theory of Henkin quantifiers [3], show that logics with Henkin quantifiers have cut-elimination, a metalogical property imperative for several Bounded Arithmetic results.

Proof Complexity. The connections between resolution-based propositional proof systems and CDCL-based SAT solving are well understood [2,26]. There are several ongoing efforts to understand the picture for QBF-solving [4]. A promising method of lifting CDCL to QBF-solving is via *dependency schemes* [18]. Recently, dependency schemes proposed for QBF were analysed and only a few were found to be sound for DQBF-solving [29]. It would be interesting to abstract out a formal proof system from dependency scheme-based DQBF-solving algorithms and see if it can be interpreted as non-uniform equivalent of the aforementioned theories of Bounded Arithmetic.

Acknowledgments. I wish to thank Anupam Das for several fruitful discussions and pointers to relevant literature on proof complexity. This work was supported by a UKRI Future Leaders Fellowship, 'Structure vs Invariants in Proofs', project reference MR/S035540/1.

References

1. Arora, S., Barak, B.: Computational Complexity. Cambridge University Press, Cambridge (2009)
2. Atserias, A., Fichte, J.K., Thurley, M.: Clause-learning algorithms with many restarts and bounded-width resolution. J. Artif. Intell. Res. **40**, 353–373 (2011). https://doi.org/10.1613/jair.3152
3. Baaz, M., Lolic, A.: A globally sound analytic calculus for Henkin quantifiers. In: Artemov, S., Nerode, A. (eds.) Logical Foundations of Computer Science, pp. 128–143. Springer, Cham (2020)
4. Beyersdorff, O., Janota, M., Lonsing, F., Seidl, M.: Chapter 31. Quantified Boolean Formulas. IOS Press (2021). https://doi.org/10.3233/faia201015
5. Blass, A., Gurevich, Y.: Henkin quantifiers and complete problems. Ann. Pure Appl. Logic **32**, 1–16 (1986). https://doi.org/10.1016/0168-0072(86)90040-0
6. Cook, S., Nguyen, P.: Logical Foundations of Proof Complexity. Cambridge University Press (2010). https://doi.org/10.1017/cbo9780511676277
7. Enderton, H.B.: Finite partially-ordered quantifiers. Math. Log. Q. **16**(8), 393–397 (1970). https://doi.org/10.1002/malq.19700160802
8. Fagin, R.: Generalized first-order spectra, and polynomial. Time recognizable sets. SIAM-AMS Proc. **7** (1974)
9. Hannula, M., Kontinen, J., Lück, M., Virtema, J.: On quantified propositional logics and the exponential time hierarchy. In: Cantone, D., Delzanno, G. (eds.) Proceedings of the Seventh International Symposium on Games, Automata, Logics and Formal Verification, GandALF 2016, Catania, Italy, 14–16 September 2016. EPTCS, vol. 226, pp. 198–212 (2016). https://doi.org/10.4204/EPTCS.226.14
10. Hemachandra, L.A.: The strong exponential hierarchy collapses. J. Comput. Syst. Sci. **39**(3), 299–322 (1989). https://doi.org/10.1016/0022-0000(89)90025-1
11. Henkin, L.: Some remarks on infinitely long formulas. J. Symb. Log. **30**(1), 167–183 (1961). https://doi.org/10.2307/2270594
12. Hintikka, J.: Quantifiers vs. quantification theory. Dialectica **27**(3–4), 329–358 (1973)
13. John Walkoe, W.: Finite partially-ordered quantification. J. Symb. Log. **35**(4), 535–555 (1970). https://doi.org/10.2307/2271440

14. Kent, C.F., Hodgson, B.R.: An arithmetical characterization of NP. Theor. Comput. Sci. **21**(3), 255–267 (1982). https://doi.org/10.1016/0304-3975(82)90076-7
15. Krajíček, J.: Proof Complexity. Cambridge University Press (2019). https://doi.org/10.1017/9781108242066
16. Krynicki, M., Mostowski, M.: Henkin Quantifiers, pp. 193–262. Springer, Dordrecht (1995). https://doi.org/10.1007/978-94-017-0522-6_7
17. Lipton, R.J.: Model theoretic aspects of computational complexity. In: 19th Annual Symposium on Foundations of Computer Science (SFCS 1978), pp. 193–200 (1978). https://api.semanticscholar.org/CorpusID:16784399
18. Lonsing, F., Biere, A.: Depqbf: a dependency-aware QBF solver: system description. J. Satisfiability Boolean Model. Comput. **7**(2-3), 71–76 (2010). https://doi.org/10.3233/sat190077
19. Lorenzen, P., Lorenz, K. (eds.): Dialogische Logik. Wissenschaftliche Buchgesellschaft, [Abt. Verl.], Darmstadt (1978)
20. Mocas, S.E.: Separating classes in the exponential-time hierarchy from classes in PH. Theor. Comput. Sci. **158**(1), 221–231 (1996). https://doi.org/10.1016/0304-3975(95)00078-X
21. Mostowski, M.: An extension of the logic with branched quantifiers. In: 8th International Congress of Logic, Methodology and Philosophy of Science, Abtracts, vol. 5, pp. 261–263 (1987)
22. Mostowski, M.: Arithmetic with the Henkin quantifier and its generalizations. In: Gaillard, F., Richard, D. (eds.) Seminaire du Laboratoire Logique, Algorithmique Et Informatique Clermontois, pp. 1–25 (1991)
23. Mostowski, M.: Branched quantifiers. DziaŁ Wydawnictw Filii Uniwersytetu Warszawskiego (1991)
24. Peterson, G., Reif, J., Azhar, S.: Lower bounds for multiplayer noncooperative games of incomplete information. Comput. Math. Appl. **41**(7), 957–992 (2001). https://doi.org/10.1016/S0898-1221(00)00333-3
25. Peterson, G.L., Reif, J.H.: Multiple-person alternation. In: 20th Annual Symposium on Foundations of Computer Science (SFCS 1979), pp. 348–363 (1979). https://doi.org/10.1109/SFCS.1979.25
26. Pipatsrisawat, K., Darwiche, A.: On the power of clause-learning sat solvers as resolution engines. Artif. Intell. **175**(2), 512–525 (2011). https://doi.org/10.1016/j.artint.2010.10.002. https://www.sciencedirect.com/science/article/pii/S0004370210001669
27. Pudlak, P.: Logical Foundations of Mathematics and Computational Complexity. 2013th edn. Springer Monographs in Mathematics. Springer, Cham (2013)
28. Voss, C.: Descriptive complexity and some unusual quantifiers. Master's thesis, MIT (2016)
29. Wimmer, R., Scholl, C., Wimmer, K., Becker, B.: Dependency schemes for DQBF. In: Creignou, N., Le Berre, D. (eds.) Theory and Applications of Satisfiability Testing - SAT 2016, pp. 473–489. Springer, Cham (2016)

Monotone Modal Logic Beyond Distributivity

Yiwen Ding[1], Krishna Manoorkar[1(✉)], Alessandra Palmigiano[1,3], and Ruoding Wang[1,2]

[1] Vrije Universiteit Amsterdam, Amsterdam, The Netherlands
krishna.manoorkar@gmail.com
[2] Xiamen University, Xiamen, China
[3] Department of Mathematics and Applied Mathematics,
University of Johannesburg, Johannesburg, South Africa

Abstract. In this paper, based on duality-theoretic techniques, we introduce a sound and complete neighborhood semantics based on polarities for basic lattice-based monotone modal logic, and show that monotone modal operators in lattice-based monotone modal logic can be represented as the composition of suitable normal modal operators via a multi-type representation of complete lattices with monotone operators.

Keywords: Lattice-based modal logic · Monotone modal logic · Polarity-based semantics · Neighborhood semantics

1 Introduction

Non-normal modal logics, especially monotone modal logics, have been considered since the inception of the modern development of modal logic [4,21,22] particularly motivated by a philosophical logic perspective; indeed, it has been argued that, in various settings, certain epistemic [27], deontic [4], or game-theoretic [25] phenomena need not satisfy the characterizing properties of normal modal operators. More recently, monotone modal logic has been studied in theoretical computer science [20]. Also from a purely mathematical logic perspective, monotone modal logic has been used as a tool to study general lattices [15,26].

The study of monotone modal operators on a non-classical (distributive) propositional base has recently started gaining traction [12]. This latter development aligns with a line of research aimed at developing the theory of lattice-based modal logics, and more generally, the logics of lattice expansions [8,11,13] both from a semantic and a proof-theoretic perspective [1,2,18]. This line of research is conceptually motivated by the interpretation of lattice-based logics as the logics of categories and concepts [9], which made it possible to develop a theory unifying rough set theory and formal concept analysis [6]. In particular, the key of interpreting lattice-based logics as the logics of categories and concepts is the definition of a uniform relational semantics based on *polarities* (a.k.a. *formal concepts*) [7,10].

In the present paper, we consider the basic *lattice-based monotone modal logic*, i.e. the logic resulting from augmenting the logic of general (i.e. not necessarily distributive) lattices with monotone modal operators. By dualizing its algebraic semantics given by lattices expanded with monotone operators, we introduce a complete polarity-based neighborhood semantics for it. Because of the page limit of the ICLA 2025 conference proceeding, we omitted some proofs in this paper and refer them to a ResearchGate version [14].

Structure of the Paper. In Sect. 2, we provide necessary preliminaries and explain the notations for the paper. In Sect. 3, we define polarity-based neighborhood frames and establish their relationship with lattices with monotone modal operators. In Sect. 4, we define polarity-based semantics for lattice-based monotone modal logic. In Sect. 5, we provide a multi-type translation of lattice-based monotone modal logic into normal lattice-based modal logic. In Sect. 6, we conclude and provide directions for future research.

2 Preliminaries

In this section, we provide preliminaries on lattice-based monotone modal logic, monotone modal logic, and its multi-type representation.

2.1 Lattice-Based Monotone Modal Logic

Let Prop be a countable set of propositional variables. The language $\mathcal{L}$ of the lattice-based monotone modal logic is defined by the following recursion:

$$\phi ::= p \mid \bot \mid \top \mid \phi \wedge \phi \mid \phi \vee \phi \mid \Delta\phi \mid \nabla\phi,$$

where $p \in$ Prop. Given any $\phi, \psi \in \mathcal{L}$, $\phi \vdash \psi$ is an $\mathcal{L}$-sequent. A *lattice-based monotone modal logic* is a set of $\mathcal{L}$-sequents containing the following axioms:

$$p \vdash p, \quad p \vdash \top \quad \bot \vdash p, \quad p \vdash p \vee q, \quad q \vdash p \vee q, \quad p \wedge q \vdash p, \quad p \wedge q \vdash q,$$

and closed under the following inference rules:

$$\frac{\phi \vdash \chi \quad \chi \vdash \psi}{\phi \vdash \psi} \quad \frac{\phi \vdash \psi}{\phi(\chi/p) \vdash \psi(\chi/p)} \quad \frac{\chi \vdash \phi \quad \chi \vdash \psi}{\chi \vdash \phi \wedge \psi} \quad \frac{\phi \vdash \chi \quad \psi \vdash \chi}{\phi \vee \psi \vdash \chi} \quad \frac{\phi \vdash \psi}{\nabla\phi \vdash \nabla\psi} \quad \frac{\phi \vdash \psi}{\Delta\phi \vdash \Delta\psi}.$$

The smallest such set **L** is called the *basic lattice-based monotone modal logic*. For any sequent $\phi \vdash \psi$ in **L**, we denote it as $\phi \vdash_{\mathbf{L}} \psi$. Note that, even though both ∇ and Δ are general monotone operators algebraically, as it will be made clear later, the main difference between them lies in the way they are interpreted semantically, i.e. they are interpreted using different types of neighborhood functions defined on polarities. In the following part, we describe a natural algebraic semantics for **L**.

A *monotone lattice algebra* (MLA) is a tuple $\mathbb{A} = (\mathbb{L}, \nabla, \Delta)$ where $\mathbb{L}$ is a bounded lattice and ∇ and Δ are monotone operators on it. An MLA-*valuation* on $\mathbb{A}$ is a homomorphic assignment $V : \mathcal{L} \to \mathbb{A}$ from the set of $\mathcal{L}$-formulas into $\mathbb{A}$. An MLA $\mathbb{A}$ is said to be *complete* if the lattice $\mathbb{L}$ is complete. An MLA-*model* is a tuple $\mathbb{M} = (\mathbb{A}, V)$ where $\mathbb{A}$ is an MLA and V is an MLA-valuation. For any MLA-model $\mathbb{M} = (\mathbb{A}, V)$, and any $\mathcal{L}$-sequent $\phi \vdash \psi$, $\mathbb{M} \models \phi \vdash \psi$ iff $V(\phi) \leq V(\psi)$. $\mathbb{A} \models \phi \vdash \psi$ iff for any MLA-valuation V on $\mathbb{A}$, $(\mathbb{A}, V) \models \phi \vdash \psi$. An MLA-model is said to be *complete* if its underlying MLA is complete.

From the Lindenbaum-Tarski construction it is straightforward to prove that **L** is sound and complete w.r.t. the class of all MLA. That is, for any $\mathcal{L}$-sequent $\phi \vdash \psi$, $\phi \vdash_\mathbf{L} \psi$ iff for any MLA $\mathbb{A}$, $\mathbb{A} \models \phi \vdash \psi$. In fact, we can prove a stronger completeness result for **L**.

Let $\mathbb{M} = (\mathbb{A}, V)$ be an MLA-model such that $\mathbb{M} \not\models \phi \vdash \psi$. Let $\mathbb{A}^\delta = (\mathbb{L}^\delta, \nabla^\sigma, \Delta^\sigma)$ be the *canonical extension* (see [17], for more details on canonical extensions) of $\mathbb{A}$. It follows from the general theory of canonical extensions that $\mathbb{A}^\delta$ is a complete MLA, and there is a natural embedding $e : \mathbb{A} \hookrightarrow \mathbb{A}^\delta$. Let $V^\delta : \mathcal{L} \to \mathbb{A}^\delta$ be the map $e \circ V$. It is easy to check that $(\mathbb{A}^\delta, V^\delta)$ is an MLA-model. Moreover, $(\mathbb{A}^\delta, V^\delta) \not\models \phi \vdash \psi$. Thus, we have the following result.

Proposition 1. *The basic lattice-based monotone modal logic* **L** *is sound and complete w.r.t. the class of complete monotone lattice algebras.*

Polarity-Based Semantics for Lattice-Based Propositional Logic. Polarity-based semantics has been studied extensively as relational semantics for the lattice-based propositional logic and lattice-based normal modal logic [7,8,10].

Given any binary relation $R \subseteq A \times X$ and any $B \subseteq A$, $Y \subseteq X$, the maps $R^{(1)} : \mathcal{P}(A) \to \mathcal{P}(X)$ and $R^{(0)} : \mathcal{P}(X) \to \mathcal{P}(A)$ are defined as $R^{(1)}[B] := \{x \in X \mid \forall b(b \in B \Rightarrow bRx)\}$ and $R^{(0)}[Y] := \{a \in A \mid \forall y(y \in Y \Rightarrow aRy)\}$. A *polarity* or *formal context* is a tuple $\mathbb{P} = (A, X, I)$, where A, and X are sets and $I \subseteq A \times X$ is a relation on them. The maps $I^{(1)} : \mathcal{P}(A) \to \mathcal{P}(X)$ and $I^{(0)} : \mathcal{P}(X) \to \mathcal{P}(A)$ form a Galois connection between posets $(\mathcal{P}(A), \subseteq)$ and $(\mathcal{P}(X), \subseteq)$, i.e. $Y \subseteq I^{(1)}[B]$ iff $B \subseteq I^{(0)}[Y]$ for all $B \in \mathcal{P}(A)$ and $Y \in \mathcal{P}(X)$. A *formal concept* of $\mathbb{P}$ is a pair $(\llbracket c \rrbracket, (\!(c)\!))$ such that $\llbracket c \rrbracket \subseteq A$, $(\!(c)\!) \subseteq X$, $I^{(1)}[\llbracket c \rrbracket] = (\!(c)\!)$, and $I^{(0)}[(\!(c)\!)] = \llbracket c \rrbracket$. The set of all the formal concepts of $\mathbb{P}$ can be partially ordered as follows: for any formal concepts c, d, $c \leq d$ iff $\llbracket c \rrbracket \subseteq \llbracket d \rrbracket$ iff $(\!(d)\!) \subseteq (\!(c)\!)$. This poset is denoted as $\mathbb{P}^+$, which is further a complete lattice and called the *concept lattice* of $\mathbb{P}$. Given a polarity $\mathbb{P}$, we use $\mathbb{P}^+$ to denote concept lattice of $\mathbb{P}$.

A *valuation* on a polarity $\mathbb{P}$ is a map $V : \mathsf{Prop} \to \mathbb{P}^+$. A *polarity-based model* is a tuple $\mathbb{M} = (\mathbb{P}, V)$, where $\mathbb{P}$ is a polarity, and V is a valuation on it. For any polarity-based model $\mathbb{M} = (\mathbb{P}, V)$, *modal satisfaction relation* $\Vdash$ and *modal co-satisfaction relation* $\succ$ are defined inductively as follows:

$M, a \Vdash p$	iff $a \in \llbracket V(p) \rrbracket$	$M, x \succ p$	iff $x \in (\!(V(p))\!)$
$M, a \Vdash \top$	always	$M, x \succ \top$	iff $(\forall a \in A) a I x$
$M, x \succ \bot$	always	$M, a \Vdash \bot$	iff $(\forall x \in X) a I x$
$M, a \Vdash \phi \wedge \psi$	iff $M, a \Vdash \phi$ and $M, a \Vdash \psi$	$M, x \succ \phi \wedge \psi$	iff $(\forall a \in A)(M, a \Vdash \phi \wedge \psi \Rightarrow a I x)$
$M, x \succ \phi \vee \psi$	iff $M, x \succ \phi$ and $M, x \succ \psi$	$M, a \Vdash \phi \vee \psi$	iff $(\forall x \in X)(M, x \succ \phi \vee \psi \Rightarrow a I x)$

Theorem 1 (Proposition 3, [7]). *Lattice-based propositional logic is sound and complete w.r.t. the class of polarity-based models.*

2.2 Monotone Modal Logic and Its Multi-type Representation

In this subsection, we briefly recall the classical monotone modal logic and its neighborhood semantics, i.e., classical neighborhood frames and models (also called Montague-Scott semantics), detailedly studied in [24]. We also discuss the duality for neighborhood frames and Boolean algebra expansions based on [5,19]. The language $\mathcal{L}_c$ of monotone modal logic consists of Boolean connectives $\vee, \wedge$, $\neg$ along with an unary modal operator ∇. All the Boolean connectives behave as in classical propositional logic, while ∇ is a monotone operator.

A *neighborhood frame* (N-frame) is a pair $\mathbb{F} = (W, N)$, where W is an non-empty set and $N : W \to \mathcal{P}(\mathcal{P}(W))$ is a neighborhood function. A neighborhood frame $\mathbb{F} = (W, N)$ is *monotonic* if $\forall w \in W$, $X \in N(w)$ and $X \subseteq Y$ imply that $Y \in N(w)$. A *neighborhood model* (N-model) is a tuple $\mathbb{M} = (\mathbb{F}, V)$, where $\mathbb{F}$ is a N-frame and $V : \mathsf{Prop} \to \mathcal{P}(W)$ is a valuation on it. V can be homomorphically extended to a map $\overline{V}$ on all the the Boolean formulas on Prop in the standard manner. We extend the map to formulas with connective ∇ as follows: For any $\phi \in \mathcal{L}_c$, $w \in \overline{V}(\nabla \phi)$ iff $\overline{V}(\phi) \in N(w)$. For any formula ϕ, $\mathbb{M}, w \models \phi$ iff $w \in \overline{V}(\phi)$. We denote $\mathbb{M} \models \phi$ iff for all $w \in \mathbb{M}$, $\mathbb{M}, w \models \phi$, and $\mathbb{F} \models \phi$ iff for all models $\mathbb{M}$ on $\mathbb{F}$, $\mathbb{M} \models \phi$.

A *monotonic Boolean algebra expansion* (m-algebra) is a tuple $\mathbb{A} = (\mathbb{B}, f)$ where $\mathbb{B}$ is a Boolean algebra and f is a monotone operation on it. It is *perfect* if $\mathbb{B}$ is complete and atomic. Let $\mathbb{F} = (W, N)$ be a monotonic N-frame, the *complex algebra* of $\mathbb{F}$ is a tuple $\mathbb{F}^* = (\mathcal{P}(W), \nabla^{\mathbb{F}^*})$ such that $\mathcal{P}(W)$ is the powerset algebra and for any $U \subseteq W$, $\nabla^{\mathbb{F}^*}[U] := \{w \mid U \in N(w)\}$. It is easy to check that $\mathbb{F}^*$ is a perfect m-algebra. Conversely, let $\mathbb{A} = (\mathcal{P}(W), \nabla)$ be a perfect m-algebra. Then it is easy to check that $\mathbb{A}_* = (W, N_\nabla)$ such that for any $w \in W$, $N_\nabla(w) := \{U \subseteq W \mid w \in \nabla[U]\}$, is a monotonic N-frame. Thus, the following theorem shows the duality between a monotonic neighborhood frame and a perfect Boolean algebra expansion.

Theorem 2 (Proposition 2, [5]). *If $\mathbb{A}$ is a perfect m-algebra and $\mathbb{F}$ is a monotonic N-frame, then $(\mathbb{F}^*)_* \cong \mathbb{F}$ and $(\mathbb{A}_*)^* \cong \mathbb{A}$.*

Now, we introduce the relational and algebraic multi-type representation of monotone modal logics based on [5]. For any relation $R \subseteq A \times B$, $R^{-1} \subseteq B \times A$ is a relation given by $bR^{-1}a$ iff aRb.

Let $\mathbb{F} = (W, N)$ be a monotonic N-frame. We define a *supported two-sorted N-frame* $\mathbb{F}^\star = (W, \mathcal{P}(W), R_\ni, R_{\not\ni}, R_N, R_{N^c})$ such that, $R_\ni \subseteq \mathcal{P}(W) \times W$, $R_{\not\ni} \subseteq$

$\mathcal{P}(W) \times W$, $R_N \subseteq W \times \mathcal{P}(W)$, $R_{N^c} \subseteq W \times \mathcal{P}(W)$ are defined as follows: For any $w \in W$ and $U \subseteq W$,

- $UR_\ni w$ iff $w \in U$;
- $UR_{\not\ni} w$ iff $w \notin U$;
- $wR_N U$ iff $U \in N(w)$;
- $wR_{N^c} U$ iff $U \notin N(w)$;
- $R_N^{-1}(R_\ni^{-1}(U^c))^c = R_{N^c}^{-1}(R_{\not\ni}^{-1}(U)^c)^c$.

The *complex algebra* of $\mathbb{F}^*$ is defined as $\mathbb{F}^{*+} = (\mathcal{P}(W), \mathcal{P}(\mathcal{P}(W)), \Box_\ni^+, \Diamond_{\not\ni}^+, \Diamond_N^+, \Box_{N^c}^+)$ such that for any $U \in \mathcal{P}(W)$, $V \in \mathcal{P}(\mathcal{P}(W))$,

- $\Box_\ni^+ : \mathcal{P}(W) \to \mathcal{P}(\mathcal{P}(W))$ is meet-preserving such that $\Box_\ni^+ U = (R_\ni^{-1}(U^c))^c$;
- $\Diamond_N^+ : \mathcal{P}(\mathcal{P}(W)) \to \mathcal{P}(W)$ is join-preserving such that $\Diamond_N^+ V = R_N^{-1} V$;
- $\Diamond_{\not\ni}^+ : \mathcal{P}(W) \to \mathcal{P}(\mathcal{P}(W))$ is join-preserving such that $\Diamond_{\not\ni}^+ U = R_{\not\ni}^{-1} U$;
- $\Box_{N^c}^+ : \mathcal{P}(\mathcal{P}(W)) \to \mathcal{P}(W)$ is meet-preserving such that $\Box_{N^c}^+ V = (R_{N^c}^{-1}(V^c))^c$.

Let $\mathbb{F}^* = (\mathcal{P}(W), \nabla^{\mathbb{F}^*})$ be the complex algebra of $\mathbb{F} = (W, N)$. It is easy to check that $\Diamond_N^+ \Box_\ni^+ U = \Box_{N^c}^+ \Diamond_{\not\ni}^+ U = \nabla^{\mathbb{F}^*} U$ for any $U \subseteq W$.

The following definition is motivated by the above observations. A *supported heterogeneous m-algebra* is a multi-type algebra $\mathbb{A} = (\mathbb{C}, \mathbb{D}, \Box_\ni^{\mathbb{A}}, \Diamond_{\not\ni}^{\mathbb{A}}, \Diamond_N^{\mathbb{A}}, \Box_{N^c}^{\mathbb{A}})$, where $\mathbb{C}$ and $\mathbb{D}$ are Boolean algebras, and $\Box_\ni^{\mathbb{A}} : \mathbb{C} \to \mathbb{D}, \Box_{N^c}^{\mathbb{A}} : \mathbb{D} \to \mathbb{C}$ are finitely meet-preserving operators and $\Diamond_{\not\ni}^{\mathbb{A}} : \mathbb{C} \to \mathbb{D}, \Diamond_N^{\mathbb{A}} : \mathbb{D} \to \mathbb{C}$ are finitely join-preserving operators, and $\Diamond_N^{\mathbb{A}} \Box_\ni^{\mathbb{A}} a = \Box_{N^c}^{\mathbb{A}} \Diamond_{\not\ni}^{\mathbb{A}} a$ for all $a \in \mathbb{C}$. Let $\mathbb{A}_\bullet = (\mathbb{C}, \nabla^{\mathbb{A}_\bullet})$ be such that for any $a \in \mathbb{C}$, $\nabla^{\mathbb{A}_\bullet} a = \Diamond_N^{\mathbb{A}} \Box_\ni^{\mathbb{A}} a = \Box_{N^c}^{\mathbb{A}} \Diamond_{\not\ni}^{\mathbb{A}} a$. It is easy to check that $\nabla^{\mathbb{A}_\bullet}$ is a monotone operator and hence $\mathbb{A}_\bullet$ is an m-algebra.

Conversely, let $\mathbb{A} = (\mathbb{C}, \nabla^{\mathbb{A}})$ be a perfect m-algebra such that $\mathbb{A} \cong \mathbb{F}^*$ for some monotonic N-frame $\mathbb{F}$. Let $\mathbb{A}^\bullet := \mathbb{F}^{*+}$. The following lemma provides the representation of perfect m-algebras by supported heterogeneous m-algebras.

Lemma 1 (Proposition 14, [5]). *If $\mathbb{A}$ is a perfect m-algebra, then $\mathbb{A} \cong (\mathbb{A}^\bullet)_\bullet$.*

This multi-type representation shows that the monotone map in perfect m-algebras can be decomposed as a combination of normal $\Box$ and $\Diamond$ operators. Thus, it is possible to define a translation of monotone modal logic into many-sorted normal modal logic (see [5], for more details)[1].

3 Polarity-Based Neighborhood Frames

In this section, we generalize the duality between monotone perfect m-algebras and neighborhood frames, described in Sect. 2.2, to a duality between complete MLAs and a structure we call polarity-based neighborhood frames. This duality serves both as the motivation and the mathematical foundation for providing a sound and complete polarity-based neighborhood semantics for the lattice-based monotone modal logic described in Sect. 2.1.

---

[1] Similar translation of monotone modal logic into many-sorted normal modal logic using more syntactic methods has been studied previously by multiple researchers [16,19]. However, the algebraic presentation given here aligns more with our methodology for non-distributive setting.

Let $\mathbb{A} = (\mathbb{L}, \Delta, \nabla)$ be a complete MLA, and $\mathbb{L}_+ = (L_A, L_X, \leq)$ be a polarity, where $L_A = L_X$ are copies of the domain of $\mathbb{L}$, and $\leq$ is the partial order of $\mathbb{L}$. By Birkhoff's representation theorem, $(\mathbb{L}_+)^+ \cong \mathbb{L}$, where the isomporhism maps any $c \in \mathbb{L}$ to the concept $\widehat{c} = (\{l \in L_A \mid l \leq c\}, \{l \in L_X \mid c \leq l\})$.

Let L^+ be the set of all concepts of $\mathbb{L}_+$. We define a Δ-type neighborhood function $N_\Delta : L_A \to \mathcal{P}(L^+)$, and a ∇-type neighborhood function $N_\nabla : L_X \to \mathcal{P}(L^+)$ as follows: For any $c \in \mathbb{L}$, $a \in L_A$, $x \in L_X$,

$$\widehat{c} \in N_\Delta(a) \text{ iff } a \leq \Delta(c), \quad \widehat{c} \in N_\nabla(x) \text{ iff } \nabla(c) \leq x.$$

Note that for any $a \in L_A$ (resp. $x \in L_X$), $N_\Delta(a)$ (resp. $N_\nabla(x)$) is an upward-closed (resp. downward-closed) subset of L^+. For any complete MLA $\mathbb{A} = (\mathbb{L}, \Delta, \nabla)$, we denote the tuple $(\mathbb{L}_+, N_\Delta, N_\nabla)$ defined above by $\mathbb{A}_+$.

Lemma 2. *Let $\mathbb{A} = (\mathbb{L}, \Delta, \nabla)$ be any complete MLA and $\mathbb{A}_+ = (\mathbb{L}_+, N_\Delta, N_\nabla)$. Let $N_\Delta^+, N_\nabla^+ : (\mathbb{L}_+)^+ \to (\mathbb{L}_+)^+$ be two maps such that for any $\widehat{c} \in (\mathbb{L}_+)^+$,*

$$N_\Delta^+(\widehat{c}) = (\{a \in L_A \mid \widehat{c} \in N_\Delta(a)\}, \{a \in L_A \mid \widehat{c} \in N_\Delta(a)\}^\uparrow),$$
$$N_\nabla^+(\widehat{c}) = (\{x \in L_X \mid \widehat{c} \in N_\nabla(x)\}^\downarrow, \{x \in L_X \mid \widehat{c} \in N_\nabla(x)\}).$$

Then, $((\mathbb{L}_+)^+, N_\Delta^+, N_\nabla^+) \cong \mathbb{A}$.

Proof. See [14, A.1].

Given any polarity $\mathbb{P}$, we use P^+ to denote the set of all formal concepts of $\mathbb{P}$ from now. The following definition is motivated by the above lemma.

Definition 1. *A polarity-based neighborhood frame (PN-frame) is a tuple $\mathbb{F} = (\mathbb{P}, N_\Delta, N_\nabla)$, where $\mathbb{P} = (A, X, I)$ is a polarity, and $N_\Delta : A \to \mathcal{P}(P^+)$ and $N_\nabla : X \to \mathcal{P}(P^+)$ are maps, such that for any $c \in P^+$, $\{a \in A \mid c \in N_\Delta(a)\}$ and $\{x \in X \mid c \in N_\nabla(x)\}$ are Galois-stable. A PN-frame is said to be monotone if for any $a \in A$ (resp. $x \in X$), the set $N_\Delta(a)$ (resp. $N_\nabla(x)$) is upward (resp. downward) closed.*

It is straightforward to check that, for any complete MLA $\mathbb{A}$, $\mathbb{A}_+$ is a monotone PN-frame. Conversely, given a monotone PN-frame $\mathbb{F} = (\mathbb{P}, N_\Delta, N_\nabla)$, it is easy to check that $\mathbb{F}^+ = (\mathbb{P}^+, N_\Delta^+, N_\nabla^+)$ is a complete MLA. The Galois-stable condition ensures that N_∇^+ and N_Δ^+ are well-defined. The monotonicity of N_∇^+ and N_Δ^+ follows from the monotonicity of $\mathbb{F}$. Thus, the Lemma 2 can be restated in the following form which shows that the monotone PN-frames provide a representation for complete MLAs.

Lemma 3. *For any complete MLA $\mathbb{A}$, $(\mathbb{A}_+)^+ \cong \mathbb{A}$.*

Relation to Classical Neighborhood Frames and Enriched Formal Contexts. Let $\mathbb{F} = (W, N)$ be an N-frame and $\mathbb{F}^* = (\mathcal{P}(W), \nabla^{\mathbb{F}^*})$ be its complex algebra. We define its *lifted* PN-frame $\overline{\mathbb{F}} = (\mathbb{P}, N_\Delta, N_\nabla)$, where $\mathbb{P} = (W_A, W_X, \neq)$ such that W_A and W_X are two copies of W, and $\neq \subseteq W_A \times W_X$ is the 'not equal to' relation, and the maps $N_\Delta : W_A \to \mathcal{P}(P^+)$ and $N_\nabla : W_X \to \mathcal{P}(P^+)$ are defined as follows: For any $w \in W$, $N_\Delta(w) = \{(Y, Y^c) \mid Y \in N(w)\}$, and $N_\nabla(w) = \{(B, B^c) \mid B \notin N(w)\}$. Note that all the concepts of $\overline{\mathbb{F}}$ are of the form (B, B^c) for some $B \subseteq W$.

Lemma 4. *For any N-frame* $\mathbb{F} = (W, N)$, $(\mathbb{P}^+, N_\Delta^+) \cong (\mathbb{P}^+, N_\nabla^+) \cong \mathbb{F}^*$.

Proof. See [14, A.2].

This lemma shows that the classical neighborhood frames can be lifted to the equivalent PN-frames. Hence, it can be seen as a natural generalization of classical neighborhood frames. The following lemma characterizes the class of PN-frames which define the normal modal operations.

Lemma 5. *Let* $\mathbb{F} = (A, X, I, N_\Delta, N_\nabla)$ *be a monotone PN-frame. Then,*
1. N_Δ^+ *is meet-preserving (resp. completely meet-preserving) iff for any* $a \in A$, $N_\Delta(a)$ *is closed under finite (resp. arbitrary) meets.*
2. N_Δ^+ *is join-preserving (resp. completely join-preserving) iff for any* $x \in X$, $N_\Delta(x)$ *is closed under finite (resp. arbitrary) joins.*

Proof. We only prove item 1, and the proof for item 2 is similar. Suppose that $N_\Delta(a)$ is closed under finite (resp. arbitrary) meets for any $a \in A$. Let $\{c_j \mid j \in \mathcal{J}\}$ be a finite (resp. arbitrary) set of concepts in P^+.

$$\begin{aligned}[\![N_\Delta^+(\bigwedge_{j \in \mathcal{J}} c_j)]\!] &= \{a \in A \mid \bigwedge_{j \in \mathcal{J}} c_j \in N_\Delta(a)\} \quad \text{(By def. of } N_\Delta^+\text{)}\\ &= \{a \in A \mid \forall j(c_j \in N_\Delta(a))\}\\ &= \bigcap_{j \in \mathcal{J}}\{a \in A \mid c_j \in N_\Delta(a)\}\\ &= \bigcap_{j \in \mathcal{J}}[\![N_\Delta^+(c_j)]\!]\\ &= [\![\bigwedge_{j \in \mathcal{J}} N_\Delta^+(c_j)]\!].\end{aligned}$$

The second equation follows from the fact that $N_\Delta(a)$ is an up-set closed under finite (resp. arbitrary) meets. Thus, $N_\Delta^+(\bigwedge_{j \in \mathcal{J}} c_j) = \bigwedge_{j \in \mathcal{J}} N_\Delta^+(c_j)$.

Conversely, suppose that N_Δ^+ is meet-preserving (resp. completely meet-preserving) and let $a \in A$. By the definition of N_Δ^+, there is $\bigwedge_{j \in \mathcal{J}} c_j \in N_\Delta(a)$ iff $a \in [\![N_\Delta^+(\bigwedge_{j \in \mathcal{J}} c_j)]\!]$. On the other hand, as N_Δ^+ is meet-preserving (resp. completely meet-preserving), $a \in [\![N_\Delta^+(\bigwedge_{j \in \mathcal{J}} c_j)]\!]$ iff $a \in [\![\bigwedge_{j \in \mathcal{J}} N_\Delta^+(c_j)]\!]$ iff $a \in [\![N_\Delta^+(c_j)]\!]$ for all $j \in \mathcal{J}$. By the definition of N_Δ^+ again, this is equivalent to $c_j \in N_\Delta(a)$ for all $j \in \mathcal{J}$. Therefore, there is $\bigwedge_{j \in \mathcal{J}} c_j \in N_\Delta(a)$ iff $c_j \in N_\Delta(a)$ for all $j \in \mathcal{J}$. Hence proved.

Let $\mathbb{F} = (\mathbb{P}, N_\Delta, N_\nabla)$ be a monotone PN-frame such that the maps N_Δ^+ and N_∇^+ are completely meet-preserving and completely join-preserving, respectively. Let $\mathbb{F}' = (\mathbb{P}, R_\square, R_\lozenge)$ be an *enriched formal context* [6, Definition 2.5], such that $R_\square \subseteq A \times X$ and $R_\lozenge \subseteq X \times A$ are defined as follows: For any $a \in A$, $x \in X$,

$$aR_\square x \text{ iff } x \in ([\![\bigwedge N_\Delta(a)]\!]) \quad \text{and} \quad xR_\lozenge a \text{ iff } a \in [\![\bigvee N_\nabla(x)]\!].$$

For any enriched formal context $\mathbb{F}$, its *complex algebra* is a tuple $\mathbb{F}^+ = (\mathbb{P}^+, [R_\square], \langle R_\lozenge \rangle)$, where for any $c \in \mathbb{P}^+$, $[R_\square]c = (R_\square^{(0)}[[\![c]\!]], (R_\square^{(0)}[[\![c]\!]])^\uparrow)$, and $\langle R_\lozenge \rangle c = ((R_\lozenge^{(0)}[[\![c]\!]])^\downarrow, R_\lozenge^{(0)}[[\![c]\!]])$.

Lemma 6. *Let* $\mathbb{F} = (\mathbb{P}, N_\Delta, N_\nabla)$ *be a monotone PN-frame such that the maps* N_Δ^+ *and* N_∇^+ *are completely meet-preserving and completely join-preserving, respectively. Then,* $\mathbb{F}'^+ = \mathbb{F}^+$.

Proof. See [14, A.4].

Thus, PN-frames also naturally generalize the enriched formal contexts (used as semantics for normal lattice-based modal logic, see [6,7] for details).

4 Polarity-Based Neighborhood Semantics for Lattice-Based Monotone Modal Logic

In this section, we introduce the polarity-based neighborhood semantics based on polarity-based neighborhood frames and show that they provide the sound and complete semantics for basic lattice-based monotone modal logic **L**.

Definition 2. *A polarity-based neighborhood model (PN-model) is a tuple* $\mathbb{M} = (\mathbb{F}, V)$, *where* $\mathbb{F} = (\mathbb{P}, N_\Delta, N_\nabla)$ *is a PN-frame and* $V : \mathsf{Prop} \to \mathbb{F}^+$ *is a valuation on* $\mathbb{F}$. *Each valuation* V *can be extended homomorphically to a map* $\overline{V} : \mathcal{L} \to \mathbb{F}^+$ *over all* $\mathcal{L}$*-formulas. The satisfaction relation* $\Vdash$ *and co-satisfaction relation* $\succ$ *between a PN-model* $\mathbb{M} = (A, X, I, N_\Delta, N_\nabla)$ *and an* $\mathcal{L}$*-formula* ϕ *are defined as follows: For any* $a \in A$, $x \in X$, $\mathbb{M}, a \Vdash \phi$ *iff* $a \in [\![\overline{V}(\phi)]\!]$, *and* $\mathbb{M}, x \succ \phi$ *iff* $x \in (\!(\overline{V}(\phi))\!)$. *Given any* $\mathcal{L}$*-sequent* $\phi \vdash \psi$, $\mathbb{M} \models \phi \vdash \psi$ *iff* $[\![\overline{V}(\phi)]\!] \subseteq [\![\overline{V}(\psi)]\!]$ *iff* $(\!(\overline{V}(\psi))\!) \subseteq (\!(\overline{V}(\phi))\!)$.

The following lemma follows immediately from the above definition.

Lemma 7. *Let* $\mathbb{M} = (A, X, I, N_\Delta, N_\nabla, V)$ *be a PN-model. For any* $\mathcal{L}$*-formula* ϕ, *let* $[\![\phi]\!] = \{a \in A \mid \mathbb{M}, a \Vdash \phi\}$, *and* $(\!(\phi)\!) = \{x \in X \mid \mathbb{M}, x \succ \phi\}$. *Then,*
1. *For all the connectives except* ∇ *and* Δ, $\Vdash$ *and* $\succ$ *are defined as in the lattice-based propositional logic (cf. Sect. 2.1).*
2. $\mathbb{M}, a \Vdash \Delta\phi$ *iff* $([\![\phi]\!], (\!(\phi)\!)) \in N_\Delta(a)$, *and* $\mathbb{M}, x \succ \Delta\phi$ *iff* $\forall a \in A(a \Vdash \Delta\phi \Rightarrow aIx)$.
3. $\mathbb{M}, x \succ \nabla\phi$ *iff* $([\![\phi]\!], (\!(\phi)\!)) \in N_\nabla(x)$, *and* $\mathbb{M}, a \Vdash \nabla\phi$ *iff* $\forall x \in X(x \succ \nabla\phi \Rightarrow aIx)$.
4. *Moreover, if* $\mathbb{M}$ *is monotone, then*
    (i) $\mathbb{M}, a \Vdash \Delta\phi$ *iff* $\forall c \in \mathbb{F}^+((\forall x \in X(x \in (\!(c)\!)) \Rightarrow \mathbb{M}, x \succ \phi) \Rightarrow c \in N_\Delta(a))$,
    (ii) $\mathbb{M}, x \succ \nabla\phi$ *iff* $\forall c \in \mathbb{F}^+((\forall a \in A(a \in [\![c]\!] \Rightarrow \mathbb{M}, a \Vdash \phi) \Rightarrow c \in N_\nabla(x))$.

It is easy to check that all the axioms and rules of lattice-based monotone modal logic are valid on monotone PN-models. Thus, lattice-based monotone modal logic is *sound* w.r.t. the class of monotone PN-models. The following lemma follows trivially from Definition 2.

Lemma 8. *Let* $\mathbb{M} = (\mathbb{F}, V)$ *be a monotone PN-model with* $\mathbb{F} = (A, X, I, N_\Delta, N_\nabla)$, *and* $\mathbb{M}^+ = (\mathbb{F}^+, V^+)$, *where* $V^+ : \mathcal{L} \to \mathbb{F}^+$ *is equal to* $\overline{V}$. *Then,* $\mathbb{M}^+$ *is a complete MLA-model. Moreover, for any* $\phi, \psi \in \mathcal{L}$, $a \in A$, *and* $x \in X$,
1. $\mathbb{M}, a \Vdash \phi$ *iff* $a \in [\![V^+(\phi)]\!]$,
2. $\mathbb{M}, x \succ \phi$ *iff* $x \in (\!(V^+(\phi))\!)$,
3. $\mathbb{M} \models \phi \vdash \psi$ *iff* $\mathbb{M}^+ \models \phi \vdash \psi$.

The following lemma which provides the correspondence in the converse direction follows straightforwardly from Lemma 3 and Definition 2.

Lemma 9. *Let* $\mathbb{M} = (\mathbb{A}, V)$ *be a complete MLA-model. Let* $\mathbb{M}_+ = (\mathbb{A}_+, V_+)$, *where* $V_+ :$ Prop $\to (\mathbb{A}_+)^+ \cong \mathbb{A}$ *is defined by* $V_+(p) = \widehat{V(p)}$. *Then,* $\mathbb{M}_+$ *is a monotone PN-model. Moreover, for any* $\phi, \psi \in \mathcal{L}$, $a \in L_A$, $x \in L_X$,
1. $a \leq V(\phi)$ *iff* $\mathbb{M}_+, a \Vdash \phi$. 2. $V(\phi) \leq x$ *iff* $\mathbb{M}_+, x \succ \phi$.
3. $\mathbb{M} \models \phi \vdash \psi$ *iff* $\mathbb{M}_+ \models \phi \vdash \psi$.

Lemmas 8, and 9 along with the completeness of **L** w.r.t. MLA-models shown in Proposition 1, together imply that the logic **L** is complete w.r.t. the class of monotone PN-models. In conclusion, we get the following result.

Theorem 3. *Basic lattice-based monotone modal logic* **L** *is sound and complete w.r.t. the class of monotone PN-models.*

5 Multi-type Representation of Lattice-Based Monotone Modal Logic

In this section, we generalize the methodology used for multi-type representation of classical monotone modal logic to obtain a multi-type representation of lattice-based monotone modal logic. This multi-type representation allows us to translate the monotone operators Δ and ∇ into the combination of normal multi-type modal operators $\Box$ and $\Diamond$.

5.1 Normal Decomposition of Monotone Modal Operators

Let $(\mathbb{P}, N_\Delta, N_\nabla)$ be a monotone PN-frame. Let $\mathcal{P}(P^+)$ be the power-set lattice (i.e. negation-free fragment of the power-set algebra) of P^+. Let $\mathbb{P}^\star = (P_A^+, P_X^+, I_{\neq})$, where $P_A^+ = P_X^+ = P^+$, and for any $w \in P_A^+$, $v \in P_X^+$, $wI_{\neq}v$ iff $w \neq v$. Note that $\mathbb{P}^{\star+} \cong \mathcal{P}(P^+)$. Let $R_{N_\Delta} \subseteq A \times P^+$, $R_{N_\nabla} \subseteq X \times P^+$, $R_{\ni_\Delta} \subseteq P^+ \times A$, and $R_{\ni_\nabla} \subseteq P^+ \times X$ be the binary relations defined as follows: For any $a \in A$, $x \in X$, and $c \in P^+$,

$$aR_{N_\Delta}c \text{ iff } c \in N_\Delta(a), \qquad cR_{\ni_\Delta}a \text{ iff } a \in [\![c]\!],$$
$$xR_{N_\nabla}c \text{ iff } c \in N_\nabla(x), \qquad cR_{\ni_\nabla}x \text{ iff } x \in (\!(c)\!).$$

Note that for any $c \in P^+$, the sets $R_{\ni_\Delta}^{(1)}[c] = [\![c]\!]$, $R_{N_\Delta}^{(0)}[c] = \{a \mid c \in N_\Delta(a)\}$, $R_{\ni_\nabla}^{(1)}[c] = (\!(c)\!)$, and $R_{N_\nabla}^{(0)}[c] = \{x \mid c \in N_\nabla(x)\}$ are Galois-stable[2]. Moreover, any subset of P^+ is Galois-stable in $\mathbb{P}^\star$.

Hence, we can define the following maps: For any $c \in P^+$, $B \in \mathcal{P}(P^+)$,
1. $\Diamond_{\ni_\Delta} : P^+ \to \mathcal{P}(P^+)$, and $\Diamond_{N_\nabla} : \mathcal{P}(P^+) \to P^+$ are the normal join-preserving maps, such that $\Diamond_{\ni_\Delta}(c) = R_{\ni_\Delta}^{(0)}[[c]]$, and $\Diamond_{N_\nabla}(B) = ((R_{N_\nabla}^{(0)}[B])^\downarrow, R_{N_\nabla}^{(0)}[B])$.
2. $\Box_{N_\Delta} : \mathcal{P}(P^+) \to P^+$, and $\Box_{\ni_\nabla} : P^+ \to \mathcal{P}(P^+)$ are the normal meet-preserving maps, such that $\Box_{N_\Delta}(B) = (R_{N_\Delta}^{(0)}[B], (R_{N_\Delta}^{(0)}[B])^\uparrow)$, and $\Box_{\ni_\nabla}(c) = R_{\ni_\nabla}^{(0)}[(\!(c)\!)]$.

---
[2] Galois-stability of $R_{N_\Delta}^{(0)}[c]$, and $R_{N_\nabla}^{(0)}[c]$ follows from items 2 and 3 of Definition 1.

The following lemma shows that we can decompose the operator N_Δ^+ as the combination of operators $\Diamond_{\ni_\Delta}$ and $\Box_{N_\Delta}$, and the operator N_∇^+ as the combination of operators $\Box_{\ni_\nabla}$ and $\Diamond_{N_\nabla}$.

Lemma 10. *Let* $(\mathbb{P}, N_\Delta, N_\nabla)$ *be a monotone PN-frame. Then for any* $c \in P^+$, $N_\Delta^+(c) = \Box_{N_\Delta}\Diamond_{\ni_\Delta}c = (R_{N_\Delta}^{(0)}[R_{\ni_\Delta}^{(0)}[[c]]], R_{N_\Delta}^{(0)}[R_{\ni_\Delta}^{(0)}[[c]]]^\uparrow)$, *and* $N_\nabla^+(c) = \Diamond_{N_\nabla}\Box_{\ni_\nabla}c = (R_{N_\nabla}^{(0)}[R_{\ni_\nabla}^{(0)}[(c)]]^\downarrow, R_{N_\nabla}^{(0)}[R_{\ni_\nabla}^{(0)}[(c)]])$.

Proof. See [14, A.5].

The above lemma along with Lemma 2, shows that any monotone map on a concept lattice can be decomposed into a combination of normal join-preserving and normal meet-preserving maps. Conversely, it is clear that for any set of compatible relations $R_{\ni_\Delta} \subseteq P^+ \times A$, $R_{N_\Delta} \subseteq A \times P^+$, $R_{\ni_\nabla} \subseteq P^+ \times X$, and $R_{N_\nabla} \subseteq X \times P^+$, the maps $\Box_{N_\Delta}\Diamond_{\ni_\Delta} : \mathbb{P}^+ \to \mathbb{P}^+$ and $\Diamond_{N_\nabla}\Box_{\ni_\nabla} : \mathbb{P}^+ \to \mathbb{P}^+$ are monotone. Therefore, we get the following result.

Proposition 2. *Let* $\mathbb{P}$ *be any polarity. Then,*
1. *A map* $\Delta : \mathbb{P}^+ \to \mathbb{P}^+$ *is monotone iff there exist relations* $R_{\ni_\Delta} \subseteq P^+ \times A$ *and* $R_{N_\Delta} \subseteq A \times P^+$ *such that for any* $c \in \mathbb{P}^+$, $\Delta(c) = (R_{N_\Delta}^{(0)}[R_{\ni_\Delta}^{(0)}[[c]]], R_{N_\Delta}^{(0)}[R_{\ni_\Delta}^{(0)}[[c]]]^\uparrow)$.
2. *A map* $\nabla : \mathbb{P}^+ \to \mathbb{P}^+$ *is monotone iff there exist relations* $R_{\ni_\nabla} \subseteq P^+ \times X$ *and* $R_{N_\nabla} \subseteq X \times P^+$ *such that for any* $c \in \mathbb{P}^+$, $\nabla(c) = (R_{N_\nabla}^{(0)}[R_{\ni_\nabla}^{(0)}[(c)]]^\downarrow, R_{N_\nabla}^{(0)}[R_{\ni_\nabla}^{(0)}[(c)]])$.

This multi-type representation for MLAs motivates us to define the two-sorted polarity-based monotone frames and the two-sorted heterogeneous multi-type monotone lattice algebras as follows.

Definition 3. *A two-sorted polarity-based monotone frame (2-PN frame) is a tuple* $\mathbb{F} = (\mathbb{P}, \mathbb{P}^\star, R_{N_\Delta}, R_{N_\nabla}, R_{\ni_\Delta}, R_{\ni_\nabla})$, *where*
1. $\mathbb{P} = (A, X, I)$ *is a polarity, and* $\mathbb{P}^\star = (P^+, P^+, I_{\neq})$,
2. *Relations* $R_{\ni_\Delta} \subseteq P^+ \times A$, $R_{N_\Delta} \subseteq A \times P^+$, $R_{\ni_\nabla} \subseteq P^+ \times X$, *and* $R_{N_\nabla} \subseteq X \times P^+$ *are compatible, i.e., for any* $c \in P^+$, $a \in A$, $x \in X$, *the sets* $R_{\ni_\Delta}^{(0)}[a]$ *(resp.* $R_{N_\Delta}^{(0)}[c]$, *resp.* $R_{\ni_\nabla}^{(0)}[x]$, *resp.* $R_{N_\nabla}^{(0)}[c]$*) and* $R_{\ni_\Delta}^{(1)}[c]$ *(resp.* $R_{N_\Delta}^{(1)}[a]$, *resp.* $R_{\ni_\nabla}^{(1)}[c]$, *resp.* $R_{N_\nabla}^{(1)}[x]$*) are Galois-stable sets in the polarities* $\mathbb{P}^\star$ *(resp.* $\mathbb{P}$, *resp.* $\mathbb{P}^\star$, *resp.* $\mathbb{P}$*), and* $\mathbb{P}$ *(resp.* $\mathbb{P}^\star$, *resp.* $\mathbb{P}$, *resp.* $\mathbb{P}^\star$*), respectively.*

Definition 4. *A heterogeneous multi-type monotone lattice algebra (HM-MLA) is a tuple* $\mathbb{H} = (\mathbb{L}, \mathcal{P}(\mathbb{L}), \Box_{N_\Delta}, \Diamond_{N_\nabla}, \Diamond_{\ni_\Delta}, \Box_{\ni_\nabla})$, *where*
1. $\mathbb{L}$ *is a lattice, and* $\mathcal{P}(\mathbb{L})$ *is the power-set lattice of elements of* $\mathbb{L}$,
2. $\Box_{N_\Delta} : \mathcal{P}(\mathbb{L}) \to \mathbb{L}$, *and* $\Box_{\ni_\nabla} : \mathbb{L} \to \mathcal{P}(\mathbb{L})$ *(resp.* $\Diamond_{N_\nabla} : \mathcal{P}(\mathbb{L}) \to \mathbb{L}$, *and* $\Diamond_{\ni_\Delta} : \mathbb{L} \to \mathcal{P}(\mathbb{L})$*) are normal meet-preserving (resp. join-preserving) operators. An HM-MLA* $\mathbb{H}$ *is said to be complete if* $\mathbb{L}$ *is complete,* $\Box_{N_\Delta}, \Box_{\ni_\nabla}$ *are completely meet-preserving, and* $\Diamond_{N_\nabla}, \Diamond_{\ni_\Delta}$ *are completely join-preserving.*

Given a 2-PN frame $\mathbb{F} = (\mathbb{P}, \mathbb{P}^\star, R_{N_\triangle}, R_{N_\triangledown}, R_{\ni_\triangle}, R_{\ni_\triangledown})$, its *complex algebra* is the complete HM-MLA $\mathbb{F}^+ = (\mathbb{P}^+, \mathbb{P}^{\star +}, \Box_{N_\triangle}, \Diamond_{N_\triangledown}, \Diamond_{\ni_\triangle}, \Box_{\ni_\triangledown})$, where the maps $\Box_{N_\triangle} : \mathbb{P}^{\star +} \to \mathbb{P}^+$, $\Diamond_{N_\triangledown} : \mathbb{P}^{\star +} \to \mathbb{P}^+$, $\Diamond_{\ni_\triangle} : \mathbb{P}^+ \to \mathbb{P}^{\star +}$, and $\Box_{\ni_\triangledown} : \mathbb{P}^+ \to \mathbb{P}^{\star +}$ are as follows: for any $c \in \mathbb{P}^+$, $B \in \mathbb{P}^{\star +}$,

1. $\Box_{N_\triangle}(B) = (R_{N_\triangle}^{(0)}[B], (R_{N_\triangle}^{(0)}[B])^\uparrow)$, 2. $\Diamond_{N_\triangledown}(B) = ((R_{N_\triangledown}^{(0)}[B])^\downarrow, R_{N_\triangledown}^{(0)}[B])$,
3. $\Diamond_{\ni_\triangle}(c) = ((R_{\ni_\triangle}^{(0)}[\![c]\!])^\downarrow, R_{\ni_\triangle}^{(0)}[\![c]\!])$, 4. $\Box_{\ni_\triangledown}(c) = (R_{\ni_\triangledown}^{(0)}[\![c]\!], (R_{\ni_\triangledown}^{(0)}[\![c]\!])^\uparrow)$.

Conversely, given a complete HM-MLA $\mathbb{H} = (\mathbb{L}, \mathcal{P}(L), \Box_{N_\triangle}, \Diamond_{N_\triangledown}, \Diamond_{\ni_\triangle}, \Box_{\ni_\triangledown})$, its *dual 2-PN frame* is a frame $\mathbb{H}_+ = (\mathbb{P}, \mathbb{P}^\star, R_{N_\triangle}, R_{N_\triangledown}, R_{\ni_\triangle}, R_{\ni_\triangledown})$, where $\mathbb{P} = \mathbb{L}_+$, and $R_{N_\triangle} \subseteq L_A \times P^+$, $R_{\ni_\triangle} \subseteq P^+ \times L_A$, $R_{N_\triangledown} \subseteq L_X \times P^+$, and $R_{\ni_\triangledown} \subseteq P^+ \times L_X$ are maps, such that for any $a \in L_A$, $x \in L_X$, $c \in L$, $\widehat{c} \in P^+$,

(i) $aR_{N_\triangle}\widehat{c}$ iff $a \leq \Box_{N_\triangle}(\{c\}^c)$, (ii) $\widehat{c}R_{\ni_\triangle}a$ iff $\Diamond_{\ni_\triangle}a \leq c$.
(iii) $xR_{N_\triangledown}\widehat{c}$ iff $\Diamond_{N_\triangledown}(\{c\}) \leq x$, (iv) $\widehat{c}R_{\ni_\triangledown}x$ iff $c \leq \Box_{\ni_\triangledown}x$.

The following lemma shows that 2-PN frames give a representation of HM-MLAs.

Lemma 11. *For any complete HM-MLA $\mathbb{H}$, $(\mathbb{H}_+)^+ = \mathbb{H}$.*

Proof. Suppose $\mathbb{H} = (\mathbb{L}, \mathcal{P}(L), \Box_{N_\triangle}, \Diamond_{N_\triangledown}, \Diamond_{\ni_\triangle}, \Box_{\ni_\triangledown})$ is a complete HM-MLA. By definition, $(\mathbb{H}_+)^+ = (\mathbb{P}^+, \mathbb{P}^{\star +}, \Box'_{N_\triangle}, \Diamond'_{N_\triangledown}, \Diamond'_{\ni_\triangle}, \Box'_{\ni_\triangledown})$, where $\mathbb{L}_+ = \mathbb{P} = (A, X, I)$. Thus, $\mathbb{P}^+ \cong \mathbb{L}$, and $\mathbb{P}^{\star +} \cong \mathcal{P}(L)$, where the isomorphism from $\mathcal{P}(L)$ to $\mathbb{P}^{\star +}$ is given by $B \mapsto \widehat{B} = (\{\widehat{c} \mid c \in B\}, \{\widehat{c} \mid c \in B\}^c)$. We show that these lattice homomorphisms are HM-MLA-homomorphisms (i.e. they preserve all the HM-MLA operators). That is, for any $c \in \mathbb{L}$, $B \in \mathcal{P}(L)$,

$\widehat{\Box_{N_\triangle}(B)} = \Box'_{N_\triangle}(\widehat{B})$, $\widehat{\Box_{\ni_\triangledown}(c)} = \Box'_{\ni_\triangledown}(\widehat{c})$, $\widehat{\Diamond_{N_\triangledown}(B)} = \Diamond'_{N_\triangledown}(\widehat{B})$, $\widehat{\Diamond_{\ni_\triangle}(c)} = \Diamond'_{\ni_\triangle}(\widehat{c})$.

We give the proofs for $\Box_{N_\triangle}$ and $\Box_{\ni_\triangledown}$. The proofs for $\Diamond_{N_\triangledown}$ and $\Diamond_{\ni_\triangle}$ are dual.

$$\begin{aligned}
[\![\Box'_{N_\triangle}(\widehat{B})]\!] &= R_{N_\triangle}^{(0)}[\widehat{B}] \\
&= \{a \in L_A \mid \forall \widehat{d} \in ([\widehat{B}])(aR_{N_\triangle}d)\} \\
&= \{a \in L_A \mid \forall d \notin B(aR_{N_\triangle}\widehat{d})\} \quad \text{(by } ([\widehat{B}]) = B^c) \\
&= \{a \in L_A \mid \forall d \notin B(a \leq \Box_{N_\triangle}(\{d\}^c))\} \text{ (by def. of } R_{N_\triangle}) \\
&= \{a \in L_A \mid a \leq \bigwedge_{d \notin B} \Box_{N_\triangle}(\{d\}^c)\} \\
&= \{a \in L_A \mid a \leq \Box_{N_\triangle}(\bigwedge_{d \notin B}\{d\}^c)\} \\
&= \{a \in L_A \mid a \leq \Box_{N_\triangle}(B)\} \\
&= [\![\widehat{\Box_{N_\triangle}(B)}]\!]
\end{aligned}$$

Therefore, $\widehat{\Box_{N_\triangle}(B)} = \Box'_{N_\triangle}(\widehat{B})$.

$$\begin{aligned}
[\![\Box'_{\ni_\triangledown}(\widehat{c})]\!] &= R_{\ni_\triangledown}^{(0)}[[\widehat{c}]] \\
&= \{\widehat{d} \in P^+ \mid \forall y \in ([\widehat{c}])(\widehat{d}R_{\ni_\triangledown}y)\} \\
&= \{\widehat{d} \in P^+ \mid \forall y \in ([\widehat{c}])(d \leq \Box_{\ni_\triangledown}(y))\} \text{ (by def. of } R_{\ni_\triangledown}) \\
&= \{\widehat{d} \in P^+ \mid d \leq \Box_{\ni_\triangledown}(c)\} \\
&= [\![\widehat{\Box_{\ni_\triangledown}(c)}]\!]
\end{aligned}$$

Therefore, $\Box'_{\ni_\triangledown}(\widehat{c}) = \widehat{\Box_{\ni_\triangledown}(c)}$.

5.2 Multi-type Lattice-Based Normal Modal Logic

In this subsection, we define the multi-type lattice-based normal modal logic $\mathbf{L}_{MT}$ based on the models defined on 2-PN frames (called 2-PN models). We prove that this logic is sound and complete w.r.t. the class of algebraic models based on HM-MLAs (called HM-MLA models) by establishing a duality between 2-PN models and HM-MLA models.

Let Prop be a set of propositional variables. The multi-type language $\mathcal{L}_{MT}$ of $\mathbf{L}_{MT}$ is defined by the following recursion:

$$T_1 \ni \phi ::= p \mid \top \mid \bot \mid \phi \vee \phi \mid \phi \wedge \phi \mid \Box_{N_\Delta} \alpha \mid \Diamond_{N_\nabla} \alpha,$$
$$T_2 \ni \alpha ::= \Diamond_{\exists_\Delta} \phi \mid \Box_{\exists_\nabla} \phi.$$

A *heterogeneous multi-type monotone modal algebra model* (HM-MLA model) is a tuple $\mathbb{M} = (\mathbb{H}, V)$, where $\mathbb{H}$ is an HM-MLA and V is an *HM*-MLA *valuation*, which is a homomorphic assignment from $\mathcal{L}_{MT}$ to $\mathbb{H}$. A HM-MLA model is *complete* if its HM-MLA frame is complete. The satisfaction and validity of $\mathcal{L}_{MT}$-sequents on $\mathbb{M}$ is defined in a standard manner.

Multi-type Frames and Models. A *2-PN model* ($\mathcal{L}_{MT}$-*model*) is a tuple $\mathbb{M} = (\mathbb{F}, V)$, where $\mathbb{F} = (\mathbb{P}, \mathbb{P}^\star, R_{N_\Delta}, R_{N_\nabla}, R_{\exists_\Delta}, R_{\exists_\nabla})$ is a 2-PN frame and $V : \mathrm{Prop} \to \mathbb{F}^+$ is a valuation assigning a concept to every propositional variable. The *satisfaction* relation $\Vdash$ and *co-satisfaction* relation $\succ$ for $\mathbb{M}$ are defined inductively. The definition for propositional variables, constants, and propositional connectives $\vee$ and $\wedge$ are the same as in Sect. 2.1. The definition for remaining connectives is as follows: For any $a \in A$, $x \in X$, $d \in P^+$, $\phi \in T_1$, and $\alpha \in T_2$,

1. $\mathbb{M}, a \Vdash \Box_{N_\Delta} \alpha$ iff for any $d \in P^+$, $\mathbb{M}, d \succ \alpha \Rightarrow aR_{N_\Delta}d$, and $\mathbb{M}, x \succ \Box_{N_\Delta} \alpha$ iff for all $a \in A$, $\mathbb{M}, a \Vdash \Box_{N_\Delta} \alpha \Rightarrow aIx$.
2. $\mathbb{M}, d \succ \Diamond_{\exists_\Delta} \phi$ iff for any $a \in A$, $\mathbb{M}, a \Vdash \phi \Rightarrow dR_{\exists_\Delta}a$, and $\mathbb{M}, d \Vdash \Diamond_{\exists_\Delta}\phi$ iff for any $d' \in P^+$, $\mathbb{M}, d' \succ \Diamond_{\exists_\Delta}\phi \Rightarrow dI_{\neq}d'$.
3. $\mathbb{M}, x \succ \Diamond_{N_\nabla} \alpha$ iff for any $d \in P^+$, $\mathbb{M}, d \Vdash \alpha \Rightarrow xR_{N_\nabla}d$, and $\mathbb{M}, a \Vdash \Diamond_{N_\nabla}\alpha$ iff for all $x \in X$, $\mathbb{M}, x \succ \Diamond_{N_\nabla}\alpha \Rightarrow aIx$.
4. $\mathbb{M}, d \Vdash \Box_{\exists_\nabla}\phi$ iff for any $x \in X$, $\mathbb{M}, x \succ \phi \Rightarrow dR_{\exists_\nabla}x$, and $\mathbb{M}, d \succ \Box_{\exists_\nabla}\phi$ iff for any $d' \in P^+$, $\mathbb{M}, d' \Vdash \Box_{\exists_\nabla}\phi \Rightarrow d'I_{\neq}d$.
5. For any $\phi, \psi \in T_1$, $\mathbb{M} \models \phi \vdash \psi$ iff for any $a \in A$, $\mathbb{M}, a \Vdash \phi$ implies $\mathbb{M}, a \Vdash \psi$ iff for any $x \in X$, $\mathbb{M}, x \succ \psi$ implies $\mathbb{M}, x \succ \phi$. We say $\mathbb{F} \models \phi \vdash \psi$ iff for any valuation V on $\mathbb{F}$, $(\mathbb{F}, V) \models \phi \vdash \psi$.
6. For any $\alpha, \beta \in T_2$, $\mathbb{M} \models \alpha \vdash \beta$ iff for any $d \in P^+$, $\mathbb{M}, d \Vdash \alpha$ implies $\mathbb{M}, d \Vdash \beta$ iff for any $d \in P^+$, $\mathbb{M}, d \succ \beta$ implies $\mathbb{M}, d \succ \alpha$. We say $\mathbb{F} \models \alpha \vdash \beta$ iff for any valuation V on $\mathbb{F}$, $(\mathbb{F}, V) \models \alpha \vdash \beta$.

Lemma 12. *Let $\mathbb{M} = (\mathbb{F}, V)$ be a 2-PN model where $\mathbb{F} = (\mathbb{P}, \mathbb{P}^\star, R_{N_\Delta}, R_{N_\nabla}, R_{\exists_\Delta}, R_{\exists_\nabla})$, and $\mathbb{P} = (A, X, I)$. Let $\overline{V}$ be a homomorphic extension of V. Then, for any $a \in A$, $x \in X$, $d \in P^+$, $\phi, \psi \in T_1$, and $\alpha \in T_2$,*
1. *$\mathbb{M}, a \Vdash \phi$ iff $a \in [\![\overline{V}(\phi)]\!]$.*   2. *$\mathbb{M}, x \succ \phi$ iff $x \in (\!(\overline{V}(\phi))\!)$.*
3. *$\mathbb{M}, d \Vdash \alpha$ iff $d \in \overline{V}(\alpha)$.*   4. *$\mathbb{M}, d \succ \alpha$ iff $d \in \overline{V}(\alpha)$.*
5. *$\mathbb{M} \models \phi \vdash \psi$ iff $[\![\overline{V}(\phi)]\!] \subseteq [\![\overline{V}(\psi)]\!]$ iff $(\!(\overline{V}(\psi))\!) \subseteq (\!(\overline{V}(\phi))\!)$.*

Proof. See [14, A.7].

Let $\mathbb{H}$ be a complete HM-MLA and $V : \mathcal{L}_{MT} \to \mathbb{H}$ be a valuation on it. The dual 2-PN model for the algebraic model $\mathbb{M} = (\mathbb{H}, V)$ is $\mathbb{M}_+ = (\mathbb{H}_+, V_+)$, where $\mathbb{H}_+$ is a 2-PN frame and $V_+ : \mathsf{Prop} \to (\mathbb{H}_+)^+$ is a valuation defined by $V_+(p) = \widehat{V(p)}$. Conversely, given a 2-PN model $\mathbb{M} = (\mathbb{F}, V)$, we define its dual algebraic model $\mathbb{M}^+ = (\mathbb{F}^+, V^+)$, where $V^+ : \mathcal{L}_{MT} \to \mathbb{F}^+$ homomorphically extends V. The following lemmas establish a duality between complete HM-MLA models and 2-PN models.

Lemma 13. *Let $\mathbb{M} = (\mathbb{F}, V)$ be a monotone 2-PN model, where $\mathbb{F}$ is a 2-PN frame with $\mathbb{P} = (A, X, I)$, and $V : \mathsf{Prop} \to \mathbb{F}^+$. Let $\mathbb{M}^+ = (\mathbb{F}^+, V^+)$, where $V^+ : \mathcal{L}_{MT} \to \mathbb{F}^+$ is the homomorphic extension of V. Then, $\mathbb{M}^+$ is a complete HM-MLA model. Moreover, for any $a \in A$, $x \in X$, $d \in P^+$, $\phi, \psi \in \mathcal{L}_{MT}$,*
1. $\mathbb{M}, a \Vdash \phi$ *iff* $a \in [\![V^+(\phi)]\!]$. 2. $\mathbb{M}, x \succ \phi$ *iff* $x \in (\![V^+(\phi)]\!)$.
3. $\mathbb{M}, d \Vdash \phi$ *iff* $d \in V^+(\phi)$. 4. $\mathbb{M}, d \succ \phi$ *iff* $d \in V^+(\phi)$.
5. $\mathbb{M} \models \phi \vdash \psi$ *iff* $\mathbb{M}^+ \models \phi \vdash \psi$.

Proof. It is trivially true that $\mathbb{M}^+$ is a complete HM-MLA model. The remaining parts follow from the fact that $V^+ = \overline{V}$, and Lemma 12.

Lemma 14. *Let $\mathbb{M} = (\mathbb{H}, V)$ be a complete HM-MLA model. Let $\mathbb{M}_+ = (\mathbb{H}_+, V_+)$, where $V_+ : \mathsf{Prop} \to (\mathbb{H}_+)^+ \cong \mathbb{H}$ is defined by $V_+(p) = \widehat{V(p)}$. Then, $\mathbb{M}_+$ is a 2-PN model. Moreover, for any $\phi, \psi \in \mathcal{L}_{MT}$, $a \in L_A$, $x \in L_X$ and $d \in L$,*
1. $a \leq V(\phi)$ *iff* $\mathbb{M}_+, a \Vdash \phi$. 2. $V(\phi) \leq x$ *iff* $\mathbb{M}_+, x \succ \phi$.
3. $d \in V(\phi)$ *iff* $\mathbb{M}_+, d \Vdash \phi$. 4. $d \in V(\phi)$ *iff* $\mathbb{M}_+, d \succ \phi$.
5. $\mathbb{M} \models \phi \vdash \psi$ *iff* $\mathbb{M}_+ \models \phi \vdash \psi$.

Proof. See [14, A.8].

These lemmas imply the following theorem.

Theorem 4. *The class of HM-MLA models is sound and complete w.r.t. $\mathcal{L}_{MT}$.*

5.3 Translation from $\mathcal{L}$ to $\mathcal{L}_{MT}$

In this section, we define a syntactic translation from $\mathcal{L}$ to $\mathcal{L}_{MT}$ and prove that it is a valid translation. Let $T : \mathcal{L} \to \mathcal{L}_{MT}$ be the map defined by the following recursion: For any $\phi, \psi \in \mathcal{L}$,

$T(p) = p$ $T(\phi \vee \psi) = T(\phi) \vee T(\psi)$ $T(\phi \wedge \psi) = T(\phi) \wedge T(\psi)$
$T(\top) = \top$ $T(\Delta \phi) = \Box_{N_\Delta} \Diamond_{\exists_\Delta} T(\phi)$ $T(\nabla \phi) = \Diamond_{N_\nabla} \Box_{\exists_\nabla} T(\phi)$ $T(\bot) = \bot$.

Let $\mathbb{F} = (\mathbb{P}, N_\Delta, N_\nabla)$ be a monotone PN-frame. Its corresponding multi-type 2-PN frame is $\mathbb{F}^\star = (\mathbb{P}, \mathbb{P}^\star, R_{N_\Delta}, R_{N_\nabla}, R_{\exists_\Delta}, R_{\exists_\nabla})$, where the relations $R_{N_\Delta} \subseteq A \times P^+$, $R_{N_\nabla} \subseteq X \times P^+$, $R_{\exists_\Delta} \subseteq P^+ \times A$, and $R_{\exists_\nabla} \subseteq P^+ \times X$ are defined as in Sect. 5.1. We show that T defines a valid translation from $\mathcal{L}$ to $\mathcal{L}_{MT}$.

Theorem 5. *Let* $\mathbb{M} = (\mathbb{F}, V)$ *be a monotone* PN*-model. Let* $\mathbb{M}^\star$ *be the 2-PN model* $(\mathbb{F}^\star, V^+)$. *Then for any* $\phi, \psi \in \mathcal{L}$,
1. $\mathbb{M}, a \Vdash \phi$ *iff* $\mathbb{M}^\star, a \Vdash T(\phi)$,   2. $\mathbb{M}, x \succ \phi$ *iff* $\mathbb{M}^\star, x \succ T(\phi)$,
3. $\mathbb{M} \models \phi \vdash \psi$ *iff* $\mathbb{M}^\star \models T(\phi) \vdash T(\psi)$

Proof. See [14, A.9].

6 Conclusion and Future Works

In this paper, we define the polarity-based neighborhood models and show that they provide a class of sound and complete models for the lattice-based monotone modal logic using duality-theoretic methods. We also show that similar to the classical case, monotone modal operator in lattice-based monotone modal logic can be decomposed into a combination of normal modal operators via multi-type representation of complete lattices with monotone operators.

This paper opens several directions for future research. The multi-type representation of the monotone operators in terms of normal operators opens the possibility to provide the completeness results for several extensions of lattice-based monotone modal logic via canonicity. It would also be interesting to define the model-theoretic notions like p-morphisms and bi-simulations on polarity-based neighborhood frames and study their properties. Finally, the classical neighborhood frames are used as the semantics for some non-monotonic logics like the conditional logic [3,23]. It would be interesting to see if we can generalize this idea to the non-distributive setting.

References

1. van der Berg, I., De Domenico, A., Greco, G., Manoorkar, K.B., Palmigiano, A., Panettiere, M.: Labelled calculi for lattice-based modal logics. In: Indian Conference on Logic and Its Applications, pp. 23–47. Springer, Cham (2023)
2. van der Berg, I., De Domenico, A., Greco, G., Manoorkar, K.B., Palmigiano, A., Panettiere, M.: Labelled calculi for the logics of rough concepts. In: Indian Conference on Logic and Its Applications, pp. 172–188. Springer, Cham (2023)
3. Chellas, B.F.: Basic conditional logic. J. Philos. Logic 133–153 (1975)
4. Chellas, B.F.: Modal Logic: An Introduction. Cambridge University Press, Cambridge (1980)
5. Chen, J., Greco, G., Palmigiano, A., Tzimoulis, A.: Non normal logics: semantic analysis and proof theory. In: Iemhoff, R., Moortgat, M., de Queiroz, R. (eds.) WoLLIC 2019. LNCS, vol. 11541, pp. 99–118. Springer, Heidelberg (2019). https://doi.org/10.1007/978-3-662-59533-6_7
6. Conradie, W., et al.: Rough concepts. Inf. Sci. **561**, 371–413 (2021)
7. Conradie, W., Frittella, S., Palmigiano, A., Piazzai, M., Tzimoulis, A., Wijnberg, N.M.: Toward an epistemic-logical theory of categorization. EPTCS **251** (2017)
8. Conradie, W., Palmigiano, A.: Algorithmic correspondence and canonicity for non-distributive logics. Ann. Pure Appl. Logic **170**(9), 923–974 (2019)

9. Conradie, W., Palmigiano, A., Robinson, C., Wijnberg, N.: Non-distributive logics: from semantics to meaning. In: Rezus, A. (ed.) Contemporary Logic and Computing, Landscapes in Logic, vol. 1, pp. 38–86. College Publications (2020). arXiv preprint arXiv:2002.04257
10. Conradie, W., Frittella, S., Palmigiano, A., Piazzai, M., Tzimoulis, A., Wijnberg, N.M.: Categories: how i learned to stop worrying and love two sorts. In: Väänänen, J., Hirvonen, Å., de Queiroz, R. (eds.) WoLLIC 2016. LNCS, vol. 9803, pp. 145–164. Springer, Heidelberg (2016). https://doi.org/10.1007/978-3-662-52921-8_10
11. Conradie, W., Palmigiano, A.: Constructive canonicity of inductive inequalities. Logical Methods Comput. Sci. **16**(3), 1–39 (2020)
12. De Groot, J.: Positive monotone modal logic. Stud. Logica. **109**, 829–857 (2021)
13. De Rudder, L., Palmigiano, A.: Slanted canonicity of analytic inductive inequalities. ACM Trans. Comput. Logic (TOCL) **22**(3), 1–41 (2021)
14. Ding, Y., Manoorkar, K., Palmigiano, A., Wang, R.: Monotone modal logic beyond distributivity (2025). https://doi.org/10.13140/RG.2.2.20699.68646
15. Frittella, S., Palmigiano, A., Santocanale, L.: Dual characterizations for finite lattices via correspondence theory for monotone modal logic. J. Log. Comput. **27**(3), 639–678 (2017)
16. Gasquet, O., Herzig, A.: From classical to normal modal logics. Proof Theory Modal Logic 293–311 (1996)
17. Gehrke, M., Harding, J.: Bounded lattice expansions. J. Algebra **238**, 345–371 (2001)
18. Greco, G., Jipsen, P., Liang, F., Palmigiano, A., Tzimoulis, A.: Algebraic proof theory for le-logics. ACM Trans. Comput. Log. **25**(1), 1–37 (2024)
19. Hansen, H.H.: Monotonic modal logics (2003)
20. Hansen, H.H., Kupke, C.: A coalgebraic perspective on monotone modal logic. Electron. Notes Theor. Comput. Sci. **106**, 121–143 (2004)
21. Lemmon, E.J.: New foundations for Lewis modal systems1. J. Symb. Logic **22**(2), 176–186 (1957)
22. Lewis, D.: Counterfactuals. Wiley, Hoboken (2013)
23. Nute, D.: Topics in Conditional Logic, vol. 20. Springer, Cham (1980)
24. Pacuit, E.: Neighborhood Semantics for Modal Logic. Springer, Cham (2017)
25. Parikh, R.: The logic of games and its applications. In: North-Holland Mathematics Studies, vol. 102, pp. 111–139. Elsevier (1985)
26. Santocanale, L.: A duality for finite lattices (2009)
27. Vardi, M.: On the complexity of epistemic reasoning. In: Proceedings of the Fourth Annual Symposium on Logic in Computer Science, pp. 243–244. IEEE Computer Society (1989)

Recognizing Numbers

Pranshu Gaba[1] and Arnab Sur[2]

[1] Tata Institute of Fundamental Research, Mumbai, India
pranshu.gaba@tifr.res.in
[2] Chennai Mathematical Institute, Chennai, India
arnabs@cmi.ac.in

Abstract. The use of monoids in the study of word languages recognized by finite-state automata has been quite fruitful. In this work, we look at the same idea of "recognizability by finite monoids" for other monoids. In particular, we attempt to characterize recognizable subsets of various additive and multiplicative monoids over integers, rationals, reals, and complex numbers. While these recognizable sets satisfy properties such as closure under Boolean operations and inverse morphisms, they do not enjoy many of the nice properties that recognizable word languages do.

Keywords: Recognizability · Additive monoids · Multiplicative monoids · Algebraic automata theory

1 Introduction

Regular (or recognizable) subsets of the free monoid (i.e., the set of finite words over a finite alphabet) are one of the most well-studied classes of languages. Many seemingly unrelated notions—recognizability by finite-state automata, recognizability by finite monoids or congruences of finite index, rationality, monadic-second order definability—all coincide for free monoids [4,8,9]. The algebraic notion of recognizability by morphisms into finite monoids allows us to use tools from the structure theory of finite monoids which has been particularly useful to characterize certain subclasses of the recognizable languages. For instance, we have Schützenberger's celebrated result that the star-free languages (languages described by rational expressions using only the union, concatenation, and complement operations) are exactly those recognized by aperiodic monoids (monoids which do not contain a group). Further, there are algebraic characterizations of piecewise and locally testable languages as well [4,8].

Related Work. There has been a lot of research to define algebraic recognizability for structures other than sets of finite words like sets of trees [5], infinite words [7], timed words [2,6], and data words [1,2]. The goal has been to show their equivalence with recognizability by automata models or definability by logics for these structures. However these "recognizable sets" are not recognized by finite monoids but by other algebras. The survey [10] provides a great overview of algebraic recognizability of all the structures mentioned above.

Indeed, recognizability by finite monoids seems to be best suited only for finite words. However, we want to characterize the recognizable (by finite monoids) subsets of as many monoids as we can to figure out why they may or may not be interesting. For instance, the recognizable subsets of non-negative reals under addition ($\mathbb{R}_{\geq 0}$) are just $\emptyset, \mathbb{R}_{\geq 0}, \{0\}$, and $\mathbb{R}_{>0}$ [3]. This fact indicates that one must look at other algebraic structures to define recognizability for timed languages.

Contributions. In this work, we will primarily look at monoids of numbers: integers, rationals, reals, and complex numbers, each with addition and multiplication operations. For the additive monoids, we bring together known results and point out generalizations to arbitrarily divisible monoids. To the best of our knowledge, the results for multiplicative monoids are new.

For recognizability by finite monoids, monoids of numbers may seem very specific as only some of their more general facts like infinite generators or commutativity lend the characterizations we find, but these monoids illustrate them very well. The additive monoids of the naturals and the integers are the exceptions since they are special cases of the free monoid and the free group respectively.

Outline. In Sect. 2, we recall the definitions and notations that we shall be using for our work. In Sect. 3, we characterize the additive monoids $\mathbb{Z}, \mathbb{Q}, \mathbb{R}, \mathbb{C}, \mathbb{Z}_{\geq 0}$, $\mathbb{Q}_{\geq 0}$, and $\mathbb{R}_{\geq 0}$, Along the way, we show the triviality of recognizable subsets of a class of monoids that are *arbitrarily divisible*. In Sect. 4, we characterize the multiplicative monoids $\mathbb{Z}, \mathbb{Q}, \mathbb{R}$, and $\mathbb{C}$, where we also show a useful property satisfied by recognizable subsets of (countable) infinitely generated free monoids.

2 Preliminaries

We recall some definitions and notations used in this text.

Sets of Numbers. We denote the set of integers by $\mathbb{Z}$, the set of rational numbers by $\mathbb{Q}$, the set of real numbers by $\mathbb{R}$, and the set of complex numbers by $\mathbb{C}$. In addition, for $\mathbb{X} \in \{\mathbb{Z}, \mathbb{Q}, \mathbb{R}\}$, we denote non-negative numbers by $\mathbb{X}_{\geq 0}$, positive numbers by $\mathbb{X}_{>0}$, and non-zero numbers by $\mathbb{X} \setminus \{0\}$. In particular, we also sometimes denote the natural numbers, that is the non-negative integers, by $\mathbb{N}$.

Monoids. A *monoid* $(M, \cdot_M, 1_M)$ is a set M with an associative binary operation $\cdot_M$ and a unit element 1_M for the binary operation, that is, $m \cdot_M (1_M) = m = (1_M) \cdot_M m$ for all $m \in M$. We say that a monoid $(M, \cdot_M, 1_M)$ is finite (resp. infinite) if the underlying set M is finite (resp. infinite). We omit the operation and the unit element when it is clear from context, that is, we simply use the symbol of the set M to denote the monoid $(M, \cdot_M, 1_M)$. We also omit the binary operation when clear from context, that is, we represent $m_1 \cdot_M m_2$ by $m_1 m_2$ and

$m \cdot_M m$ by m^2. We extend the monoid operation to sets $X, Y \subseteq M$ by defining $X \cdot_M Y = \{x \cdot_M y \mid x \in X, y \in Y\}$. The asterate (or Kleene-star) operation is then defined as
$$X^* = \bigcup_{i \geq 0} X^i$$
where $X^0 = \{1_M\}$ and $X^i = X \cdot_M X^{i-1}$ for $i > 0$. Some examples of monoids are groups and free monoids. A *group* $(G, \cdot_G, 1_G)$ is a monoid in which every element has an inverse, that is, for all $g \in G$, there exists a unique element $g^{-1} \in G$ such that $g \cdot_G g^{-1} = 1_G = g^{-1} \cdot_G g$. The *order* of a group G is the number of elements in G if G is finite, or else, if G is infinite, then the order of G is also infinite. Moreover, if G is a finite group, then for every element $g \in G$, the *order of g* is the smallest positive integer i such that $g^i = 1_G$.

Idempotents. An element e of a monoid M is *idempotent* if $e^2 = e$. In particular, the unit element 1_M of a monoid M is idempotent. The unit element 1_G of a group G is the unique idempotent in the group G. A monoid M is *idempotent* if all elements m of M are idempotent.

Zero. An element z of a monoid M is a *zero* of M if for all $m \in M$, we have that $zm = mz = z$. If a monoid contains a zero, then the zero is unique. Given a monoid M without a zero element, let $M_0 = M \cup \{0_M\}$ be the monoid M appended with a zero element. If M has a zero element, we denote the zero element by 0_M and it is the case that $M_0 = M$. In what follows, we assume that the zero is distinct from the unit, for otherwise the monoid is trivial.

Roots. Given an element m of a monoid M and a positive integer k, an element n of M is called a kth *root* of m in M if $n^k = m$. We denote by m_k a kth root of m. In general, a kth root of m may not exist, or if it exists, it need not be unique.

Morphisms. A *morphism* φ from a monoid $(M, \cdot_M, 1_M)$ to a monoid $(N, \cdot_N, 1_N)$ is a map $\varphi \colon M \to N$ that maps the unit element of M to the unit element of N and preserves the monoid operations. Formally, we have that φ satisfies $\varphi(1_M) = 1_N$ and for all $m_1, m_2 \in M$, we have that $\varphi(m_1 \cdot_M m_2) = \varphi(m_1) \cdot_N \varphi(m_2)$.

Recognizable Sets. Given monoids M and N, we say a subset S of M is *recognized* by the monoid N if there exists a morphism $\varphi \colon M \to N$ and a subset $T \subseteq N$, such that $\varphi^{-1}(T) = S$. We say S is *recognizable* if it is recognized by a *finite* monoid. We denote the set of all recognizable subsets of M by Rec(M). We also assume that the recognizing morphisms are surjective. Recognizable sets exhibit nice closure properties, which we summarize in Proposition 1. The results follow since inverse maps are well-behaved with respect to Boolean operations.

Proposition 1 ([4,8,9]). *Let M be a monoid, and let $S_1, S_2 \in$ Rec(M) be recognizable subsets of M. Then, $S_1 \cup S_2, S_1 \cap S_2, M \setminus S_1 \in$ Rec(M). That is, recognizable sets of a monoid are closed under unions, intersections, and complements.*

3 Additive Monoids

In this section, we look at the additive monoids of numbers. We list in Table 1 the recognizable subsets of the various monoids considered. The rest of the section is dedicated to showing these results, as well as some additional properties that apply to a wider class of monoids.

Table 1. Recognizable subsets of additive monoids $(\mathbb{X}, +, 0)$

$\mathbb{X}$	$\mathrm{Rec}(\mathbb{X})$	$\mathbb{X}$	$\mathrm{Rec}(\mathbb{X})$
$\mathbb{Z}$	periodic sets	$\mathbb{Z}_{\geq 0}$	ultimately periodic sets
$\mathbb{Q}$	$\{\emptyset, \mathbb{Q}\}$	$\mathbb{Q}_{\geq 0}$	$\{\emptyset, \mathbb{Q}_{\geq 0}, \{0\}, \mathbb{Q}_{>0}\}$
$\mathbb{R}$	$\{\emptyset, \mathbb{R}\}$	$\mathbb{R}_{\geq 0}$	$\{\emptyset, \mathbb{R}_{\geq 0}, \{0\}, \mathbb{R}_{>0}\}$
$\mathbb{C}$	$\{\emptyset, \mathbb{C}\}$		

3.1 Non-negative Integers $(\mathbb{Z}_{\geq 0}, +, 0)$

A subset X of $\mathbb{Z}_{\geq 0}$ is *ultimately periodic* if there exists a positive integer p (called a period) and a positive integer n_0 such that for all integers $n > n_0$, we have that $n \in X$ if and only if $n + p \in X$. The recognizable subsets of $(\mathbb{Z}_{\geq 0}, +, 0)$ are precisely the ultimately periodic sets.

To see this, observe that $\mathbb{Z}_{\geq 0}$ is isomorphic to the free monoid generated by a unary alphabet. The recognizable subsets of free monoid with one generator are exactly the unary regular languages. Thus, the recognizable subsets of $(\mathbb{Z}_{\geq 0}, +, 0)$ are the ultimately periodic sets. A detailed exposition is found in [4].

3.2 Integers $(\mathbb{Z}, +, 0)$

A subset X of $\mathbb{Z}$ is *periodic* if there exists a positive integer p (called a period) such that for all $n \in \mathbb{Z}$, we have that $n \in X$ if and only if $n + p \in X$. The recognizable subsets of $(\mathbb{Z}, +, 0)$ are the periodic sets. We show this now.

The monoid $(\mathbb{Z}, +, 0)$ is a group, and since surjective morphisms map groups to groups, we have that any recognizable subset of $\mathbb{Z}$ must be recognized by a finite group. Let G be a finite group and let $\varphi \colon \mathbb{Z} \to G$ be a morphism. This morphism is completely determined by $\varphi(1)$. Indeed, if $\varphi(1) = g$ for some $g \in G$, then we have $\varphi(0) = 1_G$, we have $\varphi(n) = g^n$ for all positive integers n, and we have $\varphi(n) = (g^{-1})^{|n|}$ for all negative integers n, where g^{-1} is the inverse of g. If g has order p in G, then for all subsets H of G, we have that $\varphi^{-1}(H)$ is a periodic set with period p.

3.3 Non-negative Rationals ($\mathbb{Q}_{\geq 0}, +, 0$)

To find the recognizable subsets of non-negative rationals, we show and use an observation that is applicable for a wider class of monoids. We define a property of monoids, namely monoids that are *arbitrarily divisible* and characterize their recognizable subsets. The special case of non-negative reals under addition was proven in [3]. Here, we generalize it for monoids with this property.

Definition 1. A monoid M is *arbitrarily divisible* if for all $m \in M$ and positive integer k, there exists a kth root of m in M, that is, there exists $m_k \in M$ such that $(m_k)^k = m$.

The following proposition shows that the image $\varphi(M)$ of an arbitrarily divisible monoid M under a morphism φ is an idempotent monoid.

Proposition 2. *If M is an arbitrarily divisible monoid, N is a finite monoid, and $\varphi : M \to N$ is a morphism, then $\varphi(M)$ is an idempotent monoid.*

Proof. Let $m \in M$. We want to show that for all $m \in M$ that $\varphi(m)$ is an idempotent element, that is, $\varphi(m)^2 = \varphi(m)$. If $\varphi(m) = 1_N$, then we are done since the unit element 1_N is an idempotent. Thus, the only interesting case is when $\varphi(m) \neq 1_N$. Let $n \in N$ be the image of a $(|N|!)^{\text{th}}$ root of m, that is, $\varphi(m_{|N|!}) = n$. Consider the sequence:

$$n, n^2, n^3, \ldots n^{|N|+1}$$

By the pigeonhole principle, there exist at least two elements in this sequence that are equal, say $n^i = n^j$ where $1 \leq i < j \leq |N|+1$. There is an idempotent n^k where $i \leq k < j$ and $(j-i) \mid k$ [4]. Let $k' = \frac{|N|!}{k}$. Then we have $\varphi(m) = n^{|N|!} = n^{k \cdot k'} = (n^k)^{k'} = n^k$, which is an idempotent. $\square$

Remark 1. Let M be an arbitrarily divisible monoid. From Proposition 2, it follows that for all $m \in M$, it is the case that $\varphi(m)$ is idempotent, and thus for all positive integers p, we have that $\varphi(m^p) = \varphi(m)$. Moreover, for every $m \in M$, for all positive integers q, let m_q denote a qth root of m, that is, $(m_q)^q = m$. Then, again, from Proposition 2, we have that $\varphi(m_q)$ is an idempotent, and thus $\varphi(m_q) = \varphi(m_q)^q = \varphi((m_q)^q) = \varphi(m)$. Thus, for all positive integers p, q, we have $\varphi((m_q)^p) = \varphi(m)^p = \varphi(m)$.

Note that $(\mathbb{Q}_{\geq 0}, +, 0)$ is an arbitrarily divisible monoid since for every non-negative rational number r and every positive integer k, there exists a non-negative rational number that is a kth root of r, namely $\frac{r}{k}$. From Remark 1, it follows that for all positive rational numbers r, and for all morphisms to finite monoids, $\varphi : \mathbb{Q}_{\geq 0} \to M$, it is the case that $\varphi(r) = \varphi(1)$. Thus, we have that the image of a surjective φ has at most two elements $\varphi(0)$ and $\varphi(1)$, and we get that $\text{Rec}(\mathbb{Q}_{\geq 0}, +) = \{\emptyset, \{0\}, \mathbb{Q}_{> 0}, \mathbb{Q}_{\geq 0}\}$, since these sets are $\varphi^{-1}(\emptyset)$, $\varphi^{-1}(\{\varphi(0)\})$, $\varphi^{-1}(\{\varphi(1)\})$, and $\varphi^{-1}(\{\varphi(0), \varphi(1)\})$ respectively.

3.4 Rationals $(\mathbb{Q}, +, 0)$ and Beyond

Now, we look at the monoids $(\mathbb{Q}, +, 0)$, $(\mathbb{R}, +, 0)$, and $(\mathbb{C}, +, 0)$. These are all arbitrarily divisible groups. The following proposition shows that the only recognizable subsets of arbitrarily divisible groups are the trivial sets.

Proposition 3. *If M is an arbitrarily divisible group then $Rec(M) = \{\emptyset, M\}$.*

Proof. We shall show that all morphisms $\varphi : M \to N$, where N is a finite monoid, maps every element m in M to the unit 1_N in N. To see this, observe that since M is a group, $\varphi(M)$ is also a group. Let k be the order of the group $\varphi(M)$. Since M is arbitrarily divisible, there exists m_k in M, a kth root of m. Thus, we have $\varphi(m) = \varphi((m_k)^k) = \varphi(m_k)^k = 1_N$. □

3.5 Non-negative Reals $(\mathbb{R}_{\geq 0}, +, 0)$

Let $\varphi : \mathbb{R}_{\geq 0} \to M$ be a surjective morphism to a finite monoid. Since $(\mathbb{R}_{\geq 0}, +, 0)$ is a commutative arbitrarily divisible monoid, by 2 M is a commutative idempotent monoid. Moreover by Remark 1, for each $q \in \mathbb{Q}_{>0}$ we have $\varphi(q) = \varphi(1)$.

Let r be a positive irrational number. There exists $p, q \in \mathbb{Q}_{>0}$ such that $p < r < q$. Then, $\varphi(q) = \varphi(r)\varphi(q-r) = \varphi(2r)\varphi(q-r) = \varphi(r)\varphi(q)$. We also have that $\varphi(r) = \varphi(p)\varphi(r - p) = \varphi(2p)\varphi(r - p) = \varphi(r)\varphi(p) = \varphi(r)\varphi(q) = \varphi(q)$. Thus every irrational maps to $\varphi(1)$ as well.

Therefore, there are only two possible choices of φ, either for all $r \in \mathbb{R}_{\geq 0}$, $\varphi(r) = 1_M$, or for all $r \in \mathbb{R}_{>0}, \varphi(r) = \varphi(1)$ and $\varphi(0) = 1_M$. It follows that $Rec(\mathbb{R}_{\geq 0}, +, 0) = \{\emptyset, \{0\}, \mathbb{R}_{>0}, \mathbb{R}_{\geq 0}\}$.

4 Multiplicative Monoids

We are interested in finding the recognizable subsets of the multiplicative monoids $\mathbb{Z}$, $\mathbb{Q}$, $\mathbb{R}$, and $\mathbb{C}$. We will first recall some observations that will simplify our study. The first observation relates recognizable subsets of monoids with a zero element to recognizable subsets of monoids without a zero element.

Proposition 4. *Let M be a monoid without a zero. Then $Rec(M_0) = Rec(M) \cup \{R \cup \{0_M\} \mid R \in Rec(M)\}$.*

Proof. If $\varphi : M_0 \to N$ is a morphism to a finite monoid N, then $\varphi(0_M) = 0_N$. Indeed, for any $n \in N$ there exists $m \in \varphi^{-1}(n)$ (since we assume all recognizing morphisms are surjective), and $\varphi(0_M) \cdot n = \varphi(0_M) = n \cdot \varphi(0_M)$. Let $X = \varphi^{-1}(P)$ for some $P \subseteq N$ and $\psi : M \to N$ be the morphism defined by $\psi(m) = \varphi(m)$ for all $m \in M$. Then if $0_M \in X$ then $X \setminus \{0\}$ is recognized by ψ and if $0_M \notin X$ then X is recognized by ψ. □

This helps us reduce our problem to finding the recognizable subsets of the multiplicative monoids $\mathbb{Z}\setminus\{0\}$, $\mathbb{Q}\setminus\{0\}$, $\mathbb{R}\setminus\{0\}$, and $\mathbb{C}\setminus\{0\}$. The next observation is concerned with the recognizable subsets of a direct product of monoids.

Theorem 1 (Mezei [8]). *Let $M_1, \ldots, M_n$ be monoids and let $M = M_1 \times \cdots \times M_n$. A subset of M is recognizable if and only if it is a finite union of subsets of the form $R_1 \times \cdots \times R_n$, where each R_i is a recognizable subset of M_i.*

Remark 2. The monoids $(\mathbb{Z} \setminus \{0\}, \times, 1), (\mathbb{Q} \setminus \{0\}, \times, 1), (\mathbb{R} \setminus \{0\}, \times, 1)$ are all isomorphic to the direct products of $(\mathbb{Z}_{>0}, \times, 1), (\mathbb{Q}_{>0}, \times, 1), (\mathbb{R}_{>0}, \times, 1)$ respectively with $\mathbb{Z}_2$. Note that $\text{Rec}(\mathbb{Z}_2) = 2^{\{0,1\}}$. Theorem 1 thus gives us the recognizable subsets of these monoids from the recognizable subsets of the corresponding "positive" monoids.

Remark 3. Note that $(\mathbb{R}_{>0}, \times, 1)$ and $(\mathbb{C} \setminus \{0\}, \times, 1)$ are both arbitrarily divisible groups and hence by Proposition 3 their recognizable subsets are trivial. By Remark 2 we have

$$\text{Rec}((\mathbb{R}, \times, 1)) = \{\emptyset, \{0\}, \mathbb{R}_{<0}, \mathbb{R}_{\leq 0}, \mathbb{R}_{>0}, \mathbb{R}_{\geq 0}, \mathbb{R} \setminus \{0\}, \mathbb{R}\}$$
$$\text{Rec}((\mathbb{C}, \times, 1)) = \{\emptyset, \{0\}, \mathbb{C} \setminus \{0\}, \mathbb{C}\}.$$

In the rest of this section, we work towards finding the recognizable subsets of the multiplicative monoids $\mathbb{Z}_{>0}$ and $\mathbb{Q}_{>0}$. We shall look at recognizable subsets of (countable) infinitely generated free monoids. Indeed naturals and rationals with multiplication are merely the commutative infinitely generated free monoid and the commutative infinitely generated free group respectively. We show that a recognizing morphism can be "factorized" into two morphisms via a finitely generated free monoid that also recognizes the set.

Lemma 1. *Let A be an infinite alphabet and $\varphi : A^* \to M$ be a morphism to a finite monoid M. Then, there exists an equivalence relation $\sim$ on A of finite index k and a morphism $\psi : (A/\sim)^* \to M$ such that $\varphi = \psi \circ \pi_\sim$ where $\pi_\sim : A^* \to (A/\sim)^*$ is the projection morphism of $\sim$.*

Proof. We define the equivalence relation $\sim$, for all $a, a' \in A$, $a \sim a'$ iff $\varphi(a) = \varphi(a')$. Note that $\sim$ has a finite index. We define the morphism $\psi : (A/\sim)^* \to M$ by $\psi([a]_\sim) = \varphi(a)$ for $a \in A$. It is clearly well-defined and is a morphism. We now show that $\varphi = \psi \circ \pi_\sim$. Let $w = a_1 a_2 \ldots a_p \in A^*$. Then,

$$\psi(\pi_\sim(a_1 \ldots a_p)) = \psi([a_1]_\sim [a_2]_\sim \ldots [a_p]_\sim)$$
$$= \varphi(a_1)\varphi(a_2)\ldots\varphi(a_p) = \varphi(w).$$

□

Proposition 5. *If $X \in \text{Rec}(A^*)$ then, there exists an equivalence relation $\sim$ on A of finite index k such that $X = (\pi_\sim^{-1} \circ \pi_\sim)(X)$ where $\pi_\sim : A^* \to (A/\sim)^*$ is the projection morphism of $\sim$.*

Proof. Let X be recognized by the surjective morphism $\varphi : A^* \to M$ where M is a finite monoid. We have by Lemma 1 that there exists an equivalence relation $\sim$ on A of finite index k and a morphism $\psi : (A/\sim)^* \to M$ such that $\varphi = \psi \circ \pi_\sim$ where $\pi_\sim : A^* \to (A/\sim)^*$ is the projection morphism of $\sim$. Clearly, $X \subseteq (\pi_\sim^{-1} \circ \pi_\sim)(X)$. Assume on the contrary there is some $w \in (\pi_\sim^{-1} \circ \pi_\sim)(X) \setminus X$. Then there exists $w' \in X$ such that $\pi_\sim(w) = \pi_\sim(w')$. Then, $\varphi(w') = \psi(\pi_\sim(w')) = \psi(\pi_\sim(w)) = \varphi(w)$, and thus $w \in X$ which is a contradiction. □

We would like to mention a way to define $\mathrm{Rec}(A^*)$ in monadic-second order logic using Proposition 5. Given $X \in \mathrm{Rec}(A^*)$, let $\sim$ be the equivalence relation of finite index k given by Proposition 5. Now by Lemma 1, $\pi_\sim(X) \in \mathrm{Rec}((A/\sim)^*)$, and therefore $\pi_\sim(X)$ is a model for a monadic-second order (MSO) formula with first-order variables interpreted as positions in the word and second-order variables interpreted as sets of position, and unary predicates R_b for each $b \in A/\sim$ ($R_b(i)$ if and only if the ith letter is b) and binary predicate $<$ ($i < j$ if position i appears before position j) [8]. The set X is a model for this MSO formula if the predicates R_b are interpreted as $R_b(i)$ if and only if the ith letter is a such that $\pi_\sim(a) = b$. This is unsatisfactory since this is not a uniform characterization and much of the computational difficulty is hidden in the interpretation of the predicates R_b. Indeed, the predicates R_b may not even be computable as the following remark shows.

Remark 4. If A is a countably infinite alphabet then $\mathrm{Rec}(A^*)$ is uncountable. Consider the monoid $M_3 = \langle \{0_M, 1_M, p\} \mid p^2 = 0_M \rangle$ where 0_M is the zero, 1_M is the unit. Any $X \subseteq A$ is recognized by the morphism $\varphi : A^* \to M_3$ defined by $\varphi(a) = p$ for $a \in X$ and $\varphi(a) = 0_M$ otherwise. Thus, one cannot hope for a characterization by a finitely describable machine or logic.

Despite the large number of recognizable subsets, we show that many seemingly easy-to-compute subsets of naturals and rationals under multiplication are not recognizable.

4.1 Positive Integers $(\mathbb{N} \setminus \{0\}, \times, 1)$

Working directly with $(\mathbb{N} \setminus \{0\}, \times, 1)$ is a bit difficult, so we choose instead to work with an isomorphic monoid, the monoid of finite sequences of naturals that comes from the powers of the primes in a natural number's prime factorization.

Let σ and τ be finite sequences of non-negative integers of length m and n respectively. Without loss of generality we assume that $m \leq n$. The pointwise addition of σ and τ, $\sigma + \tau$ is the sequence of length n defined by

$$(\sigma + \tau)(i) = \begin{cases} \sigma(i) + \tau(i) & \text{if } i \leq m, \\ \tau(i) & \text{otherwise} \end{cases}$$

Finite sequences of non-negative integers form a monoid under pointwise addition with the sequence of length 0 as the unit element. We define for each $n \in \mathbb{N}$, the sequence of non-negative integers of length n, σ_n where for each $1 \leq i < n$, $\sigma_n(i) = 0$ and $\sigma_n(n) = 1$. Let σ_0 be the sequence of length 0 and $N =$

$\{\sigma_n\}_{n \in \mathbb{N}}$. N generates the monoid of finite sequences of non-negative integers under pointwise addition and thus we denote the monoid by N^*.

It is easy to see that $(\mathbb{N} \setminus \{0\}, \times, 1)$ is isomorphic to N^*. Indeed, by the fundamental theorem of arithmetic, one can associate each natural with the sequence of powers in its unique prime factorization. Thus, we shall focus on $\text{Rec}(N^*)$.

Let $\varphi : A^* \to M$ be a morphism to a finite monoid M. . Then, by Lemma 1 there exists an equivalence relation $\sim$ on N of finite index k and a morphism $\psi : (N/\sim)^* \to M$ such that $\varphi = \psi \circ \pi_\sim$ where $\pi_\sim : N^* \to (N/\sim)^*$ is the projection morphism of $\sim$.

By the commutativity of N^*, the monoid $(N/\sim)^*$ is a commutative free monoid generated by k elements and is hence isomorphic to $\mathbb{N}^k$ (k-tuples of non-negative integers) under pointwise addition [9]. Fix some enumeration $N/\sim = \{\alpha_1, \ldots \alpha_k\}$. We then have

$$\pi_\sim(\sigma) = \left(\sum_{i \in \alpha_j} \sigma(i)\right)_{j \in [k]}$$

Hereafter we shall denote elements of $\pi_\sim(N^*)$ by elements of $\mathbb{N}^k$ as defined above.

Applications of Proposition 5 Using Proposition 5 we can show certain subsets of N^* are not recognizable. We call a property $\mathcal{P}$ of $\mathbb{N}$, *trivial* if the set $P = \{n \in \mathbb{N} \mid \mathcal{P}(n)\}$ is finite or co-finite. For $\sigma \in N^*$ we denote by $|\sigma|$ the smallest natural number n_0 such that for all $n > n_0$, $\sigma(n) = 0$. Examples of non-trivial properties are parity and primality.

Proposition 6. *The following sets are not recognizable for every non-trivial property $\mathcal{P}$ of $\mathbb{N}$.*

1. $X = \{\sigma \in N^* \mid \mathcal{P}(|\sigma|)\}$,
2. $X = \{\sigma \in N^* \mid \forall i \in \mathbb{N}, \mathcal{P}(\sigma(i))\}$, and
3. $X = \{\sigma \in N^* \mid \exists i \in \mathbb{N}, \mathcal{P}(\sigma(i))\}$.

Proof. The proof template is the same for all three parts. We assume on the contrary there exists a finite monoid M and morphism $\varphi : N^* \to M$ that recognizes X. By Proposition 5 there exists an equivalence relation $\sim$ on N of finite index k such that $X = (\pi_\sim^{-1} \circ \pi_\sim)(X)$. Since $\sim$ is of finite index and $\mathcal{P}$ is not trivial, there exist natural numbers $n_1 < n_2 < n_3$ such that $\mathcal{P}(n_1)$ and $\mathcal{P}(n_3)$ but $\neg \mathcal{P}(n_2)$ holds and $\sigma_{n_1} \sim \sigma_{n_3}$. We show two sequences, $s_1 \in X$ and $s_2 \notin X$, but $\pi_\sim(s_1) = \pi_\sim(s_2)$ for each of the three parts.

1. $s_1 = \sigma_{n_2} + \sigma_{n_3} \in X$ and $s_2 = \sigma_{n_1} + \sigma_{n_2} \notin X$ but $\pi_\sim(s_1) = \pi_\sim(s_2)$.
2. Let $s_1 \in X$ and $\sigma_i \sim \sigma_j$ such that $s_1(i) = n_1$ and $s_1(j) = n_3$. We construct the sequence s_2 such that for all $l \neq i, j$, $s_2(l) = s_1(l)$, $s_2(i) = n_2$ and $s_2(j) = n_1 + n_3 - n_2$. Clearly $s_2 \notin X$ but $\pi_\sim(s_1) = \pi_\sim(s_2)$.
3. If $X \in \text{Rec}(N^*)$ then $X^c \in \text{Rec}(N^*)$. However, $X^c = \{\sigma \in N^* \mid \forall i \in \mathbb{N}, \neg \mathcal{P}(\sigma(i))\}$ and $\neg \mathcal{P}$ is also non-trivial, contradicting the observation above.

Thus sequences of even naturals (which translates to perfect squares for the monoid of naturals under multiplication), or sequences of even length (naturals divisible by exactly an even number of primes) are not recognizable despite seeming to be easy computable by some finite state automaton that, say, takes input sequences encoded in unary.

Since N^* is the free commutative monoid generated by a countably infinite set and the monoid M_3 defined in Remark 4 is also commutative, we conclude that $\text{Rec}(N^*)$ is uncountable. We cannot expect a nice machine or logic characterization for $\text{Rec}(N^*)$.

4.2 Positive Rationals $(\mathbb{Q}_{>0}, \times, 1)$

Recall that we looked at finite sequences of naturals with pointwise addition instead of positive naturals with multiplication. For this case, we shall look at the monoid of finite sequences of integers with pointwise addition instead.

Finite sequences of integers form a monoid under pointwise addition with the sequence of length 0 as the unit element. We define for each $n \in \mathbb{Z}$, the sequence of integers of length $|n|$, σ_n where for each $1 \leq i < |n|$, we have $\sigma_n(i) = 0$ and $\sigma_n(|n|) = n/|n|$. Let σ_0 be the sequence of length 0 and $Z = \{\sigma_n\}_{n \in \mathbb{Z}}$. Z generates the monoid of finite sequences of integers under pointwise addition and thus we denote the monoid by Z^*.

Let $\varphi : A^* \to M$ be a morphism to a finite monoid M. .Then, by Lemma 1 there exists an equivalence relation $\sim$ on Z of finite index k and a morphism $\psi : (Z/\sim)^* \to M$ such that $\varphi = \psi \circ \pi_\sim$ where $\pi_\sim : Z^* \to (Z/\sim)^*$ is the projection morphism of $\sim$.

Note that by the commutativity of Z^*, the monoid $(Z/\sim)^*$ is a commutative free monoid generated by k elements and is hence isomorphic to $\mathbb{Z}^k$ (k-tuples of integers) under pointwise addition. Fix some enumeration $Z/\sim = \{\alpha_1, \ldots \alpha_k\}$. We then have

$$\pi_\sim(\sigma) = \left(\sum_{i \in \alpha_j} \sigma(i)\right)_{j \in [k]}$$

Hereafter we shall denote elements of $\pi_\sim(Z^*)$ by elements of $\mathbb{Z}^k$ as defined above.

Applications of the Proposition 5. The next proposition shows that every non-empty recognizable subset X of Z^* contains sequences of length greater than p for all $p > 0$. A consequence of the following proposition is that in every recognizable subset of $(\mathbb{Q}_{>0}, \times, 1)$, the set of prime factors of the elements of the subset is infinite.

Proposition 7. *If $X \in \text{Rec}(Z^*)$ and $X \neq \emptyset$, then for all integers $p > 0$, there exists a sequence $\sigma \in X$ such that $|\sigma| > p$.*

Proof. Let $\sigma \in X$ and $p \in \mathbb{N}$. If $|\sigma| > p$ then we are done. Assume $|\sigma| \leq p$. Then there exists $n, n' > p$ such that $\sigma_n \sim \sigma_{n'}$. Note that $\sigma(n) = \sigma(n') = 0$. We construct the sequence σ' where $\sigma'(n) = 1$, $\sigma'(n') = -1$, and for all $m \neq n, n'$, $\sigma'(m) = \sigma(m)$. Clearly, $|\sigma'| > p$ and $\pi_\sim(\sigma) = \pi_\sim(\sigma')$, and thus $\sigma' \in X$. □

The proof of the following proposition follows the same idea in Proposition 6

Proposition 8. *The following sets are not recognizable for every non-trivial property $\mathcal{P}$ of $\mathbb{Z}$.*

1. $X = \{\sigma \in Z^* \mid \forall i \in \mathbb{N}, \mathcal{P}(\sigma(i))\}$, and
2. $X = \{\sigma \in Z^* \mid \exists i \in \mathbb{N}, \mathcal{P}(\sigma(i))\}$.

Remark 4 doesn't apply directly here as it did for N^* since M_3 is not a group but Z^* is. However indeed, Rec(Z^*) is uncountable. Given any arbitrary subset $X \subseteq \mathbb{Z}$ we construct the set

$$S(X) = \{\sigma \in Z^* \mid \sum_{n \in X} \sigma(|n|) = 1 \wedge \sum_{n \notin X} \sigma(|n|) = 0\}.$$

It is not difficult to see that $S(X) = S(Y)$ iff $X = Y$ for any $X, Y \subseteq \mathbb{Z}$. Indeed, each $S(X) \in \text{Rec}(Z^*)$. Define $\sigma_n \sim \sigma_m$ iff $n, m \in X$. Thus $\pi_\sim(S(X)) = (0, 1)$ which is trivially recognizable in Z^2. Thus we have the same conclusion for Rec(Z^*): we cannot expect to get a nice machine or logic characterization for Rec(Z^*).

5 Conclusion

We looked at the recognizable subsets of the additive and multiplicative monoids of integers, rationals, reals, and more. The arbitrarily divisible property of some of the monoids implied they have rather trivial recognizable sets. Hence recognizability by finite monoids cannot be the right notion of algebraic recognizability. On the other hand, for integers and rationals with multiplication we encounter a more complicated problem. They have an uncountable number of recognizable subsets and many computable subsets are not recognizable.

It would be interesting to look at monoids of matrices under multiplication or in general monoids of relations or functions over the domains considered. Our general results above do not translate well or immediately show us the recognizable subsets of these monoids since they have different structures, especially since they may not be arbitrarily divisible and their generators may not be easily identified. Further, recognizable sets of relations, or transductions, also have been extensively studied for the case of relations over free monoids [4,9]. It might be interesting to see if the observations in the present work also show up in the case of relations over rationals and reals.

References

1. Bojańczyk, M.: Nominal monoids. Theory Comput. Syst. **53**(2), 194–222 (2013)
2. Bouyer, P., Petit, A., Thérien, D.: An algebraic characterization of data and timed languages. In: Larsen, K.G., Nielsen, M. (eds.) CONCUR 2001. LNCS, vol. 2154, pp. 248–261. Springer, Heidelberg (2001). https://doi.org/10.1007/3-540-44685-0_17

3. Dima, C.: An algebraic theory of real-time formal languages. Ph.D. thesis, Université Joseph-Fourier-Grenoble I (2001)
4. Eilenberg, S.: Automata, Languages, and Machines. Academic press (1974)
5. Gécseg, F., Steinby, M.: Tree automata. arXiv preprint arXiv:1509.06233 (2015)
6. Maler, O., Pnueli, A.: On recognizable timed languages. In: Walukiewicz, I. (ed.) FoSSaCS 2004. LNCS, vol. 2987, pp. 348–362. Springer, Heidelberg (2004). https://doi.org/10.1007/978-3-540-24727-2_25
7. Perrin, D., Pin, J.: Infinite words. pure and applied mathematical series 141 (1992)
8. Pin, J.É.: Mathematical foundations of automata theory
9. Sakarovitch, J.: Elements of Automata Theory. Cambridge University Press (2009)
10. Weil, P.: Algebraic recognizability of languages. In: Fiala, J., Koubek, V., Kratochvíl, J. (eds.) Mathematical Foundations of Computer Science 2004, pp. 149–175. Springer, Heidelberg (2004)

There is Hope for Connexive Set Theories!

Santiago Jockwich

State University of Campinas, IFCH, Rua Cora Coralina 100, Campinas, Brazil
santijoxi@hotmail.com

Abstract. In this paper, we present an algebra-valued model on top of the four-element lattice $\mathbb{MC}$ which semantically captures Wansing's logic of material connexvity (MC). We show that the resulting model validates an axiom system that is classically equivalent to ZF. For this purpose, we tweak our semantic interpretation of set-membership and identity.

Keywords: Connexive logics · Connexive set theory · Algebra-valued models

Introduction

Connexive logics, first introduced by McCall in [11], aim to establish a meaningful *connection* between the antecedent and consequent of a valid conditional. These logics are characterized by four principles, known as Aristotle's theses: $\neg(\varphi \to \neg\varphi)$ (AT) and $\neg(\neg\varphi \to \varphi)$ (AT') and Boethius's theses: $(\varphi \to \psi) \to \neg(\varphi \to \neg\psi)$ (BT) and $(\varphi \to \neg\psi) \to \neg(\varphi \to \psi)$ (BT'), which ensure that conditionals adhere to certain intuitions about entailment and negation. Additionally, connexive logics reject the symmetry of the conditional: $(\varphi \to \psi) \to (\psi \to \varphi)$ (SC), preventing conditionals from collapsing into biconditionals.

Connexive logics hold unique philosophical significance as they are *contra-classical*: rather than being subsystems of classical logic, they validate principles that classical logic does not. This distinctiveness has sparked significant research interest in recent years, with explorations into their implications for philosophical logic, formal semantics, and the foundations of logic (see [3,4], and others).

Despite this progress, the field of *connexive set theory* remains underdeveloped. In particular, no models of connexive set theory have been constructed, leaving open the question of whether these theories can be semantically interpreted. This paper seeks to address this gap by constructing a model of a connexive set theory that is mathematically expressive.

Historically, three key contributions stand out. First, McCall's CC1-based class theory [10] provided an early framework, though it was not strictly a set theory. Second, Wiredu claimed that any set theory which validates AT, the Separation axiom and couple of minimal logical principles, is trivial [15]. Finally, Estrada-Gonzalez and Romero-Rodriguez [3] refuted Wiredu's result by identifying minimal requirements for triviality and proposing strategies to avoid it. They introduced the set theory ZqCC1, based on McCall's logic CC1 and Malinowski's *q-consequence*, and demonstrated its non-triviality.

This paper builds on another strategy outlined by Estrada-Gonzalez and Romero-Rodriguez: employing Wansing's material connexive logic (MC), which avoids triviality by rejecting the principle of explosion. Notably, MC has algebraic semantics, making it a promising candidate for constructing an algebra-valued model of set theory, a method successfully used in other non-classical frameworks (see [12,13]). Here, we propose building a $\mathbb{MC}$-valued model of set theory, where $\mathbb{MC}$ is the complete lattice which semantically characterizes MC.

The paper is organized as follows: Sect. 1 introduces the lattice $\mathbb{MC}$ and its properties. Section 2 defines the universe of $\mathbb{MC}$-valued functions. Section 3 constructs the $\mathbb{MC}$-valued model using the standard interpretation map and discusses its limitations. Finally, Sect. 4 explores an alternative interpretation map to address these shortcomings.

1 The Lattice $\mathbb{MC}$

Let us begin by introducing the complete distributive lattice $\mathbb{MC} = \langle \mathbf{A}; \wedge, \vee, \Rightarrow ,^*, 1, 0\rangle$, where the algebraic operations of $\mathbb{MC}$ correspond to the truth tables of the logical connectives of the connexive logic MC as introduced in (Sect. 4.5.3, [14]). More specifically, we take our universe to be $\mathbf{A} = \{1, \frac{1}{2}, \frac{3}{4}, 0\}$, where $\frac{1}{2} \perp \frac{3}{4}$ (i.e., $\frac{1}{2}$ and $\frac{3}{4}$ are incompatible). Finally, we fix $D = \{1, \frac{1}{2}\}$ as the set of designated values (Table 1).

Table 1. Operations for the four-element lattice $\mathbb{MC}$

$\Rightarrow$	1	$\frac{1}{2}$	$\frac{3}{4}$	0	$\wedge$	1	$\frac{1}{2}$	$\frac{3}{4}$	0	$\vee$	1	$\frac{1}{2}$	$\frac{3}{4}$	0	x	x^*
1	1	$\frac{1}{2}$	$\frac{3}{4}$	0	1	1	$\frac{1}{2}$	$\frac{3}{4}$	0	1	1	1	1	1	1	0
$\frac{1}{2}$	1	$\frac{1}{2}$	$\frac{3}{4}$	0	$\frac{1}{2}$	$\frac{1}{2}$	$\frac{1}{2}$	0	0	$\frac{1}{2}$	1	$\frac{1}{2}$	1	$\frac{1}{2}$	$\frac{1}{2}$	$\frac{1}{2}$
$\frac{3}{4}$	$\frac{1}{2}$	$\frac{1}{2}$	$\frac{1}{2}$	$\frac{1}{2}$	$\frac{3}{4}$	$\frac{3}{4}$	0	$\frac{3}{4}$	0	$\frac{3}{4}$	1	1	$\frac{3}{4}$	$\frac{3}{4}$	$\frac{3}{4}$	$\frac{3}{4}$
0	$\frac{1}{2}$	$\frac{1}{2}$	$\frac{1}{2}$	$\frac{1}{2}$	0	0	0	0	0	0	1	$\frac{1}{2}$	$\frac{3}{4}$	0	0	1

Furthermore, the binary operator $\Rightarrow$ satisfies the following properties, which we will use throughout this paper.

Lemma 1. *For any $a, b \in \mathbf{A}$ we have:*

(i) If $a \notin D$, then $(a \Rightarrow b) \in D$,
(ii) If $b \in D$, then $(a \Rightarrow b) \in D$,

Proof. (i) Let $a \notin D$, then for any $b \in \mathbf{A}$ we have $(a \Rightarrow b) = \frac{1}{2} \in D$. (ii) If $b \in D$, then for any $a \in \mathbf{A}$ we have $(a \Rightarrow b) \in D$. □

Notice that from item (ii), we can derive the following properties:

$$\text{If } a \leq b, \text{ then } (a \Rightarrow b) \in D, \tag{$\leq$}$$
$$a \in D \text{ and } b \notin D \text{ iff } (a \Rightarrow b) \notin D. \tag{$\not\Rightarrow$}$$

Additionally, the lattice $\mathbb{MC}$ satisfies the following properties, which also hold for Heyting and Boolean algebras.

Lemma 2. *For any $a, b, c \in \mathbf{A}$ we have:*

$$a \leq b \text{ implies } c \Rightarrow a \leq c \Rightarrow b, \tag{P2}$$
$$a \leq b \text{ implies if } (b \Rightarrow c) \in D \text{ then } (a \Rightarrow c) \in D, \tag{P3*}$$
$$((a \wedge b) \Rightarrow c) = (a \Rightarrow (b \Rightarrow c)). \tag{P4}$$

Proof. (P2) Suppose $a \leq b$ and $c \in D$. Then, we notice immediately that $(c \Rightarrow a) = a \leq b = (c \Rightarrow b)$. So, let $c \notin D$. Then, we have $(c \Rightarrow a) = \frac{1}{2} = (c \Rightarrow b)$. (P3*) Suppose $a \leq b$ and $(b \Rightarrow c) \in D$. If $c \in D$, then by ($\leq$) we get $(a \Rightarrow c) \in D$. If $c \notin D$, then $b \notin D$. Since $a \leq b$, we have also $a \notin D$. Thus, by ($\not\Rightarrow$) we get $(a \Rightarrow c) \in D$. (P4) Let $d \in \mathbf{A} \setminus \{\frac{1}{2}\}$. Then, we observe:

$$((a \wedge b) \Rightarrow c) = d \text{ iff } c = d \text{ and } (a \wedge b) \in D$$
$$\text{iff } (a \Rightarrow (b \Rightarrow c)) = d.$$

Additionally, we have

$$((a \wedge b) \Rightarrow c) = \frac{1}{2} \text{ iff } (a \wedge b) \notin D \text{ or } ((a \wedge b) \in D \text{ and } c = \frac{1}{2})$$
$$\text{iff } (a \notin D \text{ or } b \notin D) \text{ or } (b \Rightarrow c) = \frac{1}{2}$$
$$\text{iff } (a \Rightarrow (b \Rightarrow c)) = \frac{1}{2}.$$

$\square$

Note that properties P2 and P4 were introduced in [7] with the same labels, while P3* is a slight modification of P3.

2 The Universe of $\mathbb{MC}$-Valued Functions

Every algebra-valued model is composed of two elements: a domain and an interpretation function. The first component is a cumulative hierarchy consisting of *homogeneous* functions, usually denoted by $\mathbf{V}^{(\mathbb{A})}$, where $\mathbb{A}$ represents a complete distributive lattice.

Definition 21. *Let* **V** *represents the universe of all sets. The universe of* $\mathbb{MC}$*-valued functions, also denoted by* $\mathbf{V}^{(\mathbb{MC})}$*, is defined as follows:*

$$\mathbf{V}^{(\mathbb{MC})}_\alpha = \{x \,:\, x \text{ is a function and } \mathrm{ran}(x) \subseteq \mathbb{A}_{\mathsf{MC}}$$
$$\text{and there is } \xi < \alpha \text{ with } \mathrm{dom}(x) \subseteq \mathbf{V}^{(\mathbb{MC})}_\xi\} \text{ and}$$
$$\mathbf{V}^{(\mathbb{MC})} = \{x \,:\, \exists \alpha (x \in \mathbf{V}^{(\mathbb{MC})}_\alpha)\}.$$

The second component is the interpretation map, which assigns to each formula of the extended language of set theory $\mathcal{L}_\mathbb{A}$ (i.e., the language that we obtain by adding constant symbols for every member of $\mathbf{V}^{(\mathbb{MC})}$ to the language of set theory $\mathcal{L}_\in$) an element of the universe of $\mathbb{A}$.

Recent literature on algebra-valued models, such as [5], has shown that, in addition to the standard interpretation map denoted by $[\![\cdot]\!]$, there exist alternative interpretation maps which are able to give rise to mathematically expressive models. In Sect. 4, we will explore one of such maps, viz., the $[\![\cdot]\!]_{\mathsf{IN}}$-interpretation map, and explore the corresponding $\mathbb{MC}$-valued model. Prior to this, however, we begin by showing that the standard interpretation map gives rise to a model where Leibniz' indiscernibility of identicals, the bounded quantification properties, and Separation fail.

3 The Model $\mathbf{V}^{(\mathbb{MC},\,[\![\cdot]\!])}$

Let us denote with $\mathbf{V}^{(\mathbb{MC},[\![\cdot]\!])}$ the $\mathbb{MC}$-valued model that we obtain by applying the standard interpretation map $[\![\cdot]\!]$ (see Definitions 1.8–1.16 of [1]) to the universe of $\mathbb{MC}$-valued function.

Then, we can define a notion of validity for our model and introduce the notion of *a-like elements*.

Definition 31. *A formula* $\varphi \in \mathcal{L}_{\mathbb{MC}}$ *is said to be valid in* $\mathbf{V}^{(\mathbb{MC},\,[\![\cdot]\!])}$ *given a designated set* D*, whenever* $[\![\varphi]\!] \in D$. *We denote this fact by* $\mathbf{V}^{(\mathbb{MC},\,[\![\cdot]\!])} \models_D \varphi$.

Definition 32. *We say that* $x^a \in \mathbf{V}^{(\mathbb{MC})}$ *is an a-like element if*

$$x^a = \{\langle y^a, a\rangle : y \in x\},$$

where $a \in \mathbf{A}$ *and* $x \in \mathbf{V}$.

Notice that 1-like elements, which are denoted by x^1 for any $x \in \mathbf{V}$, are simply *canonical* names.

Having a closer look at the ontology of our model we notice that for any two 0-like and $\frac{3}{4}$-like elements are identical, as shown in the following lemma.

Lemma 3. *For any* $u^{\frac{3}{4}}, v^0 \in \mathbf{V}^{(\mathbb{MC})}$ *we have* $\mathbf{V}^{(\mathbb{MC},\,[\![\cdot]\!])} \models_D u^{\frac{3}{4}} = v^0$.

Proof. Follows by definition of $[\![\cdot = \cdot]\!]$. □

This then allows us to show that the *Leibniz' law of indiscernibility of identicals* fails within our model.

Lemma 4. *There exist* $u^{\frac{3}{4}}, v^0 \in \mathbf{V}^{(MC)}$ *such that*

$$\mathbf{V}^{(MC,\ [\![\cdot]\!])} \models_D u^{\frac{3}{4}} = v^0 \wedge \mathsf{Empty}(v^0) \text{ and } \mathbf{V}^{(MC,\ [\![\cdot]\!])} \not\models_D \mathsf{Empty}(u^{\frac{3}{4}}).$$

Proof. Pick any two $u^{\frac{3}{4}}, v^0 \in \mathbf{V}^{(MC)}$ and let $\mathsf{Empty}(x) := \neg \exists z (z \in x)$. Then, it follows by Lemma 3 that $[\![u^{\frac{3}{4}} = v^0]\!] \in D$ and for any $w \in \mathbf{V}^{(MC)}$ we have $[\![w \in v^0]\!] = 0$. Therefore, $[\![\neg \exists y (y \in v^0)]\!] = [\![\mathsf{Empty}(v^0)]\!] = 1$. Further, it is readily observable that for any $w \in \mathbf{V}^{(MC)}$ we have either $[\![w \in u^{\frac{3}{4}}]\!] = 0$ or $[\![w \in u^{\frac{3}{4}}]\!] = \frac{3}{4}$. Hence, $[\![\mathsf{Empty}(u^{\frac{3}{4}})]\!] \notin D$. □

Similarly, we can make formally precise our claims about the failure of *Leibniz' law of indiscernibility of identicals* regarding 1-like and $\frac{1}{2}$-like elements.

Lemma 5. *For any* $x \in \mathbf{V}$ *we have*

$$\mathbf{V}^{(MC,\ [\![\cdot]\!])} \models_D x^{\frac{1}{2}} = x^1 \wedge \mathsf{Empty}(x^{\frac{1}{2}}) \text{ and } \mathbf{V}^{(MC,\ [\![\cdot]\!])} \not\models_D \mathsf{Empty}(x^1).$$

Proof. Follows closely the proof of Lemma 4. □

The previous results allow us to show that the *bounded quantification* property for the universal and the existential quantifier fails within our model.

Formally, we state these properties as follows:

$$\bigwedge_{x \in \mathrm{dom}(u)} (u(x) \Rightarrow [\![\varphi(x)]\!]) \in D \text{ iff } [\![\forall x (x \in y \to \varphi(x))]\!] \in D, \qquad (\mathsf{BQ}_\forall)$$

$$\bigvee_{x \in \mathrm{dom}(u)} (u(x) \wedge [\![\varphi(x)]\!]) \in D \text{ iff } [\![\exists x (x \in y \wedge \varphi(x))]\!] \in D. \qquad (\mathsf{BQ}_\exists)$$

Then, we can show the following.

Lemma 6. $\mathsf{BQ}_\forall$ *and* $\mathsf{BQ}_\exists$ *fail for* $\mathbf{V}^{(MC,\ [\![\cdot]\!])}$, *i.e., there exists a formula* $\varphi(x) \in \mathcal{L}_{MC}$ *and an* $u \in \mathbf{V}^{(MC)}$ *such that:*

$$\bigwedge_{x \in \mathrm{dom}(u)} (u(x) \Rightarrow [\![\varphi(x)]\!]) \in D \text{ and } \mathbf{V}^{(MC,\ [\![\cdot]\!])} \not\models_D \forall x (x \in u \to \varphi(x)),$$

$$\bigvee_{x \in \mathrm{dom}(u)} (u(x) \wedge [\![\varphi(x)]\!]) \notin D \text{ and } \mathbf{V}^{(MC,\ [\![\cdot]\!])} \models_D \exists x (x \in u \wedge \varphi(x)).$$

Proof. ($\mathsf{BQ}_\forall$). Consider the functions: $\{\varnothing\}^{\frac{1}{2}} = \{\langle \varnothing, \frac{1}{2} \rangle\}$, $\{\varnothing\}^1 = \{\langle \varnothing, 1 \rangle\}$, $u = \{\langle \{\varnothing\}^{\frac{1}{2}}, 1 \rangle\}$ and the formula $\mathsf{Empty}(x)$. Then, it can be readily checked that: $\bigwedge_{x \in \mathrm{dom}(u)} (u(x) \Rightarrow [\![\varphi(x)]\!]) \in D$ and $[\![\forall x (x \in u \to \varphi(x))]\!] = 0$.

($\mathsf{BQ}_\exists$). Now, consider the functions $\{\varnothing\}^{\frac{3}{4}} = \{\langle \varnothing, \frac{3}{4} \rangle\}$, $\{\varnothing\}^0 = \{\langle \varnothing, 0 \rangle\}$ and $v = \{\langle \{\varnothing\}^{\frac{3}{4}}, 1 \rangle\}$ and the same formula of the previous item. Then, we have $\bigvee_{x \in \mathrm{dom}(v)} (v(x) \wedge [\![\varphi(x)]\!]) \notin D$ and $[\![\exists x (x \in y \wedge \varphi(x))]\!] = 1$. □

Thus, when dealing with bounded quantifiers, it becomes unclear which elements are contained within the domain of the element we are quantifying over. This "vagueness" can be exploited to demonstrate that the Separation axiom fails in our model. Moreover, this failure is critical, as our primary motivation was to show that there exist set theories where Separation is compatible with a connexive logic. Thus, suggesting the need for a more fine-grained notion of identity.

Theorem 33. $\mathbf{V}^{(\mathrm{MC},\ [\![\cdot]\!])} \not\models_D$ Separation.

Proof. Fix $w = \{\langle\{\varnothing\}^1, 1\rangle, \langle\{\varnothing\}^{\frac{1}{2}}, 1\rangle\}$. Suppose that for some $u \in \mathbf{V}^{(\mathrm{MC})}$ we have
$$\bigwedge_{v \in \mathbf{V}^{(\mathrm{MC})}} (\![\![v \in w \wedge \mathsf{Empty}(v)]\!] \Rightarrow [\![v \in u]\!]) \in D.$$

Consequently, $[\![\{\varnothing\}^{\frac{1}{2}} \in w \wedge \mathsf{Empty}(\{\varnothing\}^{\frac{1}{2}})]\!] \Rightarrow [\![\{\varnothing\}^{\frac{1}{2}} \in u]\!] \in D$ and thus we get $[\![\{\varnothing\}^{\frac{1}{2}} \in u]\!] \in D$. Therefore, there exists a $x \in \mathrm{dom}(u)$ such that $u(x) \in D$ and $[\![\{\varnothing\}^{\frac{1}{2}} = x]\!] \in D$. Then, it can be readily checked that $[\![\{\varnothing\}^1 = \{\varnothing\}^{\frac{1}{2}}]\!] \wedge [\![\{\varnothing\}^{\frac{1}{2}} = x]\!] \in D$ implies $[\![\{\varnothing\}^1 = x]\!] \in D$. Thus, we have $[\![\{\varnothing\}^1 \in u]\!] \in D$. But, at the same time,
$$[\![\{\varnothing\}^1 \in u]\!] \Rightarrow [\![\mathsf{Empty}(\{\varnothing\}^1) \wedge \{\varnothing\}^1 \in w]\!] = 0.$$

Thus,
$$\bigvee_{u \in \mathbf{V}^{(\mathrm{MC})}} (\bigwedge_{v \in \mathbf{V}^{(\mathrm{MC})}} [\![v \in u]\!] \Rightarrow ([\![\mathsf{Empty}(\{\varnothing\}^1) \wedge \{\varnothing\}^1 \in w]\!])) = 0,$$

which witnesses the failure of Separation. □

Finally, it is worth noting that the connection between the failure of (BQ$_\forall$) and that of Separation has already been observed in [5]. However, whether the failure of (BQ$_\exists$) similarly entails the failure of any ZFC axiom remains an open question. In light of this, we conclude this section by posing the following question.

Open Question: Does the failure of (BQ$_\exists$) imply the breakdown of some ZFC axiom?

4 The Model $\mathbf{V}^{(\mathrm{MC},\ [\![\cdot]\!]_{\mathsf{IN}})}$

Let us now introduce the $[\![\cdot]\!]_{\mathsf{IN}}$-interpretation map, originally presented in [8] for the purpose of constructing an algebra-valued model based on the three-valued lattice characterizing Priest's logic of paradox. This interpretation map offers a more *fine-grained* notion of identity, thereby avoiding the difficulties discussed in the previous section.

Definition 41 (Atomic formulas). *For any pair of elements $u, v \in \mathbf{V}^{(\mathrm{MC})}$,*

$$[\![u \in v]\!]_{\mathrm{IN}} = \bigvee_{x \in \mathrm{dom}(v)} \left(v(x) \wedge [\![x = u]\!]_{\mathrm{IN}} \right),$$

$[\![u = v]\!]_{\mathrm{IN}} = 1$ *iff*

for every $x \in \mathrm{dom}(u)$ we have $u(x) \leq [\![x \in v]\!]_{\mathrm{IN}}$,
and for every $y \in \mathrm{dom}(v)$ such that $v(y) \leq [\![y \in u]\!]_{\mathrm{IN}}$.
Otherwise; $[\![u = v]\!]_{\mathrm{IN}} = 0$.

Then, we extend the map $[\![\cdot]\!]_{\mathrm{IN}}$ to non-atomic formulas in the obvious way.

We denote with $\mathbf{V}^{(\mathrm{MC},\ [\![\cdot]\!]_{\mathrm{IN}})}$ the MC-valued model that we obtain by using the $[\![\cdot]\!]_{\mathrm{IN}}$-interpretation map.

The intuitive idea behind Definition 41 is to align our interpretation of identity more closely with our meta-theoretic notion of identity. In particular, now for any two $u, v \in \mathbf{V}^{(\mathrm{MC})}$ we have either $[\![u = v]\!]_{\mathrm{IN}} = 1$ or $[\![u = v]\!]_{\mathrm{IN}} = 0$ (our interpretation of set-membership, on the other hand, ranges over all the elements of **A**).

It can be readily checked by the reader that within $\mathbf{V}^{(\mathrm{MC},\ [\![\cdot]\!]_{\mathrm{IN}})}$ we have *less* valid identity statements compared to $\mathbf{V}^{(\mathrm{MC},\ [\![\cdot]\!])}$. For instance, the valid identity statements under the $[\![\cdot]\!]$-interpretation map in Lemma 3 and Lemma 4 are not valid anymore under the $[\![\cdot]\!]_{\mathrm{IN}}$-interpretation map. This, therefore, allows us to block the counterexamples that arose from such identity statements.

In fact, we go on to show that $\mathbf{V}^{(\mathrm{MC},\ [\![\cdot]\!]_{\mathrm{IN}})}$ validates Leibniz' law of indiscernibility of identicals (Corollary 42), $\mathrm{BQ}_\forall$ and $\mathrm{BQ}_\exists$ (Lemma 10) and an axiom system which is classically equivalent to ZF (Theorem 43). Due to constraints on the length of the manuscript some proofs will be merely sketched.

Lemma 7. *Consider any two elements $u, v \in \mathbf{V}^{(\mathrm{MC})}$. Then, we have $[\![u = v]\!]_{\mathrm{IN}} = 1$ if and only if the following conditions hold:*

(i) *if $u(x) = 1$ then $[\![x \in v]\!]_{\mathrm{IN}} = 1$, and if $v(y) = 1$ then $[\![y \in u]\!]_{\mathrm{IN}} = 1$;*
(ii) *if $u(x) = a$ where $a \in \{\frac{1}{2}, \frac{3}{4}\}$, then $[\![x \in v]\!]_{\mathrm{IN}} \geq a$ and if $v(y) = a$ where $a \in \{\frac{1}{2}, \frac{3}{4}\}$, then $[\![y \in u]\!]_{\mathrm{IN}} \geq a$.*

Proof. Follows immediately by Definition 41. □

We have, as well, the following lemma which shows that identity is a reflexive and transitive relation in our model.

Lemma 8. *For any $u, v, w \in \mathbf{V}^{(\mathrm{MC},\ [\![\cdot]\!]_{\mathrm{IN}})}$ the following holds:*

(i) $[\![u = u]\!]_{\mathrm{IN}} = 1$,
(ii) *for any $x \in \mathrm{dom}(u)$, $u(x) \leq [\![x \in u]\!]_{\mathrm{IN}}$,*
(iii) *If $[\![u = v \wedge v = w]\!]_{\mathrm{IN}} = 1$, then $[\![u = w]\!]_{\mathrm{IN}} = 1$.*

Proof. (i) and (ii) are trivial.
(iii) By induction on the domain of w. Assume that for all $z \in \mathrm{dom}(w)$ we have:

$$[\![u = v]\!]_{\mathsf{IN}} \wedge [\![v = z]\!]_{\mathsf{IN}} = 1 \text{ implies } [\![u = z]\!]_{\mathsf{IN}} = 1.$$

If $u(x_0) = 1$, then since $[\![u = v]\!]_{\mathsf{IN}} = 1$ by Lemma 7(i) we have that $[\![x_0 \in v]\!]_{\mathsf{IN}} = 1$ which can only be the case if

(1) there exists $y_0 \in \mathrm{dom}(v)$ such that $v(y_0) = 1$ and $[\![y_0 = x_0]\!]_{\mathsf{IN}} = 1$ or
(2) there exists two $y_0, y_1 \in \mathrm{dom}(v)$ such that $v(y_0) < 1, v(y_1) < 1, v(y_0) \vee v(y_1) = 1$ and $[\![y_0 = x_0]\!]_{\mathsf{IN}} = 1 = [\![y_1 = x_0]\!]_{\mathsf{IN}}$.

For case of (1), we can apply Lemma 7(i), so we have $[\![y_0 \in w]\!]_{\mathsf{IN}} = 1$ which can only be the case if

(1.1) there exists $z_0 \in \mathrm{dom}(w)$ such that $w(z_0) = 1$ and $[\![y_0 = z_0]\!]_{\mathsf{IN}} = 1$ or
(1.2) there exist two $z_0, z_1 \in \mathrm{dom}(w)$ such that $w(z_0) < 1, w(z_1) < 1, w(z_0) \vee w(z_1) = 1$ and $[\![y_0 = z_0]\!]_{\mathsf{IN}} = 1 = [\![y_0 = z_1]\!]_{\mathsf{IN}}$.

In the case of (1.1), we know that there exists $z_0 \in \mathrm{dom}(w)$ such that $w(z_0) = 1$ and by our induction hypothesis we know that $[\![x_0 = z_0]\!]_{\mathsf{IN}} = 1$, i.e., $[\![x_0 \in w]\!] = 1$. Case (1.2) follows similarly. In the case of (2), we get by applying Lemma 7(i) that

$$v(y_0) \leq [\![y_0 \in w]\!]_{\mathsf{IN}} \text{ and } v(y_1) \leq [\![y_1 \in w]\!]_{\mathsf{IN}},$$

which implies that there exists a $z_0 \in \mathrm{dom}(w)$ such that $v(y_0) \leq w(z_0)$ and $[\![y_0 = z_0]\!]_{\mathsf{IN}} = 1$ and a $z_1 \in \mathrm{dom}(w)$ such that $v(y_1) \leq w(z_1)$ and $[\![y_1 = z_1]\!]_{\mathsf{IN}} = 1$. Thus, there exist $z_0, z_1 \in \mathrm{dom}(w)$ such that

$$w(z_0) \vee w(z_1) = 1 \text{ and } [\![x_0 = z_0]\!]_{\mathsf{IN}} = 1 = [\![x_0 = z_1]\!]_{\mathsf{IN}},$$

and hence $[\![x_0 \in w]\!]_{\mathsf{IN}} = 1$.
If $u(x_0) = a$ where $a \in \{\frac{1}{2}, \frac{3}{4}\}$, then we can apply Lemma 7(ii) and thus $[\![x_0 \in v]\!]_{\mathsf{IN}} \geq a$. Thus, there exists a $y_0 \in \mathrm{dom}(v)$ such that $v(y_0) \geq a$ and $[\![y_0 = x_0]\!]_{\mathsf{IN}} = 1$, and by applying Lemma 7(ii) we get $[\![y_0 \in w]\!]_{\mathsf{IN}} \geq a$, i.e., there exists a $z_0 \in \mathrm{dom}(w)$ such that $w(z_0) \geq a$ and $[\![z_0 = y_0]\!]_{\mathsf{IN}} = 1$. Thus, there exists a $z_0 \in \mathrm{dom}(w)$ such that $w(z_0) \geq a$ and $[\![x_0 = z_0]\!]_{\mathsf{IN}} = 1$ by our induction hypothesis. Thus, $[\![x_0 \in w]\!]_{\mathsf{IN}} \geq a$. We can proceed similarly for any $z \in \mathrm{dom}(w)$. Therefore, we may conclude $[\![u = w]\!]_{\mathsf{IN}} = 1$. □

Lemma 9. *For any $u, v \in \mathbf{V}^{(\mathrm{MC})}$ and for any formula $\varphi(x) \in \mathcal{L}_{\mathrm{MC}}$, if $[\![u = v]\!]_{\mathsf{IN}} = 1$ and $[\![\varphi(u)]\!]_{\mathsf{IN}} = a$ where $a \in \mathbf{A} \setminus \{0\}$, then $[\![\varphi(v)]\!]_{\mathsf{IN}} = a$.*

Proof. By induction on the complexity of φ.

Base case (I). Let $\varphi(x) := w = x$, where $w \in \mathbf{V}^{(\mathrm{MC})}$ and $[\![u = w]\!]_{\mathsf{IN}} = 1$. Trivial.

Base case (II). Let $\varphi(x) := w \in x$, where $w \in \mathbf{V}^{(\mathrm{MC})}$ and suppose $[\![\varphi(u)]\!]_{\mathsf{IN}} = 1$. Then, we have two cases:

1. there exists a $p_0 \in \text{dom}(u)$ such that $u(p_0) = 1$ and $[\![p_0 = w]\!]_{\text{IN}} = 1$, or
2. there exist two $p_0, p_1 \in \text{dom}(u)$ such that $u(p_0) < 1, u(p_1) < 1, u(p_0) \vee u(p_1) = 1$ and $[\![p_0 = w]\!]_{\text{IN}} = 1 = [\![p_1 = w]\!]_{\text{IN}}$.

In the case of (1), since $[\![u = v]\!]_{\text{IN}} = 1$, by Lemma 7(i), we have that $[\![p_0 \in v]\!]_{\text{IN}} = 1$. This can only be the case if;

(1.1) there exists a $q_0 \in \text{dom}(v)$ such that $v(q_0) = 1$ and $[\![p_0 = q_0]\!]_{\text{IN}} = 1$, or
(1.2) there exist two $q_0, q_1 \in \text{dom}(v)$ such that $v(q_0) < 1, v(q_1) < 1, v(q_0) \vee v(q_1) = 1$ and $[\![q_0 = p_0]\!]_{\text{IN}} = 1 = [\![q_1 = p_0]\!]_{\text{IN}}$.

In both cases the desiderata follows straight forward by applying (twice) Lemma 8(iii).

Case (2). Follows by Lemma 7(ii) and Lemma 8(iii).

Now, suppose; $[\![\varphi(u)]\!]_{\text{IN}} = a$ where $a \in \{\frac{1}{2}, \frac{3}{4}\}$, i.e.,

$$\bigvee_{p \in \text{dom}(u)} (u(p) \wedge [\![p = w]\!]_{\text{IN}}) = a$$

This can only be the case if

1. there exists $p_0 \in \text{dom}(u)$ such that $u(p_0) = a$ and $[\![p_0 = w]\!]_{\text{IN}} = 1$, and
2. there exists no $\{p_i : i \in I\} \subseteq \text{dom}(u)$ such that

$$\bigvee_{i \in I} u(p_i) = 1 \text{ and } [\![p_i = w]\!]_{\text{IN}} = 1 \text{ for all } i \in I.$$

Since $[\![u = v]\!]_{\text{IN}} = 1$ we can apply Lemma 7(ii) and, thus, $[\![p_0 \in v]\!]_{\text{IN}} \geq a$, i.e., there exists $q_0 \in \text{dom}(v)$ such that $v(q_0) \geq a$ and $[\![p_0 = q_0]\!]_{\text{IN}} = 1$. Further, by Lemma 8($iii$) we obtain $[\![q_0 = w]\!]_{\text{IN}} = 1$ and therefore $[\![w \in v]\!]_{\text{IN}} \geq a$. Moreover, it can bre readily shown that $[\![w \in v]\!]_{\text{IN}} < 1$. Hence, we can conclude $[\![w \in v]\!]_{\text{IN}} = a$.

Base case (III). Let $\varphi(x) := x \in w$, where $w \in \mathbf{V}^{(\text{MC})}$. Follow a very similar proof strategy to the previous case.

Induction step: The induction steps $\wedge, \vee, \rightarrow$ and $\neg$ are straightforward calculations that follow from applying the induction hypothesis.

Case ($\exists$). Let $\varphi(x) := \exists y \, \psi(y, x)$. Suppose that we have $[\![\varphi(u)]\!]_{\text{IN}} = 1$. This can only be the case if

(1) there exists $p \in \mathbf{V}^{(\text{MC})}$ such that $[\![\psi(p, u)]\!] = 1$ or
(2) there exist $p_0, p_1 \in \mathbf{V}^{(\text{MC})}$ such that $[\![\psi(p_0, u)]\!]_{\text{IN}} \neq 1, [\![\psi(p_1, u)]\!]_{\text{IN}} \neq 1$ and $[\![\psi(p_0, u)]\!]_{\text{IN}} \vee [\![\psi(p_1, u)]\!]_{\text{IN}} = 1$.

Case (1). By our induction hypothesis we have $[\![\psi(p, v)]\!]_{\text{IN}} = 1$. Similarly, in Case (2), by our induction hypothesis: $[\![\psi(p_0, u)]\!]_{\text{IN}} \vee [\![\psi(p_1, u)]\!]_{\text{IN}} = 1$. Hence, in both cases $[\![\varphi(v)]\!]_{\text{IN}} = 1$.

Now, suppose $[\![\varphi(u)]\!]_{\text{IN}} = a$, where $a \in \{\frac{1}{2}, \frac{3}{4}\}$. It follows that there exists $p \in \mathbf{V}^{(\text{MC})}$ such that $[\![\psi(p, u)]\!]_{\text{IN}} = a$ and there does not exist any $q \in \mathbf{V}^{(\text{MC})}$ such that $[\![\psi(q,u)]\!]_{\mathsf{IN}} \in \{1, b\}$ where $a \vee b = 1$. The induction hypothesis ensures that $[\![\psi(p,v)]\!]_{\mathsf{IN}} = a$ and $[\![\psi(q,v)]\!]_{\mathsf{IN}} \notin \{1, b\}$, for all $q \in \mathbf{V}^{(\mathsf{MC})}$. Therefore, $[\![\varphi(v)]\!]_{\mathsf{IN}} = a$.

Case ($\forall$). Let $\varphi(x) := \forall y\, \psi(y, x)$. Similar to proof of the previous case. □

This result establishes the validity of Leibniz's law of indiscernibility of identicals and allows us to derive the bounded quantification properties in our model.

Corollary 42. *For any $u, v \in \mathbf{V}^{(\mathsf{MC})}$ and any formula $\varphi(x)$ in $\mathcal{L}_{\mathsf{MC}}$ having one free variable x, if*

$$\mathbf{V}^{(\mathsf{MC},\, [\![\cdot]\!]_{\mathsf{IN}})} \models u = v \wedge \varphi(u),\ \text{then}\ \mathbf{V}^{(\mathsf{MC},\, [\![\cdot]\!]_{\mathsf{IN}})} \models \varphi(v).$$

We go on to show that both $\mathsf{BQ}_\forall$ and $\mathsf{BQ}_\exists$ hold for $\mathbf{V}^{(\mathsf{MC},\, [\![\cdot]\!]_{\mathsf{IN}})}$.

Lemma 10. $\mathsf{BQ}_\forall$ *and* $\mathsf{BQ}_\exists$ *are valid in* $\mathbf{V}^{(\mathsf{MC},\, [\![\cdot]\!]_{\mathsf{IN}})}$.

Proof. $\mathsf{BQ}_\forall$: ($\Leftarrow$). This is trivial, given that the domain of u is a proper subcollection of $\mathbf{V}^{(\mathsf{MC})}$.

($\Rightarrow$) Suppose that

$$\bigwedge_{x \in \mathrm{dom}(u)} \left(u(x) \Rightarrow [\![\varphi(x)]\!]_{\mathsf{IN}} \right) \in D$$

i.e., for any $x \in \mathrm{dom}(u)$ we have $u(x) \Rightarrow [\![\varphi(x)]\!]_{\mathsf{IN}} \in D$. Additionally, suppose that $[\![w_0 \in u]\!]_{\mathsf{IN}} \in D$ for some $w_0 \in \mathbf{V}^{(\mathsf{MC})}$, i.e.,

$$[\![w_0 \in u]\!]_{\mathsf{IN}} = \bigvee_{x \in \mathrm{dom}(u)} \left(u(x) \wedge [\![x = w_0]\!]_{\mathsf{IN}} \right) \in D.$$

Thus, there exists a $x_0 \in \mathrm{dom}(u)$ such that $u(x_0) \in D$ and $[\![x_0 = w_0]\!]_{\mathsf{IN}} = 1$ and due to our initial assumption; $[\![\varphi(x_0)]\!]_{\mathsf{IN}} \in D$. Then by Corollary 42 we have $[\![\varphi(w_0)]\!]_{\mathsf{IN}} \in D$ and thus $[\![w_0 \in u]\!]_{\mathsf{IN}} \Rightarrow [\![\varphi(w_0)]\!]_{\mathsf{IN}} \in D$. On the other hand, for any $w \in \mathbf{V}^{(\mathsf{MC})}$ such that $[\![w \in u]\!]_{\mathsf{IN}} \notin D$ the desideratum follows immediately. Therefore for any $w \in \mathbf{V}^{(\mathsf{MC})}$ we have that $[\![w \in u]\!]_{\mathsf{IN}} \Rightarrow [\![\varphi(w)]\!]_{\mathsf{IN}} \in D$.

$\mathsf{BQ}_\exists$: ($\Rightarrow$). Trivial.

($\Leftarrow$) The proof is straight-forward and makes essentially use of Corollary 4.5. □

4.1 The Validity of $\overline{\mathsf{ZF}}$ in $\mathbf{V}^{(\mathsf{MC},\, [\![\cdot]\!])}$

We now demonstrate that the axiom system $\overline{\mathsf{ZF}}$, which arises from replacing the standard Extensionality axiom with the following axiom:

$$\forall x \forall y \forall z \big((z \in x \leftrightarrow z \in y) \wedge (\neg(z \in x) \leftrightarrow \neg(z \in y)) \big) \to x = y, \quad (\overline{\text{Extensionality}})$$

holds in our model. This particular formulation of the Extensionality axiom was first introduced by [9], and is needed given that Extensionality fails within our model.

To realize this, simply consider the $\mathbb{MC}$-valued functions: $u = \{\langle \emptyset, \frac{1}{2}\rangle\}$ and $v = \{\langle \emptyset, 1\rangle\}$. Then, the antecedent of $\overline{\text{Extensionality}}$ holds but clearly $[\![u = v]\!]_{\text{IN}} = 0$. However, we can readily show that $\overline{\text{Extensionality}}$, along with the remaining ZF axioms, holds in our models.

Theorem 43. $\mathbf{V}^{(\mathbb{MC},\ [\![\cdot]\!])} \models \overline{\text{ZF}}$.

Proof. All the proofs are straight-forward with exception of the Powerset and $\overline{\text{Extensionality}}$ axiom which requires extra work given the modified interpretation map. □

4.2 The Logic of $\mathbf{V}^{(\mathbb{MC},\ [\![\cdot]\!]_{\text{IN}})}$

It is well-known due to the results of [6] that the (propositional) logic associated to the algebraic structure and that of the corresponding algebra-valued model might not coincide. For example, the Heyting algebra with 5 elements in its universe, i.e., $\mathbb{H}_5$, validates intuitionistic propositional logic (IPL) whereas $\mathbf{V}^{(\mathbb{H}_5, [\![\cdot]\!])}$ validates the intermediate logic $\text{IPL} + (\varphi \to \psi \vee \psi \to \varphi)$. The crux being that not all the elements of the universe of the algebra might occur in the range of our algebra-valued model, thus, causing the mentioned mismatch.

We will see that we experience a similar mismatch in the case of $\mathbb{MC}$ and $\mathbf{V}^{(\mathbb{MC},\ [\![\cdot]\!]_{\text{IN}})}$. Specifically, we will argue that the propositional logic of $\mathbf{V}^{(\mathbb{MC},\ [\![\cdot]\!]_{\text{IN}})}$ is either MC or the three-valued logic CN, the latter being a conservative extension of MC. Notably, we have:

$$\text{CN} = \text{MC} + \varphi \vee \neg\varphi.$$

For a thorough presentation of the semantics and syntax of this logic, we refer to [2]. Most importantly, just as in the case of MC, CN can be captured semantically through algebraic semantics. Let us denote by $\mathbb{CN}$ the complete lattice with the operations defined below (Table 2).

Table 2. Operations of the three-element lattice $\mathbb{CN}$.

$\wedge$	1	$\frac{1}{2}$	0		$\vee$	1	$\frac{1}{2}$	0		$\Rightarrow$	1	$\frac{1}{2}$	0		$*$		
1	1	$\frac{1}{2}$	0		1	1	1	1		1	1	$\frac{1}{2}$	0		1	0	
$\frac{1}{2}$	$\frac{1}{2}$	$\frac{1}{2}$	0		$\frac{1}{2}$	1	$\frac{1}{2}$	$\frac{1}{2}$		$\frac{1}{2}$	1	$\frac{1}{2}$	0		$\frac{1}{2}$	$\frac{1}{2}$	
0	0	0	0		0	1	$\frac{1}{2}$	0		0	$\frac{1}{2}$	$\frac{1}{2}$	$\frac{1}{2}$		0	1	

As in the case of $\mathbb{H}_5$, it is very likely to be the case that not all elements of the universe of $\mathbb{MC}$ correspond to the value of a sentence in the language of set theory. In particular, it is unclear to the author (despite considerable effort) how to define a sentence that would be assigned the value $\frac{3}{4}$ in our model. Moreover, since the remaining elements of $\mathbb{MC}$ correspond to the value of a set-theoretic sentence we are left with the following two options:

(1.) We can define a sentence in the language of set theory which receives value $\frac{3}{4}$ in our model and thus the range of our interpretation map coincides with the universe of $\mathbb{MC}$. In technical terms, our model is *faithful* to its underlying algebraic structure and, therefore, the propositional logic of $\mathbf{V}^{(\mathbb{MC},\ [\![\cdot]\!]_{\mathsf{IN}})}$ is MC.

(2.) We can *not* define a sentence in the language of set theory which receives value $\frac{3}{4}$ in our model and thus the range of our interpretation map coincides with the universe of $\mathbb{CN}$ (since $\mathbb{CN}$ is the sublattice of $\mathbb{MC}$ which we obtain by "removing" the element $\frac{3}{4}$ from the universe of $\mathbb{MC}$.) In this case, the propositional logic of $\mathbf{V}^{(\mathbb{MC},\ [\![\cdot]\!]_{\mathsf{IN}})}$ is CN.

It follows that, in both cases, the propositional logic of $\mathbf{V}^{(\mathbb{MC},\ [\![\cdot]\!]_{\mathsf{IN}})}$ is indeed a connexive logic. However, it remains an open question whether *the law of excludded middle* is valid in our model.

5 Conclusion

Suppose you are a connexive logician, convinced that only connexive logics correctly capture how implication and negation function in natural language discourse. Many of the undesirable consequences of the material conditional disappear, and paradoxes no longer trouble you.

However, as a connexive logician, you also want to do set theory and mathematics within a logical framework consistent with your philosophical beliefs. This creates a strong motivation-*at least* for the connexive logician-to further develop connexive foundations of mathematics. You want to avoid abandoning the principles of connexive logic when engaging with mathematical theories. It seems that for this stripe of connexive logicians our model could become a useful tool to accommodate mathematical theories within a connexive framework.

Disclosure of Interests. The authors has no competing interests to declare.

References

1. Bell, J.L.: Set Theory: Boolean-Valued Models and Independence Proofs, 3rd edn. Cambridge University Press, Cambridge (2007)
2. Cantwell, J.: The logic of conditional negation. Notre Dame J. Formal Logic **49**(3), 245–260 (2008)
3. Estrada-Gonzalez, L., Romero-Rodriguez, M.A.: Another remark on connexivity and set theory. Submitted (2024)
4. Fazio D., John, G.S.: Connexive implications in substructural logics. Rev. Symb. Logic, 1–32 (2024)
5. Jockwich, S., Tarafder, S., Venturi, G.: Ideal objects for set theory. J. Philos. Log. **51**(3), 583–602 (2022)
6. Lowe, B., Passmann, R., Tarafder, S.: Constructing illoyal algebra-valued models of set theory. Algebra Universalis **82**(46) (2021)

7. Lowe, B., Tarafder, S.: Generalized algebra-valued models of set theory. Rev. Symb. Logic **8**(1), 192–205 (2015)
8. Martinez, S.J.: Algebra-valued models for lp-set theory. Aust. J. Log. **18**(7), 657–687 (2022)
9. Martinez, S.J., Tarafder, S., Venturi, G.: ZF and its interpretations. Ann. Pure Appl. Logic **175**(6), 103427 (2024)
10. McCall, S.: Connexive class logic. J. Symb. Log. **32**(1), 83–90 (1967)
11. McCall, S., McCall, S.: Non-classical propositional calculi. PhD thesis, University of Oxford (1964)
12. Takeuti, G., Titani, S.: Fuzzy logic and fuzzy set theory. Arch. Math. Logic **32**(1), 1–32 (1992)
13. Tarafder, S.: Non-classical foundations of set theory. J. Symb. Logic **87**(1), 347–376 (2022)
14. Wansing, H.: Connexive logic. In: Zalta, E.N., Nodelman, U. (eds.) The Stanford Encyclopedia of Philosophy. Metaphysics Research Lab, Stanford University, Summer 2023 edition (2023)
15. Wiredu, J.E.: A remark on a certain consequence of connexive logic for Zermelo's set theory. Stud. Logica. **33**(2), 127–130 (1974)

A Semantics of Basic Modal Language via a Rough Set Framework

Md. Aquil Khan[✉] and Ranjan

Department of Mathematics, Indian Institute of Technology Indore,
Indore 453552, India
{aquilk,phd2201241006}@iiti.ac.in

Abstract. This article introduces a new semantics for the basic modal language, motivated by rough set theory. Additionally, a sound and complete deductive system relative to two important classes of models are obtained.

Keywords: Monotonic logic · Approximation operator · Axiomatization

1 Introduction

Rough set theory, introduced by Zdzisław Pawlak in the early 1980s, is a robust mathematical framework for addressing uncertainty and vagueness in data analysis. In Pawlak's model, the knowledge about a set W of objects is encapsulated through an equivalence relation R on W. The pair (W, R) is referred to as an approximation space. Concepts, represented as subsets of the domain, are then approximated using a pair of operators known as the lower and upper approximation operators. These operators, denoted by L_R and U_R on an approximation space (W, R), are defined as follows: for any $X \subseteq W$,

$$L_R(X) := \{x \in W : R(x) \subseteq X\},$$
$$U_R(X) := \{x \in W : R(x) \cap X \neq \emptyset\},$$

where $R(x) := \{y \in W : (x, y) \in R\}$.

Over time, Pawlak's rough set model has been extensively generalized. Examples include the multi-granulation rough set model (cf. e.g., [6,7,10,17,20]), the variable precision rough set model [23], the covering-based rough set model (cf. e.g., [19,21]), and models based on neighborhood systems [16] or Bayesian principles [22]. In [12,13], a rough set model was proposed that focuses on subsets of the domain, rather than the entire domain, through the concept of a *subset approximation structure* (in brief, SAS). This structure extends the generalized approximation space (W, R) by incorporating a collection σ of non-empty subsets of W. The necessity lower approximation operator L^n, as discussed in [13], is defined such that an element w belongs to $L^n(X)$ if and only if for every

$S \in \sigma_w$, it holds that $R(w) \cap S \subseteq X$, where $\sigma_w := \{S \in \sigma : w \in S\}$. It is important to note that if $\sigma_w = \emptyset$, then $w \in L^n(Y)$ for all Y, which is an undesirable outcome. To address this, one can impose the condition that $\sigma_w \neq \emptyset$ for w to belong to the lower approximation of a set. Consequently, we define the refined lower approximation operator L^{n*} such that $w \in L^{n*}(X)$ if and only if $\sigma_w \neq \emptyset$ and $w \in L^n(X)$.

In this article, we focus on the approximation operator L^{n*} and introduce a new semantics of the basic modal language based on this operator. The article serves a dual purpose: from the rough set perspective, it provides a formal system for reasoning about the approximation operator L^{n*} and offers insights into its characterizing properties. From the modal logic perspective, the article contributes to the literature by enriching it with a new semantic framework.

At this juncture, it is worth to mention that the logic aspects of rough set theory remains an important area of research and one can find several works on the proposals of logics for structures inherited from rough set theory. We refer to survey articles [1,4,14] for a detail study on this.

The remainder of the article is structured as follows. In Sect. 2, we provide a brief overview of the subset approximation structure and the corresponding approximation operators. Section 3 introduces a new semantics for the basic modal language and demonstrates how this language can be utilized to express properties of the approximation operator L^{n*}. Section 4 presents a modal system for the classes of SASs based on tolerance and equivalence relations, along with the corresponding soundness and completeness theorems. Finally, Sect. 5 concludes the article.

2 Subset Approximation Structure

As highlighted in the introduction, since its inception, Pawlak's rough set model has undergone numerous generalizations. In [12,13], a rough set model based on the concept of a *subset approximation structure* was introduced and explored. This model is particularly useful in scenarios where it is beneficial to focus on specific subsets of the domain rather than the entire system. Let us recall the definition of the subset approximation structure. For a set W, we denote its power set by $\wp(W)$.

Definition 1 ([13]). *A subset approximation structure, (in brief, SAS) is defined as a tuple $\mathfrak{F} := (W, \sigma, R)$, where*

- *W is a non-empty set of objects,*
- *$\sigma \subseteq (\wp(W) \setminus \{\emptyset\})$,*
- *$R \subseteq W \times W$.*

The component W of a SAS (W, σ, R) is referred to as the *carrier set* of the SAS. For any $w \in W$, we use σ_w to denote the set $\{X \in \sigma : w \in X\}$.

It is evident that an SAS is an extension of a hypergraph [5], with the addition of a binary relation on the carrier set. Moreover, it is important to note that σ may not necessarily be a covering of the carrier set (cf. e.g. [21]).

In rough set literature, it is customary to study approximation operators in a general setting without imposing any specific restrictions on the underlying relation. However, equivalence and tolerance relations are particularly important as they serve as instances of *indiscernibility* and *similarity relations*, respectively, obtained from information systems (cf. [8,11]). These two types of relations are crucial for the notion of *distinguishability* in rough set theory. In this article, our study is centered on these two types of relations, specifically considering the classes S_e and S_{rs} of SASs, which are based on equivalence and tolerance relations, respectively.

The following lower and upper approximation operators, based on SASs, were proposed in [13].

Definition 2. *Let $\mathfrak{F} := (W, \sigma, R)$ be an SAS. The* necessity lower approximation *operator $L_{\mathfrak{F}}^n$, and the* possibility upper approximation *operator $U_{\mathfrak{F}}^p$ are defined as follows. Let $X \subseteq W$.*

$$L_{\mathfrak{F}}^n(X) = \{w \in W : w \in [R]_S(X) \text{ for all } S \in \sigma_w\},$$
$$U_{\mathfrak{F}}^p(X) = \{w \in W : w \in \langle R \rangle_S(X) \text{ for some } S \in \sigma_w\}, \text{ where}$$
$$[R]_S(X) = \{u \in S : R(u) \cap S \subseteq X\}, \text{ and}$$
$$\langle R \rangle_S(X) = \{u \in S : R(u) \cap S \cap X \neq \emptyset\}.$$

Unfolding the definition, we have $w \in L_{\mathfrak{F}}^n(X)$ if and only if, for all $S \in \sigma_w$, $R(w) \cap S \subseteq X$. Similarly, $w \in U_{\mathfrak{F}}^p(X)$ if and only if there exists some $S \in \sigma_w$ such that $R(w) \cap S \cap X \neq \emptyset$.

As mentioned in the introduction, if $\sigma_w = \emptyset$, then $w \in L_{\mathfrak{F}}^n(X)$ for all $X \subseteq W$. To prevent this situation, we introduce the following refined lower approximation operator:

$$L_{\mathfrak{F}}^{n*}(X) = \{w \in W : \sigma_w \neq \emptyset \ \& \ w \in [R]_S(X) \text{ for all } S \in \sigma_w\}$$

The operator $U_{\mathfrak{F}}^{n*}$, which is the dual of $L_{\mathfrak{F}}^{n*}$, is defined as follows:

$$U_{\mathfrak{F}}^{n*}(X) = \{w \in W : \sigma_w = \emptyset \text{ or, } w \in \langle R \rangle_S(X) \text{ for some } S \in \sigma_w\}.$$

Let us recall the concept of covering space and P_1 covering systems (cf. [19,21]). Note that in an SAS $\mathfrak{F} := (W, \sigma, R)$ based on a covering space (W, σ), the operator $L_{\mathfrak{F}}^{n*}$ coincides with $L_{\mathfrak{F}}^n$. Moreover, in an SAS $\mathfrak{F} := (W, \sigma, W \times W)$ based on a covering space (W, σ), we also have $L_{\mathfrak{F}}^{n*}(X) = \underline{P_1}(X) = \{w \in W : F_\sigma(w) \subseteq X\}$, where $F_\sigma(w) := \bigcup \{S \in \sigma : w \in S\} = \bigcup_{S \in \sigma_w} S$.

Remark 1. If there is no ambiguity, we will simplify the notation by omitting the subscript and/or superscript, referring to $L_{\mathfrak{F}}^{n*}(X)$ and $U_{\mathfrak{F}}^{n*}(X)$ simply as $L(X)$ (or $L_{\mathfrak{F}}(X)$), and $U(X)$ (or $U_{\mathfrak{F}}(X)$), respectively.

The following proposition outlines the key properties of the operator L.

Proposition 1. *Let $\mathfrak{F} := (W, \sigma, R)$ be an SAS, and $X, Y \subseteq W$. Then the following hold.*

(C) $L(X) \cap L(Y) = L(X \cap Y)$.
(T_0) $U(X) \subseteq U(X \cap L(W))$
(T) $L(X) \subseteq X$, if R is reflexive.
(B_0) $L(W) \cap X \subseteq L(U(X))$, if R is symmetric.
(4_0) $L(W) \subseteq L(L(W))$.

Proof. (C): Let $w \in L(X) \cap L(Y)$. Then $\sigma_w \neq \emptyset$ and for all $S \in \sigma_w$, we have $R(w) \cap S \subseteq X$ and $R(w) \cap S \subseteq Y$. This gives $R(w) \cap S \subseteq X \cap Y$. Thus, $w \in L(X \cap Y)$. For converse, assume that $w \in L(X \cap Y)$. Then $\sigma_w \neq \emptyset$ and for all $S \in \sigma_w$, we have $R(w) \cap S \subseteq X \cap Y$. So, we get $R(w) \cap S \subseteq X$ and $R(w) \cap S \subseteq Y$. Hence, $w \in L(X) \cap L(Y)$.

(T_0): Let $w \in U(X)$. Then either $\sigma_w = \emptyset$ or there exists some $T \in \sigma_w$ such that $R(w) \cap T \cap X \neq \emptyset$. If $\sigma_w = \emptyset$, it follows trivially that $w \in U(X \cap L(W))$. Now assume $\sigma_w \neq \emptyset$ and let $u \in R(w) \cap T \cap X$. Since $T \in \sigma_u$, it follows that $\sigma_u \neq \emptyset$, which implies, $u \in L(W)$. Thus, we have $u \in R(w) \cap T \cap X \cap L(W)$, establishing that $w \in U(X \cap L(W))$.

(T): Let $w \in L(X)$. Then, $\sigma_w \neq \emptyset$, and for all $S \in \sigma_w$, we have $R(w) \cap S \subseteq X$. Let $T \in \sigma_w$. Thus, $R(w) \cap T \subseteq X$. Since R is reflexive, we obtain $w \in R(w) \cap T$, and hence we get $w \in X$.

(B_0): Let $w \in L(W) \cap X$. Then $\sigma_w \neq \emptyset$. Consider an arbitrary $S \in \sigma_w$, and we aim to show that $R(w) \cap S \subseteq U(X)$.

Take any $u \in S \cap R(w)$. Since $u \in S$, we have $S \in \sigma_u$. Given that R is symmetric, $w \in R(u)$. Thus, $w \in S \cap R(u) \cap X$, which implies $u \in U(X)$. Consequently, $w \in L(U(X))$.

(4_0): Let $w \in L(W)$. Then $\sigma_w \neq \emptyset$. Consider an arbitrary $S \in \sigma_w$, and we show that $R(w) \cap S \subseteq L(W)$. Let $u \in S \cap R(w)$. We get $\sigma_u \neq \emptyset$ as $S \in \sigma_u$. Thus $u \in L(W)$. Hence $w \in L(L(w))$. □

In Sect. 4.1, it will be proved using the modal logic approach that the properties listed in Proposition 1 are, in fact, the characterizing properties of the operator L based on SASs from the classes S_e and S_{rs}.

3 A New Semantics of Basic Modal Language

In this section, we present the semantics of the basic modal language using the operator L^{n*} defined in the previous section. Recall that the basic modal language comprises a non-empty, countable set Φ of propositional variables, along with the propositional constant $\top$, logical connectives $\neg$ and $\wedge$, and the unary modal operator $\square$. The well-formed formulas (wffs) are defined inductively according to the following rules:

$$p \mid \top \mid \neg \alpha \mid \alpha \wedge \beta \mid \square \alpha,$$

where $p \in \Phi$ and α, β are wffs. Other propositional connectives such as $\vee$, $\rightarrow$, $\leftrightarrow$, and $\perp$ are defined in the standard way, while the derived modal connective $\Diamond$ is given by $\Diamond \alpha := \neg \Box \neg \alpha$. Let $\mathcal{L}$ be the set of all wffs.

We now introduce the semantics of the basic modal language based on SASs. The notion of a model is defined as follows:

Definition 3 (Model). *A model is a tuple $\mathfrak{M} := (F, V)$, where F is an SAS, and $V : \Phi \to \wp(W)$.*

Definition 4. *The truth set $[\![\alpha]\!]_{\mathfrak{M}}$ of a wff α in a model $\mathfrak{M} := (\mathfrak{F}, V)$, where $\mathfrak{F} := (W, \sigma, R)$, is defined inductively as follows:*

$$[\![\top]\!]_{\mathfrak{M}} := W \text{ and } [\![\perp]\!]_{\mathfrak{M}} := \emptyset;$$
$$[\![p]\!]_{\mathfrak{M}} := V(p), \text{ for } p \in \Phi;$$
$$[\![\neg \alpha]\!]_{\mathfrak{M}} := W \setminus [\![\alpha]\!]_{\mathfrak{M}};$$
$$[\![\alpha \wedge \beta]\!]_{\mathfrak{M}} := [\![\alpha]\!]_{\mathfrak{M}} \cap [\![\beta]\!]_{\mathfrak{M}};$$
$$[\![\Box \alpha]\!]_{\mathfrak{M}} := L_{\mathfrak{F}}([\![\alpha]\!]_{\mathfrak{M}}).$$

It is evident from the definition of the truth set that the operators $\neg$ and $\wedge$ correspond to the set-theoretic operations of complementation and intersection, respectively. Moreover, the operator $\Box$ corresponds to the lower approximation operator L. Following the standard practice in modal logic, we use the notation $\mathfrak{M}, w \models \alpha$ to denote that $w \in [\![\alpha]\!]_{\mathfrak{M}}$. Using this notation, and by unfolding the definition of the truth set for the operators $\Box$ and $\Diamond$, we obtain the following satisfiability conditions.

Proposition 2.

$\mathfrak{M}, w \models \Box \alpha \iff \sigma_w \neq \emptyset$ & *for each $S \in \sigma_w$, and for each $u \in S \cap R(w)$,*
$\mathfrak{M}, u \models \alpha.$

$\mathfrak{M}, w \models \Diamond \alpha \iff \sigma_w = \emptyset$, *or there exists an $S \in \sigma_w$ and a $v \in S \cap R(w)$ such that $\mathfrak{M}, v \models \alpha.$*

Note that the first two components (W, σ) of an SAS (W, σ, R) determine a neighborhood frame (W, ρ_σ), where $\rho_\sigma(w) := \sigma_w$ for each $w \in W$. As the semantics of basic modal logic based on neighborhood frames are well-studied in the literature (cf. e.g., [9,18]), a natural question arises: is there any connection between the proposed semantics and the existing semantics for neighborhood frames? To address this, observe that the satisfiability condition for the operator $\Box$ in an SAS $(W, \sigma, W \times W)$ reduces to the following:

$$\mathfrak{M}, w \models \Box \alpha \iff \sigma_w \neq \emptyset \ \& \text{ for each } S \in \sigma_w, \text{ and for each } u \in S, \mathfrak{M}, u \models \alpha. \tag{1}$$

From (1), it is evident that in the class of SASs where the relation R is the universal relation on the carrier set, the proposed semantics for the operator closely resemble the satisfiability conditions of the outer modality [o] [15] and the modal operator ⊡ [3], though they are not identical. Specifically, in the satisfiability condition given in (1), there is an additional constraint that $\rho_\sigma(w) \neq \emptyset$, which is absent in the satisfiability conditions for the operators [o] and ⊡.

Furthermore, in the class of models based on SASs (W, σ, R) with $\sigma = \{W\}$, we obtain the following:

$$\mathfrak{M}, w \models \Box\alpha \iff \text{for each } u \in R(w), \mathfrak{M}, u \models \alpha. \qquad (2)$$

From (2), it follows that one can identify the Kripke frame (W, R) with the SAS $(W, \{W\}, R)$. Under this identification, the proposed semantics reduce to the standard relational Kripke semantics of the basic modal language.

The notion of validity for wffs is defined in the standard way. A wff α is said to be valid in a model $\mathfrak{M}$, denoted as $\mathfrak{M} \models \alpha$, if $[\![\alpha]\!]_\mathfrak{M} = W$. A wff is termed valid in an SAS $\mathfrak{F}$, denoted by $\mathfrak{F} \models \alpha$, if $\mathfrak{M} \models \alpha$ for all models $\mathfrak{M}$ based on $\mathfrak{F}$. Similarly, a wff α is said to be valid in a class C of SASs (notation: $\mathsf{C} \models \alpha$) if $\mathfrak{F} \models \alpha$ for all $\mathfrak{F} \in \mathsf{C}$.

Let $\Gamma \cup \{\alpha\}$ be a set of wffs. We write $\mathfrak{F} \models \Gamma$ if $\mathfrak{F} \models \alpha$ for all $\alpha \in \Gamma$. For a class C of SASs, α is said to be a semantic consequence of Γ (denoted by $\Gamma \models_\mathsf{C} \alpha$) if, for all $\mathfrak{F} \in \mathsf{C}$, whenever $\mathfrak{F} \models \Gamma$, it follows that $\mathfrak{F} \models \alpha$.

The following proposition shows that the properties of the approximation operator, as listed in Proposition 1, translate into valid wffs relative to the class S_{rs}. This result also highlights how the basic modal language can be employed to express properties of the operator L. Let S be the class of all SASs.

Proposition 3. *The following hold.*

(C) $\mathsf{S} \models \Box\alpha \wedge \Box\beta \to \Box(\alpha \wedge \beta)$.
(T_0) $\mathsf{S} \models \Diamond\alpha \to \Diamond(\alpha \wedge \Box\top)$.
(T) $\mathsf{S}_{rs} \models \Box\alpha \to \alpha$.
(B_0) $\mathsf{S}_{rs} \models \Box\top \wedge \alpha \to \Box\Diamond\alpha$.
(4_0) $\mathsf{S} \models \Box\top \to \Box\Box\top$.

Proof. Follows from Proposition 1. □

4 Modal System for the Classes S_{rs} and S_e

As mentioned in Sect. 2, we focus on the classes S_{rs} and S_e and present a modal system that is sound and complete with respect to these classes. We consider the monotonic modal logic $\mathrm{MCTT}_0\mathrm{B}_0 4_0$, which is defined by the following axioms and inference rules:

All axioms of classical propositional logic	(Taut)
$\Box\alpha \land \Box\beta \to \Box(\alpha \land \beta)$	(C)
$\Box\alpha \to \alpha$	(T)
$\Diamond\alpha \to \Diamond(\alpha \land \Box\top)$	(T_0)
$\Box\top \land \alpha \to \Box\Diamond\alpha$	(B_0)
$\Box\top \to \Box\Box\top$	(4_0)
From α and $\alpha \to \beta$, infer β	(MP)
From $\alpha \to \beta$, infer $\Box\alpha \to \Box\beta$	(RM)

For brevity, throughout this article, we will use the symbol Λ to refer to the logic $MCTT_0B_04_0$. We denote by $\vdash_\Lambda \alpha$ that the wff α is a theorem of the system Λ, where the notion of a theorem in a modal system is defined in the standard manner (cf. e.g. [2]). A wff α is said to be deducible from a set Γ (denoted by $\Gamma \vdash_\Lambda \alpha$) if α is a theorem of the system Λ, or there exist wffs $\alpha_1, \alpha_2, \ldots, \alpha_k$ in Γ such that $\vdash_\Lambda (\alpha_1 \land \alpha_2 \land \cdots \land \alpha_k) \to \alpha$.

We note the following soundness theorem.

Theorem 1 (Soundness Theorem). *Let $\Gamma \cup \{\alpha\}$ be a set of wffs. If $\Gamma \vdash_\Lambda \alpha$, then $\Gamma \models_{S_{rs}} \alpha$.*

Proof. The proof follows from Proposition 3 through a standard approach. The details are omitted. □

4.1 Completeness Theorem

The completeness theorem will be proved using the step-by-step technique described in [2], with the necessary modifications to accommodate our specific requirements. Let W^Λ denote the set of all maximal Λ-consistent sets. We start with the following well-known result.

Lemma 1 (Lindenbaum's Lemma). *For every Λ-consistent set of wffs Γ, there exists a maximal Λ-consistent set Γ^+ containing Γ.*

The following existence theorem will be needed to obtain the completeness theorem.

Lemma 2 (Existence Lemma). *Let Γ be a maximal Λ-consistent set such that $\Box\top \land \Diamond\alpha \in \Gamma$. Then, there exists a maximal Λ-consistent set Δ such that $\{\beta : \Box\beta \in \Gamma\} \cup \{\alpha, \Box\top\} \subseteq \Delta$.*

Proof. Let $\Delta^* = \{\beta : \Box\beta \in \Gamma\} \cup \{\alpha, \Box\top\}$. We claim that Δ^* is an Λ-consistent set. Suppose not; then either $\vdash_\Lambda \Box\top \to \neg\alpha$, or there exist $\beta_1, \beta_2, \ldots, \beta_k$ in Δ^* such that $\vdash_\Lambda \beta_1 \land \beta_2 \land \cdots \land \beta_k \to \neg(\alpha \land \Box\top)$.

If $\vdash_\Lambda \Box\top \to \neg\alpha$, then we have $\vdash_\Lambda \Box\Box\top \to \Box\neg\alpha$. By applying Axiom 4_0, it follows that $\vdash_\Lambda \Box\top \to \Box\neg\alpha$, which implies $\Box\neg\alpha \in \Gamma$. This leads to a contradiction since $\Diamond\alpha \in \Gamma$.

Next, suppose there exist $\beta_1, \beta_2, \ldots, \beta_k \in \Delta$ such that $\vdash_\Lambda \beta_1 \wedge \beta_2 \wedge \cdots \wedge \beta_k \to \neg(\alpha \wedge \Box\top)$. This gives

$$\vdash_\Lambda \Box(\beta_1 \wedge \beta_2 \wedge \cdots \wedge \beta_k) \to \Box\neg(\alpha \wedge \Box\top)$$
$$\implies \vdash_\Lambda \Box\beta_1 \wedge \Box\beta_2 \wedge \cdots \wedge \Box\beta_k \to \neg\Diamond(\alpha \wedge \Box\top)$$
$$\implies \neg\Diamond(\alpha \wedge \Box\top) \in \Gamma$$
$$\implies \neg\Diamond\alpha \in \Gamma \text{ (using axiom } \mathsf{T}_0),$$

which is a contradiction. Therefore, we obtain Δ^* as a Λ-consistent set. By Lindenbaum's Lemma, there exists a maximal Λ-consistent set Δ such that $\Delta^* \subseteq \Delta$. □

The notion of a *network* plays a crucial role in the step-by-step technique. We define it as follows:

Definition 5 (Network). *A Λ-network is defined as a tuple $N = (W, \sigma, R, \chi)$, where*

- (W, σ, R) *is an SAS and*
- $\chi : W \longrightarrow W^\Lambda$.

Definition 6 (Coherent Network). *A Λ-network $N = (W, \sigma, R, \chi)$ is said to be coherent if it satisfies the following conditions:*

(C1) *If $\Box\beta \in \chi(w)$, then for each $S \in \sigma_w$ and for each $v \in S \cap R(w)$, we have $\beta \in \chi(v)$.*
(C2) *If $\neg\Box\top \in \chi(w)$, then $\sigma_w = \emptyset$.*

Definition 7 (Saturated Network). *A Λ-network $N = (W, \sigma, R, \chi)$ is said to be saturated if it satisfies the following conditions:*

(S1) *If $\Box\top \wedge \Diamond\beta \in \chi(w)$, then there exists a $T \in \sigma_w$ and a $v \in T \cap R(w)$ such that $\beta \in \chi(v)$.*
(S2) *If $\Box\beta \in \chi(w)$, then $\sigma_w \neq \emptyset$.*

The absence of properties (S1) and (S2) in a network lead to certain defects, which are defined as follows:

Definition 8 (Defect). *Let $N = (W, \sigma, R, \chi)$ be a Λ-network.*

(D1-Defect) *A tuple $(\Diamond\beta, w) \in \mathcal{L} \times W$ is called a D1-defect of N if $\Box\top \wedge \Diamond\beta \in \chi(w)$, and for each $T \in \sigma_w$ and each $v \in T \cap R(w)$, we have $\beta \notin \chi(v)$.*
(D2-Defect) *A tuple $(\Box\beta, w) \in \mathcal{L} \times W$ is called a D2-defect of N if $\Box\beta \in \chi(w)$, and $\sigma_w = \emptyset$.*

Definition 9 (Perfect Network). *If a Λ-network is both coherent and saturated, then it is called a* perfect network.

A Λ-network lead us to the following model.

Definition 10 (Induced Model). *Given a Λ-network $N = (W, \sigma, R, \chi)$, we obtain a model $\mathfrak{M}_N = (\mathfrak{F}_N, V_N)$, where*

- $\mathfrak{F}_N = (W, \sigma, R)$, *and*
- $V_N(p) = \{w \in W : p \in \chi(w)\}$.

The definition of a coherent and saturated network provided above differs slightly from those commonly found in the literature. Nevertheless, the proposed notion of a perfect network is sufficiently robust to yield the following truth lemma.

Theorem 2 (Truth Lemma). *Let $N = (W, \sigma, R, \chi)$ be a perfect network. Then, for each wff α, we have*

$$\mathfrak{M}_N, w \models \alpha \text{ if only if } \alpha \in \chi(w).$$

Proof. The proof proceeds by induction on the complexity of the wff α. We will explicitly prove the case where α is of the form $\Box \beta$. The remaining cases can be handled similarly.

Assume $\mathfrak{M}_N, w \models \Box \beta$. This implies that $\sigma_w \neq \emptyset$, and for each $S \in \sigma_w$ and each $v \in S \cap R(w)$, we have $\mathfrak{M}_N, v \models \beta$. By the induction hypothesis, this further implies that $\beta \in \chi(v)$. Utilizing the (C2) property of coherence, we obtain $\Box \top \in \chi(w)$. Applying the (S1) property of saturation, it follows that $\Diamond \neg \beta \notin \chi(w)$. Hence, $\Box \beta \in \chi(w)$.

Conversely, suppose $\Box \beta \in \chi(w)$. By the (S2) property, we conclude that $\sigma_w \neq \emptyset$. Additionally, by (C1) and the induction hypothesis, it follows that $\mathfrak{M}_N, v \models \beta$ for each $S \in \sigma_w$ and each $v \in S \cap R(w)$. Thus, we deduce that $\mathfrak{M}_N, w \models \Box \beta$. This completes the proof. $\square$

Due to the truth lemma, the task of proving the desired completeness theorem now reduces to constructing a suitable perfect network. This is achieved by formulating repair lemmas that systematically eliminate defects. Before presenting the repair lemmas, we first need to introduce the notion of a network extension.

Definition 11. *Let $N = (W, \sigma, R, \chi)$ and $N' = (W', \sigma', R', \chi')$ be two Λ-networks. We say that N' extends N, denoted as $N' \triangleright N$, if*

- $W \subseteq W'$,
- $\sigma_w \subseteq \sigma'_w$ for all $w \in W$,
- $R'|_{W \times W} = R$,
- $\chi'|_W = \chi$.

Proposition 4. Let $N = (W, \sigma, R, \chi)$ and $N' = (W', \sigma', R', \chi')$ be two Λ-networks such that $N' \rhd N$, and $w \in W$.

1. If $(\Diamond\alpha, w)$ is not a D1-defect of N, then $(\Diamond\alpha, w)$ is also not a defect of N'.
2. If $(\Box\alpha, w)$ is not a D2-defect of N, then $(\Box\alpha, w)$ is also not a defect of N'.

Proof. We provide the proof of Item 1. The proof of Item 2 follows in a similar manner.

Assume, for the sake of contradiction, that $(\Diamond\alpha, w)$ is a defect in N'. This would mean that $\Box\top \wedge \Diamond\alpha \in \chi'(w)$, and for each $T \in \sigma'_w$ and each $v \in T \cap R'(w)$, we have $\alpha \notin \chi'(v)$. Since $N' \rhd N$ and $w \in W$, it follows that $\Box\top \wedge \Diamond\alpha \in \chi(w)$, $\sigma_w \subseteq \sigma'_w$, and $R'|_{W \times W} = R$. Therefore, for each $T \in \sigma_w$ and for each $v \in T \cap R(w)$, we must have $\alpha \notin \chi(v)$. This contradicts the assumption that $(\Diamond\alpha, w)$ is not a D1-defect of N. $\square$

Now we are in a position to state the repair lemmas.

Theorem 3 (Repair Lemma for D1-defect). Let $(\Diamond\alpha, w)$ be a defect of a coherent Λ-network $N = (W, \sigma, R, \chi)$, where $(W, \sigma, R) \in \mathsf{S_e}$ and W is finite. Then, there exists a Λ-network $N' = (W', \sigma', R', \chi')$ such that (a) W' is finite, (b) $(W', \sigma', R') \in \mathsf{S_e}$, (c) $N' \rhd N$, (d) $(\Diamond\alpha, w)$ is not a defect of N', and (e) N' is coherent.

Proof. Since $(\Diamond\alpha, w)$ is a defect of N, it follows that $\Box\top \wedge \Diamond\alpha \in \chi(w)$, and for each $T \in \sigma_w$ and each $v \in T \cap R(w)$, we have $\alpha \notin \chi(v)$. Applying Existence Lemma 2, we obtain a maximal Λ-consistent set Δ such that

$$\{\beta : \Box\beta \in \chi(w)\} \cup \{\alpha, \Box\top\} \subseteq \Delta. \tag{3}$$

Let u be a new element, that is $W \cap \{u\} = \emptyset$. For a relation P, let us use P^+ to denote its transitive closure. Consider the network $N' = (W', \sigma', R', \chi')$, where

$$W' := W \cup \{u\}$$

$$\sigma'_x := \begin{cases} \sigma_x, & \text{if } x \in W \setminus \{w\} \\ \sigma_w \cup \{\{u, w\}\}, & \text{if } x = w \\ \{\{u, w\}\}, & \text{if } x = u \end{cases}$$

$$\sigma' = \bigcup_{x \in W'} \sigma'_x$$

$$R' := (R \cup \{(w, u), (u, w), (u, u)\})^+$$

$$\chi'(x) := \begin{cases} \chi(x), & \text{if } x \in W \\ \Delta, & \text{if } x = u. \end{cases}$$

Note that (a) W' is finite, (b) $(W', \sigma', R') \in \mathsf{S_e}$, and (c) $N' \rhd N$. Further, $(\Diamond\alpha, w)$ is not a defect of N' as $\{u, w\} \in \sigma'_w$, and $u \in \{u, w\} \cap R'(w)$ such that $\alpha \in \chi'(u) = \Delta$. So, it remains to show that N' is coherent.

(C1): Let $\Box\beta \in \chi'(x)$, and $S \in \sigma'_x$ and $v \in S \cap R'(x)$. We need to show that $\beta \in \chi'(v)$.
If $x \in W \setminus \{w\}$, then coherency of N gives $\beta \in \chi(v) = \chi'(v)$. So, let us consider the case when $x \in \{w, u\}$.
If $x = w$, and $S \in \sigma_w$, then we must have $v \in W$, and $v \in S \cap R(w)$. Therefore, coherency of N again gives us $\beta \in \chi(v) = \chi'(v)$.
Next, suppose $x = w$, $S = \{u, w\}$, and $v \in \{u, w\}$. If $v = w$, then we obtain $\beta \in \chi'(v) = \chi(w)$ as $\Box\beta \in \chi(w)$, and $\vdash_\Lambda \Box\beta \to \beta$ (Axiom T). If $v = u$, then we obtain $\beta \in \chi'(u) = \Delta$ using (3).
Finally, suppose $x = u$, then we must have $S = \{u, w\}$ and $v \in \{u, w\}$. Therefore, we need to show that $\beta \in \chi'(w) \cap \chi'(u)$. We obtain $\beta \in \chi'(u)$ as $\Box\beta \in \chi'(u)$, and $\vdash_\Lambda \Box\beta \to \beta$ (Axiom T). Next, if possible, suppose $\beta \notin \chi'(w)$, that is, $\neg\beta \in \chi'(w)$. Now, since $\vdash_\Lambda \Box\top \wedge \neg\beta \to \Box\Diamond\neg\beta$ (Axiom B_0), and $\Box\top \wedge \neg\beta \in \chi'(w)$, it follows that $\Box\Diamond\neg\beta \in \chi'(w) = \chi(w)$. Using (3), we obtain $\Diamond\neg\beta \in \Delta = \chi'(u)$. This is not possible as we have taken $\Box\beta \in \chi'(u)$.
(C2): Let $\neg\Box\top \in \chi'(x)$, and we show that $\sigma'_x = \emptyset$. Note that $x \notin \{w, u\}$ as we have $\Box\top \in \chi'(w) \cap \chi'(u)$. So, let $x \in W \setminus \{w\}$. Then, we obtain $\neg\Box\top \in \chi(x)$, and hence, using the coherency of N, we get $\sigma'_x = \sigma_x = \emptyset$. □

Theorem 4 (Repair Lemma for D2-defect). Let $(\Box\alpha, w)$ be a defect of a coherent Λ-network $N = (W, \sigma, R, \chi)$, where $(W, \sigma, R) \in \mathsf{S_e}$ and W is finite. Then, there is a network $N' = (W', \sigma', R', \chi')$ such that (a) W' is finite, (b) $(W', \sigma', R') \in \mathsf{S_e}$, (c) $N' \triangleright N$, (d) $(\Box\alpha, w)$ is not a defect of N', and (e) N' is coherent.

Proof. Since $(\Box\alpha, w)$ is a defect of N, we obtain $\Box\alpha \in \chi(w)$ and $\sigma_w = \emptyset$. Consider the Λ-network $N' = (W', \sigma', R', \chi')$, where

$$W' := W,$$

$$\sigma'_x := \begin{cases} \sigma_x, & \text{if } x \in W \setminus \{w\} \\ \{\{w\}\}, & \text{if } x = w \end{cases}$$

$$\sigma' = \bigcup_{x \in W} \sigma'_x$$

$$R' := R$$

$$\chi'(x) := \chi(x) \text{ for each } x \in W.$$

It is not difficult to observe that (a) W' is finite, (b) $(W', \sigma', R') \in \mathsf{S_e}$, and (c) $N' \triangleright N$. Further, $(\Diamond\alpha, w)$ is not a defect of N' as $\sigma'_w \neq \emptyset$. So, it remains to show that N' is coherent.
(C1): Let $\Box\beta \in \chi'(x)$, and $S \in \sigma'_x$ and $v \in S \cap R'(x)$. We need to prove that $\beta \in \chi'(v)$. If $x \neq w$, then we obtain the result using the coherency of N. So, let us consider the case when $x = w$. In this case we obtain $S = \{w\}$ and $v = w$. Since $\Box\beta \in \chi'(w)$, and $\vdash_\Lambda \Box\beta \to \beta$ (Axiom T), we get $\beta \in \chi'(w)$ as required.
(C2): Let $\neg\Box\top \in \chi'(x)$, and we show that $\sigma'_x = \emptyset$. Since $\Box\alpha \in \chi'(w)$, using $\vdash_\Lambda \Box\alpha \to \Box\top$, we obtain $\Box\top \in \chi'(w)$. Thus, $x \neq w$. Now, using the coherency of N, we obtain $\sigma'_x = \emptyset$. □

Theorem 5 (Completeness Theorem). *Let $\Gamma \cup \{\alpha\}$ be a set of wffs. If $\Gamma \models_{\mathsf{S_e}} \alpha$, then $\Gamma \vdash_\Lambda \alpha$.*

Proof. If possible, let $\Gamma \nvdash_\Lambda \alpha$. Then, $\Gamma \cup \{\neg \alpha\}$ is Λ-consistent. By Lindenbaum's Lemma, there exists a maximal Λ-consistent set Δ such that $\Gamma \cup \{\neg \alpha\} \subseteq \Delta$. Consider a Λ-network $N = (W, \sigma, R, \chi)$, where $W = \{w\}$, $R = \{(w, w)\}$ and $\chi(w) = \Delta$, and

$$\sigma = \begin{cases} \{\{w\}\}, & \text{if } \Box\beta \in \chi(w) \text{ for some wff } \beta \\ \emptyset, & \text{otherwise.} \end{cases}$$

Note that the network N is coherent and $(W, \sigma, R) \in \mathsf{S_e}$. The network N may have some defects. These defects can be removed step-by-step by the repeated applications of Repair Lemmas (Theorems 3 and 4), leading us to a perfect network $N' := (W', \sigma', R', \chi')$ such that $N' \triangleright N$ and $(W', \sigma', R') \in \mathsf{S_e}$. We omit the details as it is standard. Finally, using Truth Lemma 2, we obtain $\mathfrak{M}'_N, w \models \beta$ for all $\beta \in \Gamma \cup \{\neg \alpha\}$, a contradiction. Thus, we obtain $\Gamma \vdash_\Lambda \alpha$. □

Combining Theorems 1 and 5, we obtain the following.

Theorem 6 (Soundness & Completeness). *Let $\Gamma \cup \{\alpha\}$ be a set of wffs. Then, we have the following*

$$\Gamma \vdash_\Lambda \alpha \iff \Gamma \models_{\mathsf{S_{rs}}} \alpha \iff \Gamma \models_{\mathsf{S_e}} \alpha.$$

5 Conclusions

In this article, the semantics of the basic modal language based on the operator L^{n*} is presented, and a sound and complete deductive system for the classes $\mathsf{S_{rs}}$ and $\mathsf{S_e}$ of SASs is obtained. Although the study has been restricted to these specific classes, the approach can be extended to encompass many other classes of SASs. The extension of this work, along with other significant topics such as invariance and definability related to the proposed semantics, is left for an extended version of the article.

Acknowledgments. This work has been supported by the National Board for Higher Mathematics (NBHM) India, Research Grant No. 02011/13/2023 NBHM (R.P)/R&D II/5863.

References

1. Banerjee, M., Khan, M.A.: Propositional logics from rough set theory. In: Peters, J.F., Skowron, A., Düntsch, I., Grzymała-Busse, J., Orłowska, E., Polkowski, L. (eds.) Transactions on Rough Sets VI. LNCS, vol. 4374, pp. 1–25. Springer, Heidelberg (2007). https://doi.org/10.1007/978-3-540-71200-8_1

2. Blackburn, P., de Rijke, M., Venema, Y.: Modal Logic. Cambridge University Press, Cambridge (2001)
3. Brown, M.A.: Action and ability. J. Philos. Log. **19**(1), 95–114 (1990)
4. Demri, S., Orłowska, E.: Incomplete Information: Structure, Inference, Complexity. Springer, Heidelberg (2002)
5. Ding, Y., Liu, J., Wang, Y.: Someone knows that local reasoning on hypergraphs is a weakly aggregative modal logic. Synthese **201**(2), 46 (2023)
6. Farinas Del Cerro, L., Orłowska, E.: DAL - a logic for data analysis. Theor. Comput. Sci. **36**, 251–264 (1985)
7. Gargov, G.: Two completeness theorems in the logic for data analysis. Technical Report 581, Institute of Computer Science, Polish Academy of Sciences, Warsaw (1986)
8. Grzymała-Busse, J.W., Rzasa, W.: Local and global approximations for incomplete data. Trans. Rough Sets **VIII**, 21–34 (2008)
9. Hansen, H.H.: Monotonic modal logics. Master's thesis, Institute for Logic, Language and Computation (2003)
10. Khan, M.A., Banerjee, M.: Formal reasoning with rough sets in multiple-source approximation systems. Int. J. Approx. Reason. **49**(2), 466–477 (2008)
11. Khan, M.A., Banerjee, M.: Logics for information systems and their dynamic extensions. ACM Trans. Comput. Log. **12**(4), 29 (2011)
12. Khan, M.A., Patel, V.S.: A formal study of a generalized rough set model based on relative approximations. In: Nguyen, H.S., Ha, Q.T., Li, T., Przybyła-Kasperek, M. (eds.) IJCRS 2018. LNCS (LNAI), vol. 11103, pp. 502–510. Springer, Cham (2018). https://doi.org/10.1007/978-3-319-99368-3_39
13. Khan, M.A., Patel, V.S.: A formal study of a generalized rough set model based on subset approximation structure. Int. J. Approx. Reason. **140**, 52–74 (2022)
14. Khan, M.A., Ranjan, Talukdar, A.: A study of modal logic with semantics based on rough set theory. J. Appl. Non-Classical Logics **34**(2-3), 223–247 (2024)
15. Lewis, D.: Counterfactuals. Blackwell Publishers, Oxford (1973)
16. Lin, T.Y., Yao, Y.Y.: Neighborhoods system: measure, probability and belief functions. In: Proceedings of the 4th International Workshop on Rough Sets and Fuzzy Sets and Machine Discovery, pp. 202–207 (1996)
17. Orłowska, E.: Kripke semantics for knowledge representation logics. Studia Logica **XLIX**, 255–272 (1990)
18. Pacuit, E.: Neighborhood Semantics for Modal Logic. Springer, Heidelberg (2017)
19. Patel, V.S., Khan, M.A., Chakraborty, M.K.: Modal systems for covering semantics and boundary operator. Int. J. Approx. Reason. **135**, 110–126 (2021)
20. Pawlak, Z.: Rough Sets, Theoretical Aspects of Reasoning about Data. Kluwer Academic Publishers, Dordrecht (1991)
21. Samanta, P., Chakraborty, M.K.: Interface of rough set systems and modal logics: a survey. In: Peters, J.F., Skowron, A., Ślęzak, D., Nguyen, H.S., Bazan, J.G. (eds.) Transactions on Rough Sets XIX. LNCS, vol. 8988, pp. 114–137. Springer, Heidelberg (2015). https://doi.org/10.1007/978-3-662-47815-8_8
22. Ślęzak, D., Ziarko, W.: The investigation of the Bayesian rough set model. Int. J. Approx. Reason. **40**, 81–91 (2005)
23. Ziarko, W.: Variable precision rough set model. J. Comput. Syst. Sci. **46**, 39–59 (1993)

Modal and Intermediate Logics of Spiked Boolean Algebras

Benedikt Löwe[1,2(✉)] and Han Xiao[1]

[1] Fachbereich Mathematik, Universität Hamburg, Bundesstrasse 55, 20146 Hamburg, Germany
{benedikt.loewe,han.xiao}@uni-hamburg.de
[2] Churchill College, Lucy Cavendish College, St. Edmund's College & Department of Pure Mathematics and Mathematical Statistics, University of Cambridge, Storey's Way, Cambridge CB3 0DS, England

Abstract. We consider the class of finite spiked Boolean algebras introduced by Inamdar and show that the modal and intermediate logics associated to it are not finitely axiomatisable.

1 Introduction

One of the most interesting intermediate logics is *Medvedev logic* Med, the logic of Boolean algebras with their top element removed, or *topless Boolean algebras*. Maksimova, Skvortsov, and Shehtman proved in 1979 that Medvedev logic is not finitely axiomatisable [8]. In 2003, van Benthem, Bezhanishvili, and Gehrke introduced the logic Cheq of *chequered sets* which is a sublogic of Med and asked whether Cheq is finitely axiomatisable [10, in particular, p. 343]; this question remains open, but Fontaine proved in 2006 that Med is not finitely axiomatisable over Cheq [2,3].

In a curious twist of events, the class of topless Boolean algebras played a role in the *Modal Logic of Forcing*. In this area of research based on work by Hamkins and the first author,[1] the goal is to determine for a given class $\mathcal{C}$ of forcing notions which modal statements are provable in ZFC if the modality $\Box\varphi$ is interpreted as "in all generic extensions by a forcing in $\mathcal{C}$, φ holds". The most important remaining open question in this area of research is the determination of the *modal logic of c.c.c. forcing*, i.e., the case where $\mathcal{C}$ is the class of all partial orders with the countable chain condition. Hamkins, Leibman, and the first author proved that the modal logic of ω_1-preserving forcing is contained in S4.tBA which is the smallest modal companion of Med [5, Theorem 36] and conjectured that the modal logic of c.c.c. forcing is also contained in S4.tBA.

In 2013, Inamdar introduced a variant of the class of topless Boolean algebras, the class of *spiked Boolean algebras* and proved that the modal logic of c.c.c. forcing is contained in S4.sBA, the smallest modal companion of the logic

---

[1] The original paper of the area is [6]. Other results on modal logics of forcing classes can be found in [4,5,7,12].

of spiked Boolean algebras [7, Theorem 150]. Recently, the second author developed ideas of Inamdar to provide a technique to prove that the modal logic of c.c.c. forcing is included in S4.FPFA, the smallest modal companion of Cheq; cf. [11, Chapter 3]. Since S4.FPFA ⊆ S4.tBA (cf. § 2.7), this technique yields a proof of the conjecture by Hamkins, Leibman, and the first author under an additional technical assumption that is expected to be removed in due time [11, Theorem 3.6.3].

In this paper, we shall have a closer look at Inamdar's class of algebras and its associated logics: *Inamdar logic* Inam, the intermediate logic of spiked Boolean algebras and its smallest and largest modal companions S4.sBA and Grz.sBA, respectively. We shall show the analogues of the results by Maksimova, Skvortsov, Shehtman, and Fontaine for these logics.

2 Definitions

Throughout this paper, we shall assume that the reader knows the basic theory of modal logic and of intermediate logics, i.e., propositional logics between intuitionistic and classical logic. We write IPC for *intuitionistic propositional calculus* and S4 and Grz for the well-known modal systems. In the following, we shall consider partial orders as *intuitionistic Kripke frames*; all of our partial orders will be finite.

2.1 Thickenings

If **P** is a finite partial order, we say that **P**$^\bullet$ is a *thickening of* **P** if **P**$^\bullet$ is a finite partial pre-order (i.e., ≤ is reflexive and transitive, but not necessarily anti-symmetric), ≡ is the induced equivalence relation (i.e., $x \equiv y$ if and only if $x \leq y$ and $y \leq x$) and **P** = **P**$^\bullet$/≡. If $\mathcal{C}$ is any class of finite partial orders, we let $\mathcal{C}^\bullet$ be the class of thickenings of elements of $\mathcal{C}$; its elements are usually referred to with the prefix "pre-": e.g., if the elements of $\mathcal{C}$ are called "examples", then the elements of $\mathcal{C}^\bullet$ are called "pre-examples".

2.2 Logics

If **P** is a finite partial (pre-)order, we write IL(**P**) for the propositional logic of all formulas valid on the frame **P** and ML(**P**) for the modal logic of all formulas valid on the frame **P**; if $\Lambda \subseteq$ IL(**P**) or $\Lambda \subseteq$ ML(**P**), we call **P** a Λ-*frame*. If $\mathcal{C}$ is a class of finite partial (pre-)orders, we write IL($\mathcal{C}$) and ML($\mathcal{C}$) for the intermediate or modal logic consisting of all modal formulas valid on every element of $\mathcal{C}$, respectively; in this case, we say that the class $\mathcal{C}$ *characterises* the logic.

2.3 Terminology for Partial Orders

Let **P** = $(P, \leq)$ be a finite partial order; as usual, we write $p < q$ if $p \leq q$ and $p \neq q$. An element of **P** is called *maximal* if there are no strictly larger elements in **P**;

it is called a *top* if everything is below it. Clearly, a top is unique. We say that **P** *has a top* if there is a top in **P**. If $p < q$, we say that q is an *immediate successor* of p if there is no r such that $p < r < q$. An element of a partial order **P** is called *coatom* if is not maximal and all of its immediate successors are maximal.

A subset $U \subseteq P$ is called an *upset* if it upwards closed under $\leq$ and a *downset* if it is downwards closed under $\leq$; sets of the form $u{\uparrow} := \{p \in P\,;\, u \leq p\}$ and $u{\downarrow} := \{p \in P\,;\, u \geq p\}$ for some $u \in P$ are upsets and downsets, respectively; we call sets of the form $u{\uparrow}$ or $u{\downarrow}$ *rooted upsets* or *rooted downsets*, respectively, and we call u the *root* of the set.

If **P** and **Q** are partial orders, we write $\mathbf{P} \oplus \mathbf{Q}$ for their *linear sum*: all elements of **Q** are strictly above all elements of **P**. We define iterated finite sums by recursion via $\mathbf{P} \times 1 := \mathbf{P}$ and $\mathbf{P} \times (n+1) := (\mathbf{P} \times n) \oplus \mathbf{P}$.

2.4 Morphisms

If **P** and **Q** are partial orders and $f : P \to Q$ is a map, we call f a *morphism* if for all $p_1, p_2 \in P$, if $p_1 \leq p_2$, then $f(p_1) \leq f(p_2)$ and if $f(p_1) \leq q$, then there is some $p_2 \geq p_1$ such that $f(p_2) = q$.

Theorem 1 (Folklore). *If Λ is a modal logic characterised by a class of finite frames $\mathcal{C}$, then the class of (finite) rooted Λ-frames consists of the morphic images of rooted upsets of elements of $\mathcal{C}$ (or, equivalently, the rooted upsets of morphic images of elements of $\mathcal{C}$).*

In particular, if $\mathcal{C}$ is closed under rooted upsets (i.e., any rooted upset of an element of $\mathcal{C}$ is in $\mathcal{C}$), then the (finite) rooted Λ-frames are precisely the morphic images of elements of $\mathcal{C}$.

If $p, q \in P$, we say that they are α-*related* in **P** if q is the only immediate successor of p. We say that they are β-*related* in **P** if they have the same immediate successors. A partial order **Q** is a *reduction* of **P** if **Q** is the quotient of **P** taken by identifying two elements of P that are either α-related or β-related. A finite sequence $\mathbf{P}_0, \ldots, \mathbf{P}_n$ is called a *reduction sequence from* **P** *to* **Q** if $\mathbf{P} = \mathbf{P}_0$, $\mathbf{Q} = \mathbf{P}_n$, and for each $0 \leq i < n$, we have that $\mathbf{P}_{i+1}$ is a reduction of $\mathbf{P}_i$.

Lemma 2 (Folklore; cf. [1, Lemma 3.1.7]) *If **P** and **Q** are finite partial orders, then **Q** is a morphic image of **P** if and only if there is a reduction sequence from **P** to **Q**.*

2.5 Modal Companions

The *Gödel translation* T is the map from propositional logic to modal logic defined recursively as follows.

$$T(\bot) := \bot, \qquad T(p) := \Box p,$$
$$T(\varphi \wedge \psi) := T(\varphi) \wedge T(\psi), \qquad T(\varphi \wedge \psi) := T(\varphi) \wedge T(\psi),$$
$$T(\neg \varphi) := \Box \neg T(\varphi), \text{ and} \qquad T(\varphi \to \psi) := \Box(T(\varphi) \to T(\psi)).$$

If Λ is an intermediate logic, we say that Λ' is a *modal companion to* Λ, if $\varphi \in \Lambda$ if and only if $T(\varphi) \in \Lambda'$. Modal companions are not unique; for each $\Lambda = \mathsf{IPC} + S$, there is a largest modal counterpart $\sigma(\Lambda) := \mathsf{Grz} + \{T(\varphi)\,;\, \varphi \in S\}$ and a smallest modal counterpart $\tau(\lambda) := \mathsf{S4} + \{T(\varphi)\,;\, \varphi \in S\}$ (cf. [9, Proposition 7]).

Theorem 3 (Esakia; cf. [9, Proposition 9]). *For any class $\mathcal{C}$ of finite partial orders, we have $\mathsf{ML}(\mathcal{C}) = \sigma(\mathsf{IL}(\mathcal{C}))$.*

Theorem 4 (Zakharyaschev; cf. [9, Proposition 10]). *For any class $\mathcal{C}$ of finite partial orders, we have $\mathsf{ML}(\mathcal{C}^\bullet) = \tau(\mathsf{IL}(\mathcal{C}))$.*

Theorem 5 (Maksimova; cf. [9, Corollary 8]). *Any intermediate logic Λ is finitely axiomatisable if and only if $\sigma(\mathsf{L})$ is finitely axiomatisable.*

As a consequence, if $\tau(\Lambda)$ is finitely axiomatisable, then $\sigma(\Lambda) = \mathsf{Grz} + \tau(\Lambda)$ is, and thus, by Theorem 5, Λ is finitely axiomatisable. As a consequence, it is enough to show that the intermediate logic Λ is not finitely axiomatisable in order to have that all three logics Λ, $\sigma(\Lambda)$, and $\tau(\Lambda)$ are not finitely axiomatisable.

2.6 Boolean Algebras and Finite Partial Functions

Finite Boolean algebras can be considered as power sets of a finite set C. We identify $\varnothing$ with the top, the singletons $\{c\}$ for $c \in C$ with the coatoms, and the set C with the root of the Boolean algebra. The number of coatoms determines the Boolean algebra uniquely up to isomorphism.

Proposition 6 (Maksimova, Skvortsov, & Shehtman; cf. [8, Lemma 4]). *A partial order $\mathbf{P}$ is the morphic image of a finite Boolean algebra if and only if $\mathbf{P}$ is a finite rooted frame with a top.*

We denote the class of finite rooted frames with a top (i.e., by Proposition 6, the class of morphic images of finite Boolean algebras) by $\mathcal{R}$.

The finite partial order that we obtain by removing the top is called a *topless Boolean algebra* (note that topless Boolean algebras are *not* Boolean algebras!). The class of topless Boolean algebras is denoted by $\mathcal{T}$.

If $\mathbf{P}$ is a finite Boolean algebra with n co-atoms $\{c_i\,;\, 1 \leq i \leq n\}$, we define a *spiked Boolean algebra* by adding n additional nodes $\{d_i\,;\, 1 \leq i \leq n\}$ such that for every $b \in P$, we have $b \leq d_i$ if and only if $b \leq c_i$ (again, spiked Boolean algebras are *not* Boolean algebras). The spiked Boolean algebras for the Boolean algebras with two, four and eight elements can be seen in Fig. 1. We write $\mathcal{S}$ for the class of spiked Boolean algebras.

Fix a natural number n and a subset $A \subseteq n$. Any function $f: A \to \{0,1\}$ is called a *finite partial function*. The collection of all finite partial functions is denoted by $\mathcal{F}$ ordered by set-theoretic inclusion and called the *finite partial function algebra*.

We can give an alternative description: let $\mathbf{F}$ be the *two-fork*, i.e., the partial order consisting of three elements $\{0, L, R\}$ such that 0 is the root and L and R are incomparable maximal elements. We define by recursion $\mathbf{F}_1 := \mathbf{F}$ and $\mathbf{F}_{n+1} := \mathbf{F}_n \times \mathbf{F}$ where $\times$ denotes the usual product on partial orders. It is easy to see that $\mathbf{F}_n$ is isomorphic to the algebra of finite partial functions on n.

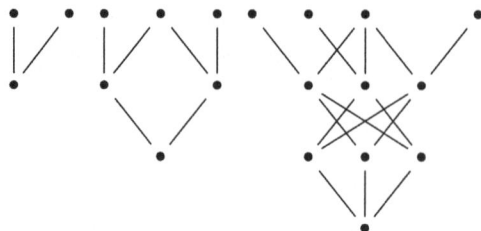

Fig. 1. The first three spiked Boolean algebras $\mathbf{S}_1$, $\mathbf{S}_2$, and $\mathbf{S}_3$.

2.7 The Logics Derived from the Classes of Algebras and Their Relationship

With the classes $\mathcal{T}$, $\mathcal{S}$, and $\mathcal{F}$, we obtain three intermediate logics and six modal logics as listed in the following table. As usual, we call $\mathsf{IL}(\mathcal{F})$ the *intermediate logic of chequered sets* Cheq and $\mathsf{IL}(\mathcal{T})$ *Medvedev logic* Med. We shall call Inam := $\mathsf{IL}(\mathcal{S})$ *Inamdar logic* in recognition of the fact that Inamdar introduced the class $\mathcal{S}$ in [7] (however, Inamdar did not consider the logic Inam, but only S4.sBA).

Since $\mathbf{F}_n$ is isomorphic to the algebra of finite partial functions on n, the set $\{\mathbf{F}_n\,;\,n\geq 1\}$ characterises Cheq, and, in particular, all morphic images of the $\mathbf{F}_n$ are Cheq-frames.

$\mathcal{C}$	$\mathsf{IL}(\mathcal{C})$	$\mathsf{ML}(\mathcal{C}) = \sigma(\mathsf{IL}(\mathcal{C}))$	$\mathsf{ML}(\mathcal{C}^\bullet) = \tau(\mathsf{IL}(\mathcal{C}))$
$\mathcal{F}$	Cheq	Grz.FPFA	S4.FPFA
$\mathcal{T}$	Med	Grz.tBA	S4.tBA
$\mathcal{S}$	Inam	Grz.sBA	S4.sBA

It is easy to see that Cheq $\subseteq$ Med $\cap$ Inam, Grz.FPFA $\subseteq$ Grz.tBA $\cap$ Grz.sBA, and S4.FPFA $\subseteq$ S4.tBA $\cap$ S4.sBA. Let p_0, p_1, p_2, q_0, and q_1 be propositional variables and define

$\varphi_0^2 := \Box(p_0 \wedge \neg p_1)$ and $\varphi_1^2 := \Box(p_1 \wedge \neg p_0)$,

$\varphi_0^3 := \Box(p_0 \wedge \neg p_1 \wedge \neg p_2)$, $\varphi_1^3 := \Box(p_1 \wedge \neg p_0 \wedge \neg p_2)$, and $\varphi_2^3 := \Box(p_2 \wedge \neg p_0 \wedge \neg p_1)$,

$\psi_\mathrm{s} := (\Diamond(\Box q_0 \wedge \Diamond\varphi_0^2 \wedge \Diamond\varphi_1^2) \wedge \Diamond(\Box q_1 \wedge \Diamond\varphi_0^2 \wedge \Diamond\varphi_1^2)) \to \Diamond\Box(q_0 \wedge q_1)$, and

$\psi_\mathrm{t} := (\Diamond\varphi_0^3 \wedge \Diamond\varphi_1^3 \wedge \Diamond\varphi_2^3) \to \Diamond(\Diamond\varphi_0^3 \wedge \Diamond\varphi_1^3 \wedge \Box\neg\varphi_2^3)$.

The formulas φ_j^i express that among i possible options, j has been made necessary and the others impossible; the formula ψ_s expresses that if it is possible to make both q_0 and q_1 separately necessarily true while still keeping the option of deciding between p_0 and p_1, then they can be made necessarily true together; finally, the formula ψ_t expresses that if you can decide between three options, you can exclude one of them but keep the other two options. The formula ψ_s is true in all spiked Boolean algebras but not in all topless Boolean algebras; the formula ψ_t is true in all topless Boolean algebras but not in all spiked Boolean algebras. As a consequence, we have that S4.FPFA$\subseteq$S4.tBA$\cap$S4.sBA$\subsetneq$S4.tBA and S4.FPFA$\subseteq$S4.tBA$\cap$S4.sBA$\subsetneq$S4.sBA. We do not know whether the first inclusions in these statements are strict, i.e., whether S4.FPFA $\subsetneq$ S4.tBA $\cap$ S4.sBA.

3 Inamdar Logic is not Finitely Axiomatisable

In this section, we shall show that all three logics associated with spiked Boolean algebras, i.e., Inam, S4.sBA, and Grz.sBA are not finitely axiomatisable. We shall use the general technique of [8], using the so-called *Chinese lanterns* defined by Maksimova, Skvortsov, and Shehtman. Given integers s and n, the *Chinese lantern* $\mathbf{L}_{s,n}$ is the set $(s+1\times 2)\cup(\{s+1\}\times n+1)$ ordered as described on the left in Fig. 2: if $\mathbf{C}_n$ is the partial order consisting of a root and exactly n immediate successors and $\mathbf{D}$ is the partial order consisting of two incomparable nodes (i.e., $d_0 \not\leq d_1$ and $d_1 \not\leq d_0$), then $\mathbf{L}_{s,n} := \mathbf{C}_n \oplus (\mathbf{D} \times s + 1)$. If we remove the node $(m, 1)$ from $\mathbf{L}_{s,n}$ we obtain the Chinese lantern $\mathbf{L}'_{s,n,m}$ depicted on the right in Fig. 2.

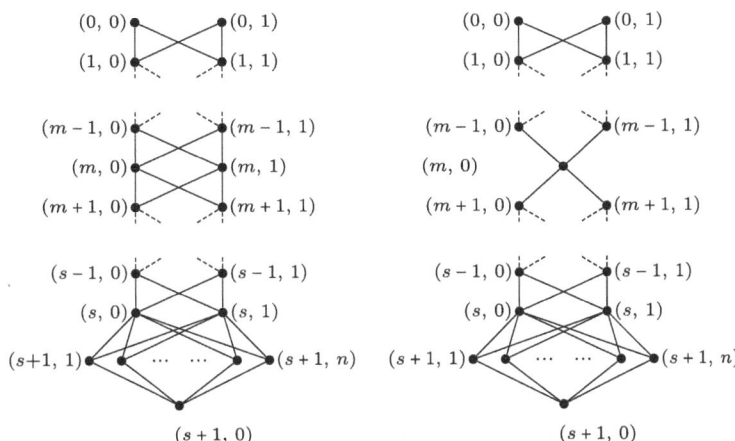

Fig. 2. The Chinese lanterns $\mathbf{L}_{s,n}$ (left) and $\mathbf{L}'_{s,n,m}$ (right).

As above, let $\mathbf{D}$ be the partial order consisting of two incomparable nodes. Let $\mathbf{P}$ be a finite partial order. We define the nth suspension of $\mathbf{P}$ by letting $\mathbf{P}^{(0)} := \mathbf{P}$ and $\mathbf{P}^{(n)} := \mathbf{P} \oplus (\mathbf{D} \times n)$ (for $n \geq 1$). The next lemma is [8, Corollary 4]; a proof can be found in [2, Claim 9].

Lemma 7 (Maksimova, Skvortsov, & Shehtman). *Assume $\mathbf{P}$ is a finite rooted frame with a top, then for any $n \geq 0$, the nth suspension $\mathbf{P}^{(n)}$ is a morphic image of a finite topless Boolean algebra.*

Using Lemma 7, it is easy to obtain the following result.

Lemma 8. Assume **P** is a finite rooted frame with a top, then for any $n \geq 0$, the nth suspension $\mathbf{P}^{(n)}$ is a morphic image of a finite spiked Boolean algebra.

Proof Sketch. There are two cases to consider. If $n = 0$, we just apply Lemma 7 to write **P** as a morphic image of a topless Boolean algebra. Consider the corresponding spiked Boolean algebra, i.e., it has additional pure spikes and a top element. Preserve the property of being a morphism by mapping all additional elements to the top element of **P**.

If $n = k+1$, then apply Lemma 7 to $\mathbf{P}^{(k)}$ to obtain a morphism from a topless Boolean algebra and extend this to $\mathbf{P}^{(k+1)} = \mathbf{P}^{(k)} \oplus \mathbf{D}$. Again, the corresponding spiked algebra has additional pure spikes and a top element. Preserve the property of being a morphism by mapping the top element to the left element of **D** and all of the spikes to the right element of **D**. □

Corollary 9. For every s, n, m, the Chinese lantern $\mathbf{L}'_{s,n,m}$ is an Inam-frame.

Proof. The downset $\mathbf{P} := (m, 0){\downarrow}$ is a finite rooted frame with a top and we have that $\mathbf{L}'_{s,n,m} = \mathbf{P}^{(m)}$. Therefore, by Lemma 8, $\mathbf{L}'_{s,n,m}$ is a morphic image of a spiked Boolean algebra and thus an Inam-frame. □

The following propositions are [8, Lemma 9] and [8, Corollary 3], respectively; for proofs, cf. [2, Proposition 11] and [2, Claim 7].[2]

Proposition 10 (Maksimova, Skvortsov, & Shehtman) For any formula φ with s variables, there is an $m \leq s$ such that

$$\mathbf{L}_{s,n} \vDash \varphi \text{ if and only if } \mathbf{L}'_{s,n,m} \vDash \varphi.$$

Proposition 11 (Maksimova, Skvortsov, & Shehtman) If the Chinese lantern $\mathbf{L}_{s,n}$ is the morphic image of a topless Boolean algebra, then $n < 2^{s+3}$.

The following lemma is the spiked analogue of Proposition 11.

Lemma 12. If the Chinese lantern $\mathbf{L}_{s,n}$ is the morphic image of a spiked Boolean algebra, then $n < 2^{s+2}$.

Proof. Let f be the morphism from a finite spiked Boolean algebra **S** to $\mathbf{L}_{s,n}$. Let C be the set of coatoms of **S**: we can think of **S** as the power set of C with additional spikes, i.e., $\varnothing$ is the top of the Boolean algebra corresponding to **S**, the singletons $\{c\}$ are the coatoms, and C is the root of **S**. Let **T** be the corresponding topless Boolean algebra, i.e., the set of non-empty subsets of C.

Clearly, both the top of **S** and its spikes have to be sent by f to maximal elements in $\mathbf{L}_{s,n}$, i.e., either $(0,0)$ or $(0,1)$. Without loss of generality, the top is

---

[2] Fontaine only proves this for the special cases $\mathbf{L}_{s,2^{s+3}}$ and $\mathbf{L}'_{s,2^{s+3},i}$, but her proof does not depend on this.

mapped to $(0,0)$. This means that at least one spike is mapped to $(0,1)$. Note furthermore that any element of **S** that is not a spike cannot be mapped to $(0,1)$ (since it is below $\varnothing$). We consider two cases.

Case 1. All spikes are mapped by f to $(0,1)$. Note that the part of $\mathbf{L}_{s,n}$ that is strictly below $(0,1)$ is just $\mathbf{L}_{s-1,n}$. This means that every coatom of **S** has a successor mapped to $(0,0)$ and a successor mapped to $(0,1)$, so it cannot be mapped to either and therefore has to be mapped to an element of $\mathbf{L}_{s-1,n}$. Thus, $f\!\restriction\!\mathbf{T}$ is a morphism from **T** onto $\mathbf{L}_{s-1,n}$, so by Proposition 11, we have that $n < 2^{(s-1)+3} = 2^{s+2}$.

Case 2. Some spikes are mapped by f to $(0,0)$ and some to $(0,1)$. In this case, let C_0 and C_1 be the sets of coatoms whose corresponding spikes are mapped to $(0,0)$ and $(0,1)$, respectively. Both sets are non-empty in this case. All coatoms in C_0 have no successor mapped to $(0,1)$, so they must be mapped by f to $(0,0)$ as well; similarly, all subsets of C_0 must be mapped to $(0,0)$.

For any $X \subseteq C$, let $X_0 := X \cap C_0$ and $X_1 := X \cap C_1$. Note that if $X_1 \neq \varnothing$, then there is some $c \in X_1 = C_1 \cap X$, so X_1 lies below the spike associated with c which is mapped to $(0,1)$. Thus $f(X_1) \neq (0,0)$. Note that we observed before that $f(X_1) \neq (0,1)$ for any $X \subseteq C$.

Claim. For each $X \subseteq C$, we have $f(X) = f(X_1)$.

[We can prove the claim by induction on the size of X. The case of $|X| = 0$, i.e., $X = X_1 = \varnothing$ is trivial. We deal with two special cases first:

(i) If $f(X_1) = (s+1, 0)$, i.e., the bottom element of the Chinese lantern, then $f(X) \leq f(X_1)$ and thus $f(X) = (s+1, 0) = f(X_1)$.
(ii) If $X_1 = \varnothing$, then $X \subseteq C_0$, so $f(X) = f(X_1) = f(\varnothing) = (0,0)$ by the above remark.

Therefore, without loss of generality, we can assume from now on that $f(X_1)$ is neither $(0,0)$, $(0,1)$, nor $(s+1, 0)$. Suppose that we have $f(X) < f(X_1)$ for some X. Because of our assumption, we know $f(X_1) \neq (s+1, 0)$. Thus, there is some x in the Chinese lantern that is incomparable with $f(X_1)$ and $f(X) \leq x$. Since the Chinese lantern is the morphic image of f, there is some element of the spiked Boolean algebra mapping to x. Note that by our assumption $(0,0) \neq x \neq (0,1)$ and therefore the preimage of x can neither be the top nor any spike. So, let $Z \subsetneq X$ such that $x = f(Z)$. Since it is a proper subset, the induction hypothesis applies to Z, so $f(Z_1) = f(Z)$. But $Z_1 \subseteq X_1$, and so $x = f(Z) = f(Z_1) \geq f(X_1)$ in contradiction to the choice of x.]

Let $\mathbf{T}_1$ be the topless Boolean algebra corresponding to the power set Boolean algebra of C_1. Our claim implies that $f\!\restriction\!\mathbf{T}_1$ is a morphism from $\mathbf{T}_1$ onto $\mathbf{L}_{s-1,n}$, so by Proposition 11, we have that $n < 2^{(s-1)+3} = 2^{s+2}$. □

Corollary 13. *For each natural number $s \geq 1$, the frame $\mathbf{L}_{s,2^{s+2}}$ is not an Inam-frame.*

Proof. Follows directly from Lemma 12. □

Theorem 14. *Inamdar logic* Inam *is not finitely axiomatisable.*

Proof. Suppose φ is a formula with s variables axiomatising the Inam-frames. By Proposition 10, we find $m \leq s$ such that $\mathbf{L}_{s,2^{s+2}} \vDash \varphi$ if and only if $\mathbf{L}'_{s,2^{s+2},m} \vDash \varphi$. By Corollary 9, $\mathbf{L}'_{s,2^{s+2},m}$ is an Inam-frame; by Corollary 13, $\mathbf{L}_{s,2^{s+2}}$ is not. Contradiction! □

Corollary 15. The modal logics S4.sBA and Grz.sBA are not finitely axiomatisable.

Proof. Follows directly from Theorem 14 and (the remark after) Theorem 5. □

4 Inamdar Logic is not Finitely Axiomatisable over Cheq

In 2006, Fontaine improved the result by Maksimova, Skvortsov, and Shehtman and proved that Med is not even finitely axiomatisable over Cheq. In this section, we provide the analogue of this result for spiked Boolean algebras. The key observation here is that all of the Chinese lanterns discussed in Sect. 3 are Cheq-frames. This was done for the Chinese lanterns $\mathbf{L}_{s,2^{s+3}}$ and $\mathbf{L}'_{s,2^{s+3},i}$ in [2, §4]; we shall consider the general case here.

We denote by $\mathbf{W} := \mathbf{D} \times 2 = \mathbf{D} \oplus \mathbf{D} = \mathbf{D}^{(1)}$ the *bowtie*, i.e., the partial order with four elements, two incomparable minimal elements and two incomparable maximal elements such that each maximal element is above each minimal element; cf. Fig. 3. We formulate the Bowtie Lemma whose proof is not difficult, but tedious: details can be found in [11, Lemma 6.2.3].

Fig. 3. The bowtie $\mathbf{W}$.

Lemma 16 (Bowtie Lemma) If $\mathbf{P}$ is any finite rooted frame, then $\mathbf{P} \oplus \mathbf{W}$ is the morphic image of some $\mathbf{F}_n$, i.e., the n-fold product of the two-fork.

Corollary 17. The Chinese lanterns $\mathbf{L}_{s,n}$ and $\mathbf{L}'_{s,n,m}$ are Cheq-frames.

Proof. We proved that $\mathbf{L}'_{s,n,m}$ is an Inam-frame in Corollary 9, so $\mathbf{L}'_{s,n,m}$ is also a Cheq-frame.

All Chinese lanterns $\mathbf{L}_{s,n}$ are of the form $\mathbf{P} \oplus \mathbf{W}$ for some finite rooted frame $\mathbf{P}$, so by the Bowtie Lemma 16, they are morphic images of some $\mathbf{F}_n$, and hence Cheq-frames. □

Theorem 18. *Inamdar logic Inam is not finitely axiomatisable over Cheq.*

Proof. Suppose φ is a formula with s variables that characterises Inam over Cheq. By Proposition 10, we find $m \leq s$ such that $\mathbf{L}_{s,2^{s+2}} \vDash \varphi$ if and only if $\mathbf{L}'_{s,2^{s+2},m} \vDash \varphi$. By Lemma 17, both $\mathbf{L}_{s,2^{s+2}}$ and $\mathbf{L}'_{s,2^{s+2},m}$ are Cheq-frames, but as in the proof of Theorem 14, the two Chinese lanterns disagree about the truth value of φ. Contradiction! □

5 Spikings

If **P** is a partial order with a top, we call **S** a *spiking of* **P** if **P** is a suborder of **S**, all elements of **P** have the same predecessors in **S** that they have in **P**, $S \setminus P$ is a finite set of maximal elements of **S** (called *spikes*) that all have only coatoms of **P** as immediate predecessors, and each coatom of **P** has at most one element of $S \setminus P$ as immediate successor. This means that a spiking is an extension of **P** with a additional maximal elements that sit on top of the coatoms of **P** in such a way that each coatom has at most one spike (and coatoms can share a spike). A spike is called *pure* if it has exactly one coatom as immediate predecessor and a partial order is called a *pure spiking* if **P** if each of the spikes is pure. Cf. Fig. 4 for an example.

Note that the top of **P** remains maximal in a spiking. The set of predecessors of all elements of **P** remains the same in **S**; the set of successors of elements of **P** remains the same unless they are coatoms. Clearly, upsets of spikings of **P** are spikings of upsets of **P**.

Theorem 19. *Let $\mathcal{C}$ be a class of partial orders closed under morphic images and rooted upsets and let $\mathcal{C}^v$ be class of spikings of elements of $\mathcal{C}$. Then $\mathcal{C}^v$ is closed under morphic images.*

Proof. By Lemma 2, our claim can be proved by showing that $\mathcal{C}^v$ is closed under reductions. Suppose that $\mathbf{S} \in \mathcal{C}^v$ is a spiking of some $\mathbf{P} \in \mathcal{C}$ and **Q** is a reduction of **S**. We aim to show that $\mathbf{Q} \in \mathcal{C}^v$.

Case 1. The reduction **Q** is produced by identifying two α-related elements $p_1 < p_2$ in **S**. Since all spikes are maximal, we have that $p_1 \in P$.

Subcase 1a. If p_2 is a (pure) spike, then identifying p_1 and p_2 is tantamount to not adding the spike in the first place, so **Q** is the spiking of **P** where the spike p_2 was not added.

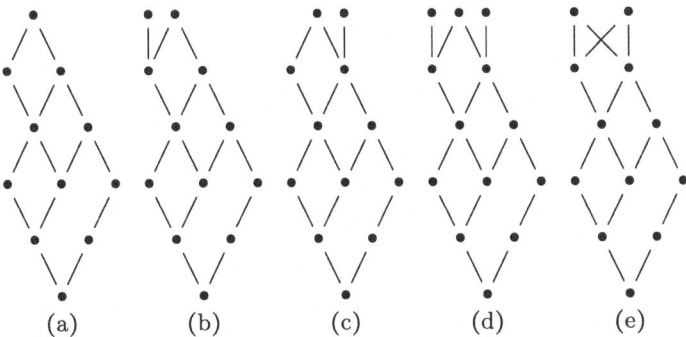

Fig. 4. A partial order (a) with five spikings (a) to (e). Of these, (a), (b), (c), and (d) are pure spikings.

Subcase 1b. If $p_2 \in P$ and is not the top element of **P**, then p_1 is not a coatom of **P** and therefore, the p_1 has precisely the same immediate successors in **P** and **S**. Thus, p_1 and p_2 were already α-related in **P**. If **Q*** is the reduction of **P** by identifying p_1 and p_2, then **Q** is a spiking of **Q*** by adding the same spikes that **S** adds to **P**. By closure of $\mathcal{C}$ under morphic images, **Q**$^* \in \mathcal{C}$, and so **Q** $\in \mathcal{C}^v$.

Subcase 1c. If p_2 is the top element of **P**, then p_1 is a coatom of **P**. Since we assumed that p_2 is the unique immediate successor of p_1 in **S**, this means that the coatom p_1 did not receive a spike in **S**. Since **P** is a partial order with a top element, all coatoms are α-related to the top element: we can reduce **P** to **Q*** by identifying p_1 and p_2 in **P**. All other coatoms of **P** remain coatoms in **Q*** and by closure of $\mathcal{C}$ under morphic images, **Q**$^* \in \mathcal{C}$. We now obtain **Q** by adding the same spikes that **S** adds to **P**.

Case 2. The reduction **Q** is produced by identifying two β-related elements p_1 and p_2. Note that in this case p_1 is maximal if and only if p_2 is maximal and p_1 is a coatom if and only if p_2 is a coatom.

Subcase 2a. Both p_1 and p_2 are maximal. This means that they are both spikes or the top element of **P**. Identifying two spikes is tantamount to adding one spike that has all of the immediate predecessors of both spikes are its immediate predecessors. Identifying a spike with the top element of **P** is tantamount to not adding the spike in the first place. In both cases, **Q** is a spiking of the **P** $\in \mathcal{C}$.

Subcase 2b. Both p_1 and p_2 are coatoms. Since **P** was a partial order with a top element, this means that p_1 and p_2 are β-related in **P**. Since they have the same successors in **S**, this means that either none of them had a spike added or there is a spike that they both share. If **Q*** is the reduction of **P** by identifying p_1 and p_2, then we can add the same spikes to **Q*** as **S** adds to **P** and obtain **Q**. By closure of $\mathcal{C}$ under morphic images, this means that **Q** $\in \mathcal{C}^v$.

Subcase 2c. Both p_1 and p_2 are neither maximal nor coatoms. Since only the immediate successors of coatoms change between **P** and **S**, this means that p_1 and p_2 are β-related in **P**. Let **Q*** be the reduction of **P** by identifying p_1 and p_2. All coatoms of **P** are preserved in **Q***, so we can add the same spikes to **Q*** as **S** adds to **P** and obtain **Q**. By closure of $\mathcal{C}$ under morphic images, this means that **Q** $\in \mathcal{C}^v$. □

Remember that the class of morphic images of Boolean algebras is the same as the class $\mathcal{R}$ of finite rooted frames with a top by Proposition 6

Corollary 20. Any rooted Inam-frame is a spiking of an element of $\mathcal{R}$.

Proof. Note that the class $\mathcal{R}$ is closed under rooted upsets, so $\mathcal{R}$ satisfies the assumptions of Theorem 19. Clearly every spiked Boolean algebra is a spiking of a Boolean algebra and hence in $\mathcal{R}^v$. If **P** is an Inam-frame, it is the morphic image of a spiked Boolean algebra. By Theorem 19, $\mathcal{R}^v$ is closed under morphic images, so **P** $\in \mathcal{R}^v$. □

References

1. Bezhanishvili, N.: Lattices of intermediate and cylindric modal logics. PhD thesis, Universiteit van Amsterdam (2006)
2. Fontaine, G.: Axiomatization of ML and Cheq. Master's thesis, Universiteit van Amsterdam (2006)
3. Fontaine, G.: ML is not finitely axiomatizable over Cheq. In: Governatori, G., Hodkinson, I.M., Venema, Y. (eds.) Advances in Modal Logic 6. Papers from the Sixth Conference on "Advances in Modal Logic," held in Noosa, Queensland, Australia, 25–28 September 2006, pp. 139–146. College Publications (2006)
4. Golshani, M., Mitchell, W.J.: On a question of Hamkins and Löwe on the modal logic of collapse forcing (2021). Preprint arXiv:1609.02633v8
5. Hamkins, J.D., Leibman, G., Löwe, B.: Structural connections between a forcing class and its modal logic. Israel J. Math. **207**(2), 617–651 (2015). https://doi.org/10.1007/s11856-015-1185-5
6. Hamkins, J.D., Löwe, B.: The modal logic of forcing. Trans. Am. Math. Soc. **360**(4), 1793–1817 (2008)
7. Inamdar, T.C.: On the modal logics of some set-theoretic constructions. Master's thesis, Universiteit van Amsterdam (2013)
8. Maksimova, L.L., Skvortsov, D.P., Shekhtman, V.B.: The impossibility of a finite axiomatization of Medvedev's logic of finitary problems. Soviet Math. Doklady **20**, 394–398 (1979)
9. Shehtman, V.: Modal counterparts of Medvedev logic of finite problems are not finitely axiomatizable. Stud. Logica. **49**, 365–385 (1990)
10. van Benthem, J., Bezhanishvili, G., Gehrke, M.: Euclidean hierarchy in modal logic. Stud. Logica. **75**, 327–344 (2003)
11. Xiao, H.: Modal and intermediate logics motivated by an open problem on c.c.c. forcing. PhD thesis, Universität Hamburg (2024)
12. Ya'ar, U.: The modal logic of σ-centered forcing and related forcing classes. J. Symb. Log. **86**(1), 1–24 (2021)

Equivalence of Deterministic Weighted Real-Time One-Counter Automata

Prince Mathew[1]((✉))[ID], Vincent Penelle[2], Prakash Saivasan[3], and A. V. Sreejith[1]

[1] Indian Institute of Technology Goa, Ponda, India
{prince,sreejithav}@iitgoa.ac.in
[2] Univ. Bordeaux, CNRS, Bordeaux INP, LaBRI, UMR 5800, Talence 33400, France
vincent.penelle@u-bordeaux.fr
[3] The Institute of Mathematical Sciences, HBNI, Mumbai, India
prakashs@imsc.res.in

Abstract. This paper introduces deterministic weighted real-time one-counter automaton (DWROCA). A DWROCA is a deterministic real-time one-counter automaton whose transitions are assigned a weight from a field. Two DWROCAS are equivalent if every word accepted by one is accepted by the other with the same weight. DWROCA is a sub-class of weighted one-counter automata with counter-determinacy. It is known that the equivalence problem for this model is in P [7]. This paper gives a simpler proof and a better polynomial-time algorithm for checking the equivalence of two DWROCAS.

Keywords: Equivalence · Weighted automata · One-counter automata

1 Introduction

In this work, we give a polynomial-time algorithm to check the equivalence of deterministic weighted real-time one-counter automata (DWROCA) where the weights are from a field (possibly infinite). These are weighted one-counter automata with the "deterministic" condition — at most one transition on a letter from a state, and the counter is modified by at most *one* on a transition. Hence, for a word over a finite alphabet, there is at most one run starting from the initial configuration that determines its accepting weight. Additionally, "real-time" indicates that there are no epsilon transitions in this system.

We say that two DWROCAS are equivalent if any word accepted by one of the automata is accepted by the other with the same weight. Consider two DWROCAS that are not equivalent and let w be a word of minimal length that distinguishes the two machines. It suffices to show that there is a polynomial bound on the length of w. Two weighted automata that "simulate" the run of the DWROCAS up to that bound can then be used to check equivalence. To prove a bound on the length of w, we give a polynomial that bounds the counter values during

---

Related Version. *Full Version:* https://arxiv.org/abs/2411.03066.

© The Author(s), under exclusive license to Springer Nature Switzerland AG 2025
C. Aiswarya et al. (Eds.): ICLA 2025, LNCS 15402, pp. 176–189, 2025.
https://doi.org/10.1007/978-3-031-89610-1_13

the run of w on both machines. Our proof strategy borrows the "belt technique" developed in the context of deterministic real-time one-counter automata by Böhm et al. [1] (also see [2,4]), which is inspired by Valiant and Paterson's work [10]. We introduce a *pumping* technique (see Theorem 1) to adapt their proof strategy for checking the equivalence of two DWROCAS.

1.1 Related Work

One-counter automata are pushdown automata with a singleton stack alphabet. It is well known that the equivalence problem for non-deterministic pushdown automata is undecidable. On the other hand, Sénizergues [8] proved that it is decidable for deterministic pushdown automata. It was later proved that there is a primitive recursive algorithm for this [9]. The equivalence problem for deterministic one-counter automata is NL-complete [3]. Studies on probabilistic pushdown automata conducted by Forejt et al. [5] showed that the equivalence problem of probabilistic pushdown automata is equivalent to the multiplicity equivalence of context-free grammars. The latter problem is not known to be decidable. The decidability of equivalence is known for some special sub-classes of probabilistic pushdown automata. The equivalence problem is at least as hard as polynomial identity testing [5] even when the alphabet contains only one letter. For the case of visibly probabilistic pushdown automata, there is a randomised polynomial time algorithm for checking equivalence [6]. However, there is a polynomial time reduction from polynomial identity testing to this problem and is hence not likely to be in P. In a recent result, Mathew et al. [7] showed that the equivalence problem for weighted one-deterministic counter automata, which is a sub-class of weighted one-counter automata over fields, is decidable in polynomial time.

Section 2 contains the basic definitions that are used in the paper. Section 3 gives a polynomial time algorithm to check the equivalence of two DWROCAS given the existence of a polynomial length word that distinguishes them and the existence of such a word is proved in Sect. 3.1. Section 4 gives a short conclusion. Proofs omitted are provided in the full version.

2 Preliminaries

We use the symbol $\mathbb{N}$ to denote the set $\{0, 1, 2, 3, \ldots\}$. An alphabet is a finite, non-empty set of letters. In this paper, we denote the alphabet by Σ. We use Σ^* to denote the set of finite length words over Σ, and for all $l \in \mathbb{N}$, we use $\Sigma^{\leq l}$ (resp. Σ^l) to denote the set of words over Σ having length less than or equal to l (resp. exactly equal to l). Given a word $w \in \Sigma^*$, we use $|w|$ to denote the length of the word w. We use the notation $[i, j]$ to denote the interval $\{i, i+1, \ldots, j\}$. Given a word $w = w_0 \cdots w_n$, we write $w_{[i \cdots j]}$ to denote the factor $w_i \cdots w_j$. Given a tuple $e = (e_1, e_2)$, we use $e|_1$ to denote e_1 and $e|_2$ to denote e_2.

2.1 Deterministic Weighted Real-Time One-Counter Automata

A deterministic weighted real-time one-counter automaton (DWROCA) over a field $\mathcal{F}$ is a tuple $\mathcal{A} = (Q, \Sigma, q_0, s_0, \delta_0, \delta_1, \eta_F)$, where Q is a finite, nonempty set of states, Σ is the input alphabet, $q_0 \in Q$ is the initial state, $s_0 \in \mathcal{F}$ is the initial weight, $\delta_0 : Q \times \Sigma \to Q \times \{0, +1\} \times \mathcal{F}$ and $\delta_1 : Q \times \Sigma \to Q \times \{-1, 0, +1\} \times \mathcal{F}$ are the transition functions, and $\eta_F : Q \to \mathcal{F}$ is a final distribution.

We use $|\mathcal{A}|$ to denote the size of $\mathcal{A}$, which we consider to be $|Q|$. A *configuration* of a DWROCA is a triple $c = (q_c, n_c, t_c)$, with $q_c \in Q$, $n_c \in \mathbb{N}$ and $t_c \in \mathcal{F}$. The configuration $(q_0, 0, s_0) \in Q \times \mathbb{N} \times \mathcal{F}$ is called the *initial configuration* of $\mathcal{A}$. A *transition* is a sextuple $\tau = (p_\tau, d_\tau, a_\tau, \text{ce}_\tau, s_\tau, q_\tau)$ where $p_\tau, q_\tau \in Q$ are states, $d_\tau \in \{0, 1\}$ specifies which among δ_0 and δ_1 is used, $a_\tau \in \Sigma$, $\text{ce}_\tau \in \{-1, 0, 1\}$ is the *counter-effect*, and $s_\tau \in \mathcal{F}$ such that $\delta_{d_\tau}(p_\tau, a_\tau) = (q_\tau, \text{ce}_\tau, s_\tau)$. Intuitively, δ_0 is used when you have a zero-test, and δ_1 otherwise. Given a transition τ and a configuration $c = (q_c, n_c, s_c)$, we denote the application of τ to c as $\tau(c) = (q_\tau, n_c + \text{ce}_\tau, s_c \times s_\tau)$ if $q_c = p_\tau$ and $d_\tau = 0$ if and only if $n_c = 0$, and is undefined otherwise. Note that the counter values always stay positive, implying that you cannot perform a decrement operation on the counter from a configuration with counter value zero.

Given a sequence of transitions $T = \tau_0 \cdots \tau_{\ell-1}$, we denote the the word labelling it as $word(T) = a_{\tau_0} \cdots a_{\tau_{\ell-1}}$, $\text{we}(T) = s_{\tau_0} \times \cdots \times s_{\tau_{\ell-1}}$ its weight-effect, and $\text{ce}(T) = \text{ce}_{\tau_0} + \cdots + \text{ce}_{\tau_{\ell-1}}$ its counter-effect. For all $0 \leq i < j \leq |\ell - 1|$, we use $T_{i\ldots j}$ to denote the sequence of transitions $\tau_i \cdots \tau_j$ and $|T|$ to denote ℓ.

We call a sequence of transitions *grounded* if there is an i such that $d_{\tau_i} = 0$ and *floating* otherwise. We denote $\min_{\text{ce}}(T) = \min_i(\text{ce}(\tau_0 \cdots \tau_i))$ the minimal counter-effect of its prefixes and $\max_{\text{ce}}(T) = \max_i(\text{ce}(\tau_0 \cdots \tau_i))$ the maximal counter-effect of its prefixes. A sequence of transitions T is *valid* if for every $i \in [0, \ell - 2]$, $q_{\tau_i} = p_{\tau_{i+1}}$. We will only consider valid sequences of transitions.

A *run* π is an alternate sequence of configurations and transitions denoted as $\pi = c_0 \tau_0 c_1 \cdots \tau_{\ell-1} c_\ell$ such that for every i, $c_{i+1} = \tau_i(c_i)$. Given a sequence of transitions T and a configuration c, we denote $T(c)$ the run obtained by applying T to c sequentially (if it is defined). The word labelling it, its length, weight-effect and counter-effect are those of its underlying sequence of transitions. A *sub-run* is a (syntactic) sub-word of a run that is also a run. The run $\pi = c_1 \tau_1 c_2 \tau_2 \ldots c_i$ is called a *simple cycle*, if $i \leq |\mathcal{A}|$, $q_{c_1} = q_{c_i}$ and for all $j, k \in [1, i-1]$, $q_{c_j} = q_{c_k}$ if and only if $j = k$.

Observe that, for a valid floating sequence of transitions, $T(c)$ is defined if and only if $n_c > -\min_{\text{ce}}(T)$, and for a valid grounded sequence of transition, $T(c)$ is defined if and only if $n_c = -\min_{\text{ce}}(T)$ and for every i, $d_{\tau_i} = 0$ if and only if $\text{ce}(\tau_0 \cdots \tau_{i-1}) = \min_{\text{ce}}(T)$. In particular, observe that if a valid floating sequence of transitions T is applicable to a configuration (q, n, t), then for every $n' \geq n$ and weight $s' \in \mathcal{F}$, it is applicable to (q, n', s').

Since the machine is deterministic, for any word w, there is at most one run labelled by w starting in a given configuration c_0. We denote this run $\pi(w, c_0)$. A run π is called an *execution* if $c_0 = (q_0, 0, s_0)$ is the initial configuration of $\mathcal{A}$. We use $\pi(w)$ to denote the execution labelled by w. A run $\pi(w, c_0) = c_0 \tau_0 c_1 \cdots \tau_{\ell-1} c_\ell$ is also represented as $c_0 \xrightarrow{w} c_\ell$. We use the notation $c_0 \rightarrow^* c_\ell$ to denote the existence of some word w such that $c_0 \xrightarrow{w} c_\ell$. The weight with which a word w is accepted by $\mathcal{A}$ along the run $\pi(w, c_0)$ is denoted by $f_\mathcal{A}(w, c_0) = s_{c_0} \times$ we$(\pi(w, c_0)) \times \eta_F(q_{c_\ell})$. We use the notation $f_\mathcal{A}(w)$ to denote $f_\mathcal{A}(w, (q_0, 0, s_0))$.

Let $\mathcal{A}$ and $\mathcal{B}$ be two DWROCAs. Consider the configurations c_1 of $\mathcal{A}$ and c_2 of $\mathcal{B}$. We say that $c_1 \equiv_l c_2$ if and only if for all $w \in \Sigma^{\leq l}$, $f_\mathcal{A}(w, c_1) = f_\mathcal{B}(w, c_2)$. We say that the configurations c_1 and c_2 are equivalent if and only if $c_1 \equiv_l c_2$ for all $l \in \mathbb{N}$, and we denote this by $c_1 \equiv c_2$. We say that $\mathcal{A}$ and $\mathcal{B}$ are equivalent if for all $w \in \Sigma^*$, $f_\mathcal{A}(w) = f_\mathcal{B}(w)$. A word $w \in \Sigma^*$ is called a *non-equivalent witness* (or simply a *witness*) for $\mathcal{A}$ and $\mathcal{B}$ if and only if $f_\mathcal{A}(w) \neq f_\mathcal{B}(w)$. A minimal witness is a non-equivalent witness with the minimal length.

Deterministic weighted automata (DWA) is a restricted form of a DWROCA where there are no counter operations. The above definitions for DWROCA are also used for DWA.

Given a word $w = w_0 \cdots w_n$, a configuration c, and a list I of disjoint intervals included in $[0, n]$: $I = [i_0, j_0], [i_1, j_1], \cdots, [i_k, j_k]$, we call w_I the word obtained by removing from w all w_ℓ with $\ell \in I$. A list of disjoint intervals I is a pumping of w from c, if in $\pi(w, c)$ all sub-runs starting at position i_d and ending in j_d for some $d \in [0, k]$ are *loops*, i.e., start and end in the same state, the minimal counter-effect of the obtained run is not smaller than the original one, i.e., $\min_{ce}(\pi(w_I, c)) \geq \min_{ce}(\pi(w, c))$ and the pumping does not introduce any new zero-tests, while still preserving the last zero-test. Informally, a pumping is obtained by removing loops in a run while not decreasing the minimum counter value reached and without introducing additional zero-tests while still preserving the last zero-test.

Theorem 1. *Given a word w and two configurations c and c', and two disjoint sequences of intervals I and J such that I and J are pumpings of w from both c and c', and $f(w, c) \neq f(w, c')$. We get that $I \uplus J$ is also a pumping of w from both c and c', and either $f(w_I, c) \neq f(w_I, c')$, or $f(w_J, c) \neq f(w_J, c')$ or $f(w_{I \uplus J}, c) \neq f(w_{I \uplus J}, c')$.*

Now, we will define some notions on DWROCA to aid us in equivalence checking.

2.2 Underlying Weighted Automata

Floating runs of a DWROCA are isomorphic to runs of the deterministic automaton obtained by ignoring counter values. We formalise this so-called notion *underlying uninitialised weighted automaton* here.

Definition 1. *The* underlying uninitialised weighted automaton *of a* DWROCA $\mathcal{A} = (Q, \Sigma, q_0, s_0, \delta_0, \delta_1, \eta_F)$ *is the uninitialised deterministic weighted automaton given by* $U(\mathcal{A}) = (Q, \Sigma, \delta'_1, \eta_F)$, *where* δ'_1 *is a transition function from* $Q \times \Sigma \to Q \times \mathcal{F}$ *and is defined as follows:*

$$\delta'_1(q_1, a) = (q_2, s) \text{ iff } \delta_1(q_1, a) = (q_2, o, s), \text{ for some } o \in \{-1, 0, +1\}.$$

The automaton $U(\mathcal{A})$ is said to be *uninitialised* because the machine has no initial state and weight. A configuration c of DWROCA $\mathcal{A}$ is said to be k-equivalent to a configuration β of an uninitialised weighted automata $\mathcal{B}$, denoted $c \sim_k \beta$, if for all $w \in \Sigma^{\leq k}$, $f_\mathcal{A}(w, c) = f_\mathcal{B}(w, \beta)$.

Given an uninitialised weighted automaton $\mathcal{B}$ and $k > 0$, we partition the configurations of DWROCA $\mathcal{A}$ into two parts: $\mathsf{EqConfig}(\mathcal{A}, \mathcal{B}, k)$ are those k-equivalent to some configuration of $\mathcal{B}$, and $\mathsf{notEqConfig}(\mathcal{A}, \mathcal{B}, k)$ are those that are not.

$$\mathsf{notEqConfig}(\mathcal{A}, \mathcal{B}, k) = \{c \in Q \times \mathbb{N} \times \mathcal{F} \mid \forall \beta \in Q \times \mathcal{F}, c \not\sim_k \beta\}.$$
$$\mathsf{EqConfig}(\mathcal{A}, \mathcal{B}, k) = \{c \in Q \times \mathbb{N} \times \mathcal{F} \mid \exists \beta \in Q \times \mathcal{F}, c \sim_k \beta\}.$$

The following lemma shows that membership in $\mathsf{EqConfig}(\mathcal{A}, \mathcal{B}, k)$ is independent of the weight of the configuration. We prove this by taking advantage of the existence of multiplicative inverses in the field and the commutativity of multiplication.

Lemma 1. *For all* $s, \bar{s} \in \mathcal{F} \setminus \{0_e\}$, $p \in Q$, $m \in \mathbb{N}$, $(p, m, \bar{s}) \in \mathsf{EqConfig}(\mathcal{A}, \mathcal{B}, k)$ *if and only if* $(p, m, \bar{s}) \in \mathsf{EqConfig}(\mathcal{A}, \mathcal{B}, k)$.

The distance of a configuration c to a set of configurations C is the length of a minimal word that takes you from c to a configuration in C and is defined as

$$\mathsf{dist}(c, C) = \min\{|w| \mid \exists c' \in C \text{ with } c \xrightarrow{w} c'\}.$$

Notice that $\mathsf{dist}(c, C) < \infty$ if and only if C is reachable from c. We denote this by $c \to^* C$. By abuse of notation, we denote $c \xrightarrow{w} C$ if there exists a configuration $c' \in C$ such that $c \xrightarrow{w} c'$. The notion of distance will play a key role in determining which parts of the run of a non-equivalent witness can be pumped out if it is not minimal.

Consider a configuration c such that $c \xrightarrow{*} \mathsf{notEqConfig}(\mathcal{A}, \mathcal{B}, k)$. The following lemma identifies a special word u such that $c \xrightarrow{u} \mathsf{notEqConfig}(\mathcal{A}, \mathcal{B}, k)$. The proof of the lemma is similar to that of the non-weighted case presented in [1,10].

Lemma 2. *Given a configuration* c *of a* DWROCA $\mathcal{A}$, *and a weighted automaton* $\mathcal{B}$, *if* $c \xrightarrow{*} \mathsf{notEqConfig}(\mathcal{A}, \mathcal{B}, k)$ *then there exists a word* $u = u_1 u_2^r u_3$ *(with* $r \geq 0$*) such that* $c \xrightarrow{u} \mathsf{notEqConfig}(\mathcal{A}, \mathcal{B}, k)$ *and the following conditions hold:*

1. *for all* $w \in \Sigma^*$, *if* $c \xrightarrow{w} \mathsf{notEqConfig}(\mathcal{A}, \mathcal{B}, k)$, *then* $|w| \geq |u|$.
2. $|u_1 u_3| \leq 3|\mathcal{A}|^3$ *and* $|u_2| \leq |\mathcal{A}|$.
3. *either* $u_2 = \epsilon$ *or* u_2 *is a simple cycle of counter loss greater than zero.*
4. $|u| \leq (\max\{|\mathcal{A}|, n_c\} + |\mathcal{A}|^2)|\mathcal{A}|$ *and the maximum counter value encountered during the run* $c \xrightarrow{u} \mathsf{notEqConfig}(\mathcal{A}, \mathcal{B}, k)$ *is less than* $\max\{|\mathcal{A}|, n_c\} + |\mathcal{A}|^2$.

3 Equivalence of DWROCA

In the remainder of the section, we fix two DWROCAS, $\mathcal{A}_1$ and $\mathcal{A}_2$. We also fix $\mathsf{K} = |\mathcal{A}_1| + |\mathcal{A}_2|$. To simplify the reasoning, we will reason on the synchronised runs on pairs of configurations. Given two DWROCAS, we consider a *configuration pair* $c = \langle \chi_c, \psi_c \rangle$ where χ_c is a configuration of $\mathcal{A}_1$ and ψ_c is a configuration of $\mathcal{A}_2$. We similarly consider *transition pairs* of $\mathcal{A}_1$ and $\mathcal{A}_2$ and consider *synchronised runs* as the application of a sequence of transition pairs to a configuration pair. Let w be a minimal word that distinguishes $\mathcal{A}_1$ and $\mathcal{A}_2$. Henceforth, we will denote by
$$\Pi = c_0 \tau_0 c_1 \cdots \tau_{L-1} c_L$$
the run pair of w from the initial configuration pair $c_0 = \langle (p_0, 0, s_0), (q_0, 0, t_0) \rangle$. We denote by $T_\Pi = \tau_0 \cdots \tau_{L-1}$ the sequence of transition pairs of this run pair. Given a run pair Π' of a word w' from c_0, we use $\mathtt{aw}(\Pi')$ to denote the tuple $(f_{\mathcal{A}_1}(w'), f_{\mathcal{A}_2}(w'))$.

Lemma 3. *There is a polynomial* $\mathrm{poly}_0 : \mathbb{N} \to \mathbb{N}$ *such that if two* DWROCAS *$\mathcal{A}_1$ and $\mathcal{A}_2$ are not equivalent, then there exists a witness w such that counter values in the execution of w is less than* $\mathrm{poly}_0(\mathsf{K})$.

This is the technically challenging part of the proof, and we will prove this later. Now, we show that the length of w is bounded by a polynomial, assuming the counter values in Π are bounded by $\mathrm{poly}_0(\mathsf{K})$.

Lemma 4. (small model property). *For any two non-equivalent* DWROCAS *$\mathcal{A}_1$ and $\mathcal{A}_2$, the length of a minimal witness is less than or equal to* $2(\mathsf{K}\mathrm{poly}_0(\mathsf{K}))^2$.

We now state the main result of our paper. We use a small model property stated in Lemma 4 to prove this theorem. This property ensures that if two DWROCAS are not equivalent, then there exists a small word that can distinguish them. The small model property helps us to reduce the equivalence problem of DWROCA to that of weighted automata by "simulating" the runs of DWROCAS up to polynomial length by two weighted automata. The naive algorithm will give us a PSPACE procedure, but there is a polynomial time procedure to do this.

Theorem 2. *There is a polynomial time algorithm that decides if two* DWROCAS *are equivalent or outputs a minimal witness.*

Since deterministic real-time one-counter automata is a DWROCA with weight 1 on its transitions, we get the following corollary.

Corollary 1. *There is a polynomial time algorithm that decides if two deterministic real-time one-counter automata are equivalent or outputs a minimal witness.*

The rest of this section is dedicated to proving Lemma 3.

3.1 Proof Strategy

A crucial idea here is to partition the set of configurations between those that behave like a weighted automata on short runs and those that do not. In Sect. 2.2, we have defined the notion of underlying uninitialised weighted automaton for a DWROCA. Comparing the configuration of a DWROCA with a configuration of a weighted automaton helps us reduce the problem to that of weighted automata. The technique is adapted from the ideas used in [1,2] where the non-weighted model is studied. We show that the addition of weights does not change the proof outline presented in [1] and can be used for the weighted model as well with suitable adaptations.

As we need to test the equivalence of configurations from $\mathcal{A}_1$ and $\mathcal{A}_2$, we will consider the disjoint union of $U(\mathcal{A}_1)$ and $U(\mathcal{A}_2)$ to have a single automaton to compare their configurations with. We use $U(\mathcal{A}_1) \cup U(\mathcal{A}_2)$ to denote this automaton and call it the underlying weighted automata. The set notUWA consists of all configurations of $\mathcal{A}_1$ and $\mathcal{A}_2$ that do not have a K-equivalent configuration in either $U(\mathcal{A}_1)$ or $U(\mathcal{A}_2)$ and is defined as follows

$$\text{notUWA} = \bigcup_{i \in \{1,2\}} \bigcap_{j \in \{1,2\}} \text{notEqConfig}(\mathcal{A}_i, U(\mathcal{A}_i), \mathsf{K}).$$

The set UWA, the complement of notUWA, contains all configurations that are K-equivalent to some configuration in the underlying weighted automata. Note that if a configuration $c \in \text{notUWA}$, then $n_c < \mathsf{K}$. The distance that will be interesting in our reasoning is the one to the set notUWA. Therefore for simplicity, we denote $\text{dist}(\xi) = \text{dist}(\xi, \text{notUWA})$ for a configuration ξ of $\mathcal{A}_1$ or $\mathcal{A}_2$.

We say that a configuration pair $c = \langle \chi_c, \psi_c \rangle$ is

- surely-equivalent: If $\chi_c \equiv_\mathsf{K} \psi_c$ and $\text{dist}(\chi_c) = \text{dist}(\psi_c) = \infty$.
- surely-nonequivalent: $\chi_c \not\equiv_\mathsf{K} \psi_c$ or $\text{dist}(\chi_c) \neq \text{dist}(\psi_c)$.
- unresolved: otherwise, i.e., $\chi_c \equiv_\mathsf{K} \psi_c$ and $\text{dist}(\chi_c) = \text{dist}(\psi_c) < \infty$.

Let us denote by $\chi_0 = (p_0, 0, s_0)$ and $\psi_0 = (q_0, 0, t_0)$ the initial configurations of $\mathcal{A}_1$ and $\mathcal{A}_2$ respectively and $\chi_0 \neq \psi_0$. The categorisation of the configuration pairs into surely-equivalent, surely-nonequivalent and unresolved allows us to solve two easy cases before concentrating on the crux of the argument. It is easy to observe that no configuration pair in Π is surely-equivalent (see Lemma 5). Our next observation (see Lemma 6) is that if a configuration pair c_j in Π is surely-nonequivalent and has counter values $\{m_j, n_j\}$, then the counter values in the rest of the execution is bounded by a polynomial in m_j, n_j and K. Hence it will remain to prove that if c_j is the first surely-nonequivalent configuration pair in Π, then m_j and n_j are bounded by a polynomial. In order to do this, we prove that any minimal run-pair composed solely of unresolved configuration pairs has all of its counter values bounded by a polynomial. This is the most challenging part of the proof.

$$\underbrace{c_0\tau_0 c_1\tau_1 c_2\cdots c_{j-1}\tau_{j-1}}_{\text{unresolved configuration pairs} \atop \text{counters poly-bounded}} \quad \underbrace{c_j = \langle \chi_{c_j}, \psi_{c_j}\rangle}_{1^{st}\text{surely-nonequivalent} \atop \text{configuration pair}} \quad \underbrace{\tau_j\cdots\tau_{L-1}c_L}_{\text{counters} \atop \text{poly-bounded}}$$

We now solve the cases when we have a surely-equivalent or surely-nonequivalent configuration pair. The proof is similar to that of the non-weighted case. The case of unresolved configuration pairs needs to be solved using a different technique and is taken care of in Sect. 3.3.

Lemma 5. *There is no surely-equivalent configuration pair in Π.*

The following lemma bounds the length of a minimal witness from a surely-nonequivalent configuration pair.

Lemma 6. *Let $c_j = \langle \chi_{c_j}, \psi_{c_j}\rangle$ be the first surely-nonequivalent configuration pair in Π, then $L - j \leq \min\{\mathrm{dist}(\chi_{c_j}), \mathrm{dist}(\psi_{c_j})\} + \mathsf{K}$.*

From Lemma 2, we know that the distance of a configuration is polynomially bounded with respect to its counter value and K. Therefore, Lemma 6 tells us that the length of a minimal witness from a surely-nonequivalent configuration pair is polynomially bounded with respect to its counter value and K. A length bound implies a bound on the counter value also, since the counters cannot increase more than the length of a word. The next two sub-sections focus on the remaining case of paths containing only unresolved configuration pairs.

3.2 Configuration Space

Each pair of configurations $c = \langle \chi, \psi \rangle$ is mapped to a point in the space $\mathbb{N} \times \mathbb{N} \times (Q \times Q) \times \mathcal{F} \times \mathcal{F}$, henceforth referred to as the *configuration space*, where the first two dimensions represent the two counter values, the third dimension $Q \times Q$ corresponds to the pair of control states, the fourth and fifth dimensions represent the weights. We partition the configuration space into initial space, belt space, and background space. This partition is indexed on two polynomials, $\mathrm{poly}_1 = 14\mathsf{K}^6$ and $\mathrm{poly}_2 = 6\mathsf{K}^4$.

- initial space: All configuration pairs $\langle (p, m, s), (q, n, t) \rangle$ such that $m, n < \mathrm{poly}_1(\mathsf{K})$.
- belt space: Let $\alpha, \beta \geq 1$ be co-prime. The belt of thickness d and slope $\frac{\alpha}{\beta}$ consists of those configuration pairs $\langle (p, m, s), (q, n, t) \rangle$ that satisfy $|\alpha.m - \beta.n| \leq d$. The belt space contains all configuration pairs $\langle (p, m, s), (q, n, t) \rangle$ outside the initial space that are inside a belt with thickness $\mathrm{poly}_2(\mathsf{K})$ and slope $\frac{\alpha}{\beta}$, where $\alpha, \beta \in [1, \mathsf{K}^2]$.
- background space: All remaining configuration pairs.

The projection of the configuration space onto the first two dimensions is depicted in Fig. 1. The polynomials poly_1 and poly_2 are chosen in such a way that all unresolved configuration pairs fall in a belt or the initial space, and all belts are disjoint. This is proved in Lemma 7. In fact, these polynomials can be used as the size of the initial space and thickness of belts in [1] since the properties to be ensured are similar. The precise polynomials are used in the lemma proving these properties.

If the maximum counter value encountered during the execution of the minimal witness is far greater than the size of the initial space, then its length inside the belt space is very long. Since the belts are disjoint outside the initial space, we just need to show that if the length of the run of the witness inside a belt is very long, then we can find a shorter witness by pumping some portion out from the current witness. This is proved in Sect. 3.3 using Lemma 8 and Lemma 9, that enables us to perform this cut.

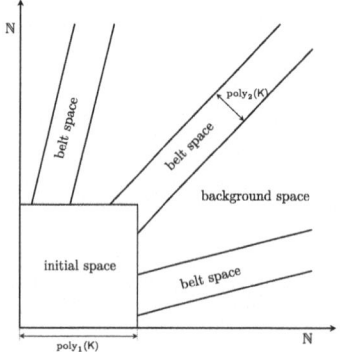

Fig. 1. Configuration space [7].   **Fig. 2.** An α-β repetition.

The partition of the unresolved configuration pairs thus helps us to show that there is a restriction on the run containing only unresolved configuration pairs in Π. The intuitive idea is that when at least one counter value is very big, the ratios between the counter value pairs on long run-pairs cannot take arbitrarily many different values. If there is a long portion of Π where the ratio between the counter value pairs repeats, we can pump a sub-run out of it and still get a witness contradicting its minimality.

The proof of the following lemma is the same as that of the non-weighted case presented in [1].

Lemma 7. *If* $c = \langle (p_c, m_c, s_c), (q_c, n_c, t_c) \rangle$ *is an unresolved configuration pair with* $\max\{m_c, n_c\} > \text{poly}_1(\mathsf{K})$ *then,*

1. *it lies in a unique belt of thickness* $\text{poly}_2(\mathsf{K})$ *and slope* α-β*, where* $\alpha, \beta \in [1, \mathsf{K}^2]$.
2. *it cannot take a transition pair to reach a configuration pair* c'*, inside another belt of thickness* $\text{poly}_2(\mathsf{K})$ *and slope* $\frac{\alpha'}{\beta'}$*, where* $\alpha', \beta' \in [1, \mathsf{K}^2]$.

By Lemma 7, the belt in which an unresolved configuration pair c lies is uniquely determined. It also ensures that the belts are disjoint outside the initial space and that no run composed only of unresolved configuration pairs can go from one belt to another without passing through the initial space.

Let $\alpha, \beta \in [1, \mathsf{K}^2]$ be co-prime and $\langle (p, m, s), (q, n, t) \rangle$, $\langle (p', m', s'), (q', n', t') \rangle$ be two configuration pairs. They are α-β related if $p = p'$ and $q = q'$ and $\alpha.m - \beta.n = \alpha.m' - \beta.n'$. Roughly speaking, two configuration pairs are α-β related if they have the same state pairs and lie on a line with slope $\frac{\alpha}{\beta}$. An α-β *repetition* is a run-pair $\bar{\pi}_1 = c_i \tau_i c_{i+1} \tau_{i+1} \cdots \tau_{j-1} c_j$ such that the configuration pairs c_i and c_j are α-β related. The projection of an α-β repetition onto the counters is depicted in Fig. 2.

The following two lemmas help us to show that if there is a witness having a very long sub-run inside a belt, then we can find a shorter witness whose sub-run inside that belt is shorter.

Lemma 8 (cut lemma). *Given a configuration pair c and a sequence of transition pairs T such that all configuration pairs of $T(c)$ are inside a belt with slope $\frac{\alpha}{\beta}$ with $\alpha, \beta \in [1, \mathsf{K}^2]$, and either $\mathsf{ce}(T)|_1 > \mathsf{K}^2 \mathrm{poly}_2(\mathsf{K})$ or $\mathsf{ce}(T)|_2 > \mathsf{K}^2 \mathrm{poly}_2(\mathsf{K})$, then there exist $i < j$ such that:*

- $T_{1\ldots i}(c)$ and $T_{1\ldots j}(c)$ are α-β related configuration pairs.
- $T_{1\ldots i} T_{j+1\ldots |T|}(c)$ is a run inside the same belt.

In the following lemma, we show that if there are long runs that stay inside a belt, one can find shorter runs inside the same belt whose last configuration pairs have identical states and counter values (but not weights).

Lemma 9 (u-turn lemma). *Let c be a configuration pair, and T a sequence of transition pairs such that:*

- $T(c)$ is completely inside some belt with slope $\frac{\alpha}{\beta}$.
- $\mathsf{ce}(T) = 0$.
- $max(\mathsf{ce}(T_{0\ldots k}) \mid 0 \le k \le |T|) > (\mathsf{K}^2 \mathrm{poly}_2(\mathsf{K}) + 1)^2$.

Then, there exist $i < j < k < \ell$ such that $T' = T_{1\ldots i} T_{j+1\ldots k} T_{\ell+1\ldots |T|}$ is such that $T'(c)$ is a run inside the same belt with $\mathsf{ce}(T') = 0$, and the ending configuration pairs of $T'(c)$ and $T(c)$ only differ by their weights.

3.3 Bounding the Counter Values in Unresolved Configuration Pairs

In this section, we look at the run-pair of the minimal witness and show that counter values appearing on portions included in belts and composed only of unresolved configuration pairs can be bounded by some polynomial. There are two cases to consider here. The first one is where we have a very long belt visit, that does not enter the background space, and the second is the case when it does. We use Theorem 1, Lemma 8 and Lemma 9 to easily show that in the first case, the length of the minimal witness can be bounded and is proved in Lemma 10. The second and the most difficult case is when we have a very long belt visit, that enters the background space from the belt space and is considered in Lemma 11. In this case, we show that if the minimal witness reaches a surely-nonequivalent configuration pair whose counter values are very large, then we can pump out some portion of the run inside a belt to reach another surely-nonequivalent configuration pair whose counter values are polynomially bounded.

Lemma 10 (belt lemma). *We consider $T = \tau_0 \cdots \tau_{L-1}$ the sequence of transition pairs in Π. Suppose there are $0 < i < k < L - 1$ such that $\tau_i \cdots \tau_k(c_i)$ is included in a belt with either $m_{c_i} = m_{c_k}$ or $n_{c_i} = n_{c_k}$ then for every $i \leq r \leq k$,*

- $\mathsf{ce}(\tau_i \cdots \tau_r)|_1 \leq 2((\mathsf{K}^2 \mathsf{poly}_2 \mathsf{K}))^2 + 1)$, *and*
- $\mathsf{ce}(\tau_i \cdots \tau_r)|_2 \leq 2((\mathsf{K}^2 \mathsf{poly}_2 \mathsf{K}))^2 + 1)$ *(Fig. 3).*

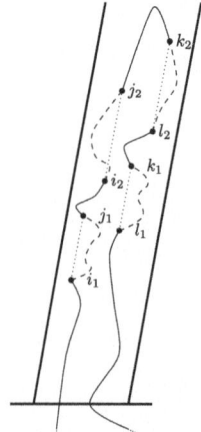
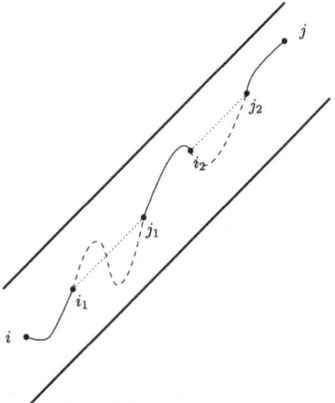

Fig. 3. A belt visit returning to the initial space.

Fig. 4. A belt visit ending in a belt.

Lemma 11. *Let c_j be the first surely-nonequivalent configuration pair in Π. Then $max\{m_{c_j}, n_{c_j}\} \leq \mathsf{poly}_1(\mathsf{K}) + \mathsf{K}(\mathsf{K}^2 \mathsf{poly}_2(\mathsf{K})) + 1$.*

Proof. Assume for contradiction that the execution of a minimal witness w reaches a surely-nonequivalent pair $c_j = \langle (p_{c_j}, m_{c_j}, s_{c_j}), (q_{c_j}, n_{c_j}, t_{c_j}) \rangle$, such that $max\{m_{c_j}, n_{c_j}\}$ is greater than $\mathsf{poly}_1(\mathsf{K}) + \mathsf{K}(\mathsf{K}^2 \mathsf{poly}_2(\mathsf{K})) + 1$. Since c_j is the first surely-nonequivalent configuration pair during the execution of a minimal witness, we know that all the previous configuration pairs are unresolved configuration pairs and lie either in the initial space or belt space. The idea here is to find an α-β repetition that is a factor of the execution inside a belt that can be cut out to obtain a shorter witness (refer Fig. 4).

Let $T = \tau_0 \cdots \tau_{L-1}$ be the sequence of transition pairs corresponding to a minimal witness from c_0. For all $0 < i \leq L$, let $c_i = \tau_0 \cdots \tau_{i-1}(c_0)$. We use $c_j = \langle \chi_{c_j}, \psi_{c_j} \rangle$, $j \leq L$, to denote the first surely-nonequivalent configuration pair during the execution of w, where $\chi_{c_j} = (p_{c_j}, m_{c_j}, s_{c_j})$ and $\psi_{c_j} = (q_{c_j}, n_{c_j}, t_{c_j})$. Let $T_1 = \tau_0 \cdots \tau_{j-1}$ and $w = v_1 v_2$, where $v_1 = word(T_{0 \cdots j-1})$ and $v_2 = word(T_{j \cdots L-1})$. Since c_j is surely-nonequivalent either (1) $\chi_{c_j} \not\equiv_K \psi_{c_j}$ or (2) $dist(\chi_{c_j}) \neq dist(\psi_{c_j})$.

Case-1: $\chi_{c_j} \not\equiv_K \psi_{c_j}$.
Since $\chi_{c_j} \not\equiv_K \psi_{c_j}$, we know that $|v_2| \leq K$. Also, since $\max\{m_j, n_j\}$ is greater than $\mathsf{poly}_1(K) + 2(K^2\mathsf{poly}_2(K)) + 1)$, by applying Lemma 8 twice, we can find $i < i_1 < j_1 \leq i_2 < j_2 < j$ such that $R(c_i) = T_{i \cdots j}(c_i)$ is a run inside a belt and $R_1 = T_{i \cdots i_1} T_{j_1+1 \cdots j}$, $R_2 = T_{i \cdots i_2} T_{j_2+1 \cdots j}$, and $R_{12} = T_{i \cdots i_1} T_{j_1+1 \cdots i_2} T_{j_2+1 \cdots j}$ are such that $R_1(c_i)$, $R_2(c_i)$ and $R_{12}(c_i)$ are runs inside in the same belt as $R(c_i)$ as shown in Fig. 4.

We can deduce that $T_1 = T_{0 \cdots i-1} R_1 T_{j+1 \cdots |T|}$, $T_2 = T_{0 \cdots i-1} R_2 T_{j+1 \cdots |T|}$ and $T_3 = T_{0 \cdots i-1} R_{12} T_{j+1 \cdots |T|}$ are such that the three runs $T_1(c_0)$, $T_2(c_0)$ and $T_{12}(c_0)$ are all valid runs that are shorter than $T(c)$ and end in configuration pairs having the same states. Let $\mathtt{aw}(T(c_0)) = (s, t)$, $\mathtt{aw}(T_1(c_0)) = (s_1, t_1)$, $\mathtt{aw}(T_2(c_0)) = (s_2, t_2)$, and $\mathtt{aw}(T_3(c_0)) = (s_3, t_3)$. We know from Theorem 1 that there exists an $i \in \{1, 2, 3\}$ such that $s_i \neq t_i$ and hence $T_i(c_0)$ contradicts the minimality of Π.

Case-2: If $dist(\chi_{c_j}) \neq dist(\psi_{c_j})$.
We show that if $\max\{m_{c_j}, n_{c_j}\} > \mathsf{poly}_1(K) + K(K^2\mathsf{poly}_2(K))) + 1$, then we can remove sub-runs to get a shorter witness contradicting the minimality of w.

Without loss of generality assume $dist(\chi_{c_j}) < dist(\psi_{c_j})$ and $\max\{m_{c_j}, n_{c_j}\} > \mathsf{poly}_1(K) + K(K^2\mathsf{poly}_2(K))) + 1$. Since c_j is the first surely-nonequivalent configuration pair in Π, for all $i \in [0, j-1]$, the configuration pairs c_i are either in the initial space or inside a belt. Let k be the smallest index such that for all $i \in [k, j-1]$, the configuration pairs c_i are inside a fixed belt. Since c_k is the first point inside a belt, $max(m_{c_k}, n_{c_k}) = \mathsf{poly}_1(K) + 1$.

Consider the run $\tau_k \cdots \tau_j(c_k)$. By the Pigeon-hole principle, there is a set of $K+1$ configuration pairs that are α-β-related to each other. Let $d_0, d_1, \ldots, d_K$ denote these configuration pairs such that $m_{d_i} < m_{d_j}$ for $i < j$. Since $dist(\chi_{c_j}) < \infty$, we know that there exists a word u such that $\chi_{c_j} \xrightarrow{u} (p', m', s') \in \mathsf{notUWA}$ with $m < K$ and $|u| = dist(\chi_{c_j})$. Let $\chi_0, \chi_1, \ldots, \chi_K$ denote the configurations, where counter values $m_{d_0}, m_{d_1}, \ldots m_{d_K}$ are encountered for the first time during the run $\chi_{c_j} \xrightarrow{u} (p', m', s')$. By the Pigeon-hole principle, there exist $r < l$, such that χ_l and χ_r have the same state. Let $u = u_1 u_2 u_3$ for some $u_1, u_2, u_3 \in \Sigma^*$ such that $\chi_{c_j} \xrightarrow{u_1} \chi_l \xrightarrow{u_2} \chi_r \xrightarrow{u_3} (p', m', s')$ and t denote the difference in counter values between χ_l and χ_r. Since χ_r and χ_l have the same state, for any $s \in \mathcal{F}$, the run $(p_{c_j}, m_{c_j} - t, s) \xrightarrow{u_1 u_3} (p, m, s'')$ is a valid run for some $s'' \in \mathcal{F}$. Therefore, for all $s \in \mathcal{F}$, $dist(p_{c_j}, m_{c_j} - t, s) < \infty$.

Note that $t = m_{d_l} - m_{d_r}$, is the same as the difference in counter values between χ_l and χ_r. Since d_l and d_r are α-β-related configuration pairs, removing the sub-run between them from the run $\tau_k \cdots \tau_j(c_k)$ will take us to the background space point $\langle \chi', \psi' \rangle$, where $\chi' = (p_{c_j}, m_{c_j} - t, s)$, for some $s \in \mathcal{F}$. Since $\langle \chi', \psi' \rangle$ is a point in the background space either $\text{dist}(\chi') = \text{dist}(\psi') = \infty$ or $\text{dist}(\chi') \neq \text{dist}(\psi')$. We already know that for all $s \in \mathcal{F}$, $\text{dist}(p_{c_j}, m_{c_j} - t, s) < \infty$, therefore $\text{dist}(\chi') \neq \text{dist}(\psi')$. Hence, $c_{j'} = \langle \chi', \psi' \rangle$ is a surely-nonequivalent configuration pair with $max\{m_{c_{j'}}, n_{c_{j'}}\} \leq \text{poly}_1(\mathsf{K}) + \mathsf{K}(\mathsf{K}^2 \text{poly}_2(\mathsf{K}))) + 1$ contradicting our initial assumption. □

Note that the technique used for proving *Case-2* is different from the one used for deterministic real-time one-counter automata by Böhm et al. [1]. Using a similar technique as mentioned above will help get a better polynomial bound on the maximum counter value encountered during the run of the minimal witness while checking the equivalence of two deterministic real-time one-counter automata.

We finally conclude the proof of our main lemma.

Proof of Lemma 3. Let $\Pi = c_0 \sigma_0 c_1 \sigma_1 \ldots c_L$ is the run of a minimal witness. From Lemma 5, we know that there is no surely-equivalent configuration pair in it. Let us assume that $c_j = \langle \chi_{c_j}, \psi_{c_j} \rangle$ is the first surely-nonequivalent configuration pair in Π. Lemma 6 ensures that $L - j \leq \min(\text{dist}(\chi_{c_j}), \text{dist}(\psi_{c_j})) + \mathsf{K}$, therefore all counter values appearing at position greater than j are bounded by $\max(n_{\chi_{c_j}}, n_{\psi_{c_j}}) + \min(\text{dist}(\chi_{c_j}), \text{dist}(\psi_{c_j}))) + \mathsf{K}$.

Lemma 10 ensures that all counter values appearing before c_j inside belts are bounded by the value $\text{poly}_1(\mathsf{K}) + 2((\mathsf{K}^2 \text{poly}_2(\mathsf{K})))^2 + 1)$, and by Lemma 11, $\max(n_{\chi_{c_j}}, n_{\psi_{c_j}}) \leq \text{poly}_1(\mathsf{K}) + \mathsf{K}(\mathsf{K}^2 \text{poly}_2(\mathsf{K}))) + 1$.

Finally, Lemma 2 ensures that $\text{dist}(\chi_{c_j})$ and $\text{dist}(\psi_{c_j})$ cannot be more than $(max\{\mathsf{K}, n_{\chi_{c_j}}, m_{\psi_{c_j}}\} + \mathsf{K}^2)\mathsf{K}$. Therefore, we conclude that there exists a polynomial $\text{poly}_0(\mathsf{K}) = \text{poly}_1(\mathsf{K}) + 2(\mathsf{K}^2 \text{poly}_2(\mathsf{K})))^2 + 1)$ such that all counter values appearing in Π are bounded by $\text{poly}_0(\mathsf{K})$. □

From Lemma 3, we get that the value of poly_0 is $\mathcal{O}(\mathsf{K}^{12})$ and from Lemma 4, we get that the the length of the minimal witness is $\mathcal{O}(\mathsf{K}^{26})$. Hence, if two DWROCAS $\mathcal{A}$ and $\mathcal{B}$ are not equivalent, then there is a word whose length is of $\mathcal{O}(\mathsf{K}^{26})$ such that $f_\mathcal{A}(w) \neq f_\mathcal{B}(w)$.

4 Conclusion and Future work

In this paper, we presented an improved polynomial time algorithm for checking the equivalence of two DWROCAS that return a word distinguishing them, if it exists. A potential research direction is to remove the "real-time" constraint in DWROCA. Note that equivalence for deterministic one-counter automata is NL-complete regardless of whether it is real-time. However, the techniques for the non-real-time case differ from those used here. Therefore, it's not guaranteed that those techniques are robust enough to the addition of weights, which makes it an interesting topic for further investigation. Finally, a polynomial time algorithm for equivalence of DWROCA also paves the way for efficient learning algorithms.

Acknowledgments. A.V. Sreejith would like to acknowledge the support by SERB for the project "Probabilistic Pushdown Automata" [MTR/2021/000788].

References

1. Böhm, S., Göller, S.: Language equivalence of deterministic real-time one-counter automata is NL-complete. In: Murlak, F., Sankowski, P. (eds.) MFCS. Lecture Notes in Computer Science, vol. 6907, pp. 194–205. Springer (2011)
2. Böhm, S., Göller, S., Jančar, P.: Bisimilarity of one-counter processes is PSPACE-Complete. In: Gastin, P., Laroussinie, F. (eds.) CONCUR. Lecture Notes in Computer Science, vol. 6269, pp. 177–191. Springer (2010)
3. Böhm, S., Göller, S., Jancar, P.: Equivalence of deterministic one-counter automata is NL-complete. In: Boneh, D., Roughgarden, T., Feigenbaum, J. (eds.) Symposium on Theory of Computing Conference, STOC 2013, Palo Alto, CA, USA, 1–4 June 2013, pp. 131–140. ACM (2013). https://doi.org/10.1145/2488608.2488626
4. Böhm, S., Göller, S., Jancar, P.: Bisimulation equivalence and regularity for real-time one-counter automata. J. Comput. Syst. Sci. **80**(4), 720–743 (2014)
5. Forejt, V., Jancar, P., Kiefer, S., Worrell, J.: Language equivalence of probabilistic pushdown automata. Inf. Comput. **237**, 1–11 (2014)
6. Kiefer, S., Murawski, A.S., Ouaknine, J., Wachter, B., Worrell, J.: On the complexity of equivalence and minimisation for q-weighted automata. Log. Methods Comput. Sci. **9**(1) (2013). https://doi.org/10.2168/LMCS-9(1:8)2013
7. Mathew, P., Penelle, V., Saivasan, P., Sreejith, A.: Weighted one-deterministic-counter automata. In: Bouyer, P., Srinivasan, S. (eds.) 43rd IARCS Annual Conference on Foundations of Software Technology and Theoretical Computer Science (FSTTCS 2023). Leibniz International Proceedings in Informatics (LIPIcs), vol. 284, pp. 39:1–39:23. Schloss Dagstuhl – Leibniz-Zentrum für Informatik, Dagstuhl (2023). https://doi.org/10.4230/LIPIcs.FSTTCS.2023.39, https://drops.dagstuhl.de/entities/document/10.4230/LIPIcs.FSTTCS.2023.39
8. Sénizergues, G.: The equivalence problem for deterministic pushdown automata is decidable. In: Degano, P., Gorrieri, R., Marchetti-Spaccamela, A. (eds.) Automata, Languages and Programming, pp. 671–681. Springer, Heidelberg (1997)
9. Stirling: Deciding DPDA equivalence is primitive recursive. In: ICALP: Annual International Colloquium on Automata, Languages and Programming (2002)
10. Valiant, Paterson: Deterministic one-counter automata. JCSS: J. Comput. Syst. Sci. **10** (1975)

Passive Learning of Fuzzy Temporal Logic Rules from Finite Traces

Sandip Paul[1] and Bornali Paul[2(✉)]

[1] Kolaghat Government Polytechnic, Purba Medinipore, West Bengal, India
[2] Department of Philosophy, Rabindra Bharati University,
Kolkata, West Bengal, India
talking2bornali@gmail.com

Abstract. This work addresses the problem of synthesizing fuzzy temporal logic rules from a set of given positive and negative examples. The examples are provided in the form of execution traces of finite length. Fuzzy Time Linear Temporal Logic over finite traces (FTL_f) is chosen as the language for rule synthesis. FTL_f is capable of capturing fuzzy temporal modalities, like, 'soon after', 'almost always', 'gradually' etc., that make the learnt rules simpler and more understandable than classical LTL representations. The proposed approach reduces the learning task to a multi-valued partial maximum satisfiability (PMaxSAT) problem. This work is useful for generating interpretable explanations of complex system behaviours.

Keywords: Passive Learning · Temporal Logic · Fuzzy Time Linear Temporal Logic · Multi-valued SAT

1 Introduction

Constructing general hypothesis from specific observations or from a set of examples is an important aspect of the human learning procedure. This process is useful for understanding the behaviour of a complex system and for providing explanations of certain behavioural patterns. This aspect of learning, known as inductive learning, is the essence of machine learning; where a system's mathematical model is inferred from a set of numerical data. Inductive learning is also employed in the form of *imitation learning* (a variant of reinforcement learning) which is used for designing autonomous robots, that performs by imitating the steps that a human would follow in that scenario. If the examples are provided in terms of logical facts and a set of logical rules (or logic program) are inferred from the examples, then the process is called *inductive reasoning* (or inductive logic programming). This inductive or passive learning finds its application in analysing and explaining the behaviour of autonomous robots [12]; process behaviour understanding or process mining [14] etc.

The task of understanding the behaviour of a reactive system, where the examples are given in terms of an infinite or finite flow of events (or state transitions), known as *traces*, is a widely studied problem. Temporal logic is mostly used to model and reason with the process-flow of such systems. In such a case, examples are provided in the form of a set of execution traces, which may be finite or infinite. The execution traces are categorized into positive (or expected) examples and negative (or unwanted) examples. A proper temporal logical rule is to be learnt from the examples that are in accordance with the positive examples and rules out the negative examples. This form of inductive learning is known as *passive learning*.

Passive learning of Linear Temporal Logic (LTL) rules has been reported in plenty of literature [5,10,13,18,22]. For this, mostly two approaches have been adopted. One approach to passive learning is to reduce the problem of learning an LTL formula into a propositional satisfiability problem; and a solution is constructed by means of a syntax directed acyclic graph (DAG) with the help of off-the-shelf SAT solver [18]. The approach proposed in [18] is further extended for incorporating noisy data [10] by using a weighted MAXSAT solver and the LTL is defined over finite traces (LTL_f). In [21] the authors employ program specification language temporal logic (PSL) in place of LTL and in [11] signal temporal logic (STL) and SMT solver are used for synthesizing system descriptions from examples. Synthesizing Metric Temporal Logic (MTL) rules has been addressed in [19], where the problem is reduced to a series of satisfiability modulo theory (SMT) problems that are solved using Linear Real Arithmetic (LRA).

Exhaustive search over all DAGs is time consuming and makes the algorithm inefficient. The problem can be resolved by partitioning the search space into small and manageable sub-parts using topology structure, called partial DAGs [20]. Another line of approaches use deterministic finite automata [22] or alternating finite automata [5] to encode the LTL formulas and construct a satisfiability problem that can be solved using SAT-solvers.

However, in some real-life cases, there are some sort of uncertainty and imprecision with respect to the time in the execution traces. For example, in the execution traces the following explanations may arise; "in all the successful traces, an event p occurs *almost always*", "event q occurs *soon* after event p". These situations cannot be expressed using the language of classical LTL. Although the aforementioned explanations are shorter and more intuitive and closer to human explanations in natural language.

To salvage this problem, we attempt to replace LTL with its fuzzy extension. Fuzzy extensions of LTL has been studied from two perspectives; 1. Fuzzy Linear Temporal Logic ($FLTL$) [15], where just connectives are interpreted as fuzzy, but the temporal modalities are not fuzzified; 2. Fuzzy Time Linear Temporal Logic (FTL) [9], where fuzzy temporal modalities are also included along with fuzzy logical connectives. Passive learning with fuzzy extensions of temporal logic has not been explored before. This is the aim of this paper.

Thus, in this work we attempt to learn FTL_f rules, which is a variation of FTL, from a set of example traces of finite length, by means of passive learning. In FTL_f the semantics is defined for finite traces (semantics of FTL is defined based on infinite-length traces). The examples are provided in the form of two disjoint sets; i.e., set of positive examples and set of negative examples.

The methodology followed here is inspired by the MAXSAT-based algorithm, proposed in [10]. However, the incorporation of FTL_f makes the proposed work much more challenging.

This work is significant for interpretable model learning, explaining system behaviours, reinforcement learning etc.

2 Preliminaries

2.1 Fuzzy Time Linear Temporal Logic on Finite Traces (FTL_f)

Temporal logic is a logical framework that is used for specifying and verifying properties of reactive systems. A popular variant of temporal logic is Linear Temporal Logic (LTL) [4], that deals with an infinite sequence of states or events, where each point in time has a unique successor. LTL can be thought of as an extension of Boolean propositional logic with temporal operators. However, LTL is unable to deal with vagueness in time, that can be expressed by some temporal operators like 'almost always', 'soon', 'gradually' etc. These modalities are necessary in situations where a formula is partially satisfied or an event occurs with slight delay than expected or a property is satisfied always except for a few points of time. To address these aspects, that may arise quite often in real-life scenarios, Fuzzy Time Linear Temporal Logic (FTL) [9] is proposed, that extends LTL with fuzzy evaluation of propositions as well as fuzzy temporal modalities. LTL, as well as FTL, are defined over an infinite sequence of events or infinite traces. Some modifications of LTL over finite traces are also proposed [6]. Here, we have presented FTL over finite traces (FTL_f). We have considered a small subset of fuzzy temporal operators that is most relevant from the perspective of passive learning.

Syntax: The formulas of FTL_f can be inductively defined as:
$$\phi = p \mid \phi \vee \psi \mid \neg \phi \mid \mathcal{O}\phi \mid \phi \mathbf{U} \psi$$
where, $p \in \mathbf{P}$ (the set of all atomic propositions), $\mathcal{O} \in \mathbf{O}$, the set of unary temporal operators. Other classical binary connectives $\wedge, \Rightarrow, \Leftrightarrow$ are defined using $\neg, \vee$ as usual.

The temporal modal operators include operators for the "future" only, not any "past" operators are included. There are a wide range of unary and binary operators included in literature [9,16]. However, in this work we have chosen a small set of fuzzy temporal operators that seem to be more intuitive for learning rules. The set $\mathbf{O}$ includes traditional temporal operators like $\mathbf{X}(Next)$, $\square(Always)$, $\Diamond(Eventually)$; as well as fuzzy temporal operators like $Soon$, $\mathbf{Gr}(Gradually)$, $\mathbf{A}\square(Almost\ Always)$ and one special operator for finite traces $\mathbf{L}(Last)$. $\mathbf{U}$ is the $Until$ operator taken from classical temporal logic. There are several other binary fuzzy temporal operators which are not considered in this work.

Semantics: Unlike the FTL defined over infinite traces, the semantics of FTL_f is defined for finite traces.

The semantics is based on a linear time structure $(\mathbb{S}, s_0, \pi, \mathbb{L}, \eta)$; where $\mathbb{S}$ is the set of states, s_0 is the initial state, π is a finite trace of the form $s_0 s_1 s_2 .. s_N \in \mathbb{S}^+$, $\mathbb{L}$ is a labeling function as $\mathbb{S} \to [0,1]^\mathbf{P}$ that assigns to each state in $\mathbb{S}$ an evaluation for each atomic proposition in $\mathbf{P}$ from $[0,1]$. $\eta : \mathbb{Z} \to [0,1]$ is a weight function or an avoiding function, that is used to define the semantics of fuzzy temporal operators. Function η accounts for the penalization imposed on the number of events that we want to ignore in evaluating the truth degree of a formula that contains an "almost" or "soon" operators.

It is customary [9,16] to take η as a decreasing function $\eta : \mathbb{Z} \to [0,1]$ and use it to define the weight functions for all the fuzzy temporal operators. For $i \in \mathbb{Z}, i \leq N$ and for a predefined positive integer $n_\eta \leq N$, $\eta(i) = 1$ for $i \leq 0$ and $\eta(i) = 0$ for $i \geq n_\eta$. The weight function corresponding to some fuzzy temporal modalities are defined as below:

- $\eta_{Soon}(i) = \eta(i)$, for any $i < N$,
- $\eta_{presently}(i) = (\eta(i))^2$, for any $i < N$,
- $\eta_{\mathbf{Gr}}(i) = \eta(n_\eta - i)$, for any $i < N$.

Now, based on a linear time structure $(\mathbb{S}, s_0, \pi, \mathbb{L}, \eta)$, as defined above, the evaluation of different FTL_f formulas along a path π is a real number in $[0,1]$. The evaluation for all the FTL_f formulas can be recursively defined for all traces of the form $\pi = s_0, s_1, ..., s_N \in \mathbb{S}^+$ or $\pi_i = s_i, s_{i+1}, ..., s_N \in \mathbb{S}^+$ as follows [16]:

1. $\|p\|(\pi_i) = \mathbb{L}(s_i)(p)$;
2. $\|\phi \vee \psi\|(\pi_i) = max\{\|\phi\|(\pi_i), \|\psi\|(\pi_i)\}$;
3. $\|\neg \phi\|(\pi_i) = 1 - \|\phi\|(\pi_i)$;
4. $\|\mathbf{X}\phi\|(\pi_i) = \|\phi\|(\pi_{i+1})$ for $i < N$;
5. $\|\Box \phi\|(\pi_i) = \bigwedge_{j \geq i}^{N} \|\phi\|(\pi_j)$;
6. $\|\Diamond \phi\|(\pi_i) = \bigvee_{j \geq i}^{N} \|\phi\|(\pi_j)$;
7. $\|\phi \mathbf{U} \psi\|(\pi_i) = \bigvee_{j \geq i}^{N} \|\psi\|(\pi_j) \wedge \bigwedge_{k \geq i}^{j} \|\phi\|(\pi_k)$;
8. $\|Soon\ \phi\|(\pi_i) = \bigvee_{j=i+1}^{min\{N, i+n_\eta\}} \|\phi\|(\pi_j).\eta_{Soon}(j-i-1)$
9. $\|Presently\ \phi\|(\pi_i) = \bigvee_{j=i+1}^{min\{N, i+n_\eta\}} \|\phi\|(\pi_j).\eta_{presently}(j-i-1)$
10. $\|\mathbf{Gr}\ \phi\|(\pi_i) = \bigvee_{j=i+1}^{min\{N, i+n_\eta\}-1} \|\phi\|(\pi_j).\eta_{\mathbf{Gr}}(j-i-1) \vee \bigvee_{j=min\{N, i+n_\eta\}}^{N} \|\phi\|(\pi_j)$
11. $\|A\Box\ \phi\|(\pi_i) = \bigvee_{j \in I_N} \bigvee_{H \subseteq I_N, |H|=N-j} \bigwedge_{h \in H} \|\phi\|(\pi_{i+h}).\eta(j)$; where $I_N = \{1, 2, ..., N\}$
12. $\|\mathbf{L}\ \phi\|(\pi) = \|\phi\|(\pi_N)$

The evaluation of classical propositional operators ($\neg, \vee$) and classical temporal operators ($\Box, \Diamond, \mathbf{X}, \mathbf{U}$) are standard. The operator *Soon* extends the $\mathbf{X}$(next) operator by ignoring a time delay of at most n_η. *Presently* imposes a more severe penalty for the ignored time instants. *Almost always* ($A\Box$) avoids at most n_η evaluations over the path π_i and penalizes the ignored instants according to the weighing function. The unary operator $\mathbf{L}$ denotes evaluation in the very last state of the finite trace.

The interpretation of fuzzy time temporal operators and evaluation of fuzzy time temporal formulas can be found in details in [9].

2.2 Multi-valued Satisfiability Problem

A multi-valued variable, x, is a variable that takes on values from a finite domain D_x, such that $|D_x| \geq 2$. For a multi-valued variable x and a value v, a multi-valued literal is of the form $x = v$ or $x \neq v$ so that $v \in D_x$.

A multi-valued clause, c, is defined as disjunction of multi-valued literals. A weighted multi-valued clause is a pair (c, w); where w is the weight of the clause c. The weight w can be infinity (then it is a hard clause), or can be any finite integer (then it is a soft clause).

Given a set of variables $X = \{x_1, x_2, ...x_n\}$, let D_X denotes the union of all the domains of the individual variables, i.e., $D_X = \{D_{x_1} \cup D_{x_2} \cup ... \cup D_{x_n}\}$. Now, a multi-valued assignment $\alpha : X \rightarrow D_X$ is said to be *complete* if it is defined for each variable in X; is said to be *admissible* if $\alpha(x_i) \in D_{x_i}$. An admissible assignment α satisfies $x_j = i$ if $\alpha(x_i) = i$; otherwise α satisfies $x_j \neq i$.

Given a set of multi-valued variables X and a set of weighted multi-valued clauses defined over X, the multi-valued weighted partial maximum satisfiability (WPMaxSAT) problem is to decide an admissible assignment that satisfies all the hard constraints and maximizes the total weight of the satisfied soft clauses. Some more terminologies are important in this regard. A weighted MaxSat (WMaxSAT) is a WPMaxSAT without any hard constraints. A partial MaxSAT (PMaxSAT) is a WPMaxSAT when all the soft constraints are of equal weight. Different many-valued SAT (MV-SAT) solvers have been proposed [2,3,7,8,17] and the search for efficient MVSAT solvers is still a very active area of research.

3 Problem Statement

This work addresses the problem of learning from a set of given positive and negative examples. This problem broadly falls under inductive reasoning, more specifically *passive learning* [1]. In this paper, we aim to infer description of complex system behavior in the form of FTL_f rules, by which we can find the best possible explanation of how two sets of execution traces differ. The rule is learnt based on a set of examples, given as finite execution traces. The set of examples, S, consists of two disjoint sets of traces; positive examples π^+ and negative examples π^-.

Now, given a set $S = \pi^+ \cup \pi^-$, and two thresholds $0 \leq \alpha \leq \beta \leq 1$, an FTL_f formula ϕ is said to be satisfied by a positive trace $\pi \in \pi^+$ if $\|\phi\|(\pi) \geq$

β and is said to be rejected by a negative trace $\pi' \in \pi^-$ if $\|\phi\|(\pi') \leq \alpha$. A formula is said to be consistent with a set of examples S, if it is satisfied by all the positive examples and is rejected by all the negative examples. But, in many practical cases, it becomes impossible to satisfy (or reject) all positive (or negative) examples. So, the constraint can be loosen up and we aim to find a formula that satisfies (and rejects) maximum number of possible positive (and negative) example traces. Formally, a *loss* can be defined. Let $\pi^\phi_{unsat} = \{\pi^+|\pi^+ \text{does not satisfy } \phi\} \cup \{\pi^-|\pi^- \text{does not reject } \phi\}$. Then the loss for a given set of examples S and a formula ϕ, is defined as

$$L^S_{\Phi_{FTLf}} = \left|\pi^\phi_{unsat}\right| / \left|\pi^+ \cup \pi^-\right|.$$

In other words, the loss is the fraction of examples that is not consistent with the derived formula ϕ.

Also, a restriction has to be imposed on the size of the learnt formula. The size of a formula is defined as the number of its subformulas. We attempt to learn the formula of minimum size. This imposed minimality eliminates the possibility of overfitting. Moreover, smaller formulas are easier for humans to comprehend than large ones.

So the task can be formally stated as: "Given a set of positive (π^+) and negative (π^-) example traces (finite) and a threshold **k**, construct an FTL_f formula of minimal size and having a maximum size of **T**, that gives $L^S_{\Phi_{FTLf}} \leq \mathbf{k}$".

4 Proposed Procedure for Passive Learning

Now, that the problem undertaken is clearly stated in a formal way, we focus on a feasible solution. The approach taken in this paper follows the one given by Neider et al. [18]. The basic idea is to reduce the problem into a satisfiability problem, so that any existing off-the-shelf SAT solvers can be used to achieve the solution. Since the aim is to find a FTL_f formula, the reduction is done to obtain a multi-valued satisfiability problem instead of a Boolean SAT problem. Moreover, assuming that it is impossible in a real-life scenario to have a formula that can accurately explain all the examples, we allow a certain relaxation and aim for maximum possible satisfiability. Thus, we reduce the passive learning problem to a multi-valued partial MAXSAT (PMaxSAT) problem.

The key step of the proposed approach is construction of a formula Φ^S_n, for a given integer n and set of example traces S; and assigning weights to the clauses of Φ^S_n, so that,

1. The size of Φ^S_n is polynomial in n,
2. Φ^S_n contains enough information to construct an FTL_f formula of size n from it,
3. By making appropriate multi-valued assignments to the variables in Φ^S_n, the incurred loss corresponding to the sample S and derived FTL_f formula is minimized.

The size n is increased by 1 in each step (starting from n=1). If for a particular n, the corresponding FTL_f formula offers a loss value less than the predefined threshold **k**, then the algorithm terminates. Otherwise the process is repeated untill we reach an upper limit of n, which is **T**. If until then no satisfactory FTL_f formula is achieved, the algorithm returns 'Unsatisfiable'; denoting that the extraction of a generalized behavior from the set of example traces is impossible for the given resources. The complete algorithm is shown in Fig. 1.

The proposed approach is based on syntactic representation of FTL_f formulas using a syntax tree, that is unique to each formula. The syntax tree is transformed into a syntax DAG by sharing common subformulas. The idea is demonstrated for an FTL_f formula $(\neg p \wedge A\Box\ r) \vee (Soon\ q \wedge r\mathbf{U}s)$ in Fig. 2. One special feature of the syntax DAG is that, the number of subformulas is exactly equal to the number of nodes in the DAG. To simplify the operations, each of the nodes of the syntax DAG (having n nodes) is assigned with a unique identifier $i \in \{1, ..., n\}$; such that the root node has identifier n; the identifier of each parent node is higher than the identifiers of its children and identifier 1 is always assigned to a node having a propositional symbol.

In the proposed algorithm, a syntax DAG with n nodes is used to represent the unknown FTL_f formula, Φ_{FTLf}, consisting of exactly n subformulas. The formula Φ_n^S captures the structural feature of the syntax DAG in its clauses. Also, the variables of Φ_n^S are constrained properly and appropriate weights are assigned to the clauses of Φ_n^S, such that a proper multi-valued assignment to the variables of Φ_n^S minimizes the loss with respect to Φ_{FTLf} and set of examples S.

Algorithm 1: Passive Learning Algorithm for Learning FTL_f Rules

Input: A set of finite traces $\pi^+ \cup \pi^-$, Threshold k
Output: An FTL_f formula with at most T nodes that makes the *loss* $\leq k$ or 'Unsatisfiable'

1. $n \leftarrow 0$;
2. **Repeat**
3. $\qquad n \leftarrow n + 1$
4. $\qquad$ Construct formula φ_n^S
5. $\qquad$ Assign infinite weights to hard clauses and weight of 1 to all the soft clauses
6. $\qquad$ Find satisfying assignments using multi-valued partial MAXSAT
7. $\qquad$ Calculate the loss
8. $\qquad$ If loss < k:
9. $\qquad\qquad$ construct FTL_f formula φ_{FTLf} from satisfying assignment of φ_n^S
10. $\qquad\qquad$ return φ_{FTLf} and exit
11. $\qquad$ **Else:**
12. $\qquad\qquad$ continue
13. **Until n = T**
14. **return 'Unsatisfiable'**

Fig. 1. Passive Learning Algorithm

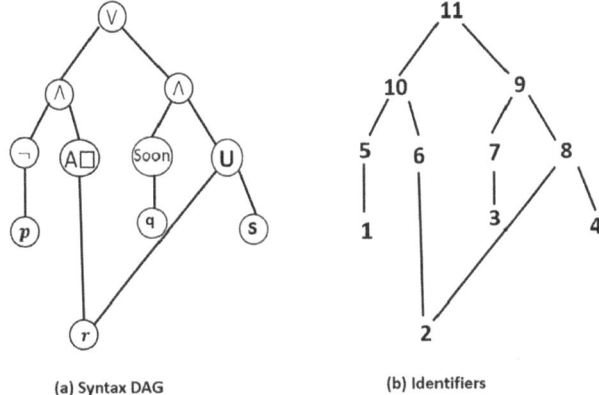

Fig. 2. (a) Syntax DAG for the formula $(\neg p \wedge A\Box\, r) \vee (\text{Soon } q \wedge r\mathbf{U}s)$, (b) Identifiers of the syntax DAG

4.1 Construction of Φ_n^S

For a given set of example traces ($S = \pi^+ \cup \pi^-$) and a positive integer n; Φ_n^S will be designed to 1) encode the syntax DAG for an FTL_f formula and 2) its semantic evaluation over the elements of S. Before going into the construction procedure, some points are needed to be elaborated.

A set of the form $\mathscr{T} = \{0, \frac{1}{m}, \frac{2}{m}..., 1\}$ is taken to be the domain of all multi-valued variables and also as set of truth-values for semantic evaluation of FTL_f formulas. The logical connectives $\wedge, \vee$ are interpreted as min and max respectively. The multiplication operations in Sect. 2.1 is taken as Lukaseiwicz t-norm ($a \otimes b = max\{a+b-1, 0\}$), to ensure closure of all operations in the set $\mathscr{T}$.

If we consider three propositional symbols $p, q, r \in \mathbf{P}$, an example trace (be it positive example or negative example) of length 4 may look like $\pi^1 = \{\{p : 0.1, q : 0.9, r : 1\}, \{p : 0.1, q : 1, r : 1\}, \{p : 0.2, q : 0, r : 0.1\}, \{p : 0.2, q : 0, r : 0.1\}\}$. It can be seen that each state of a particular finite trace $\pi \in S$ gives a mapping $\mathbf{P} \to \mathscr{T}$. This will be depicted as the labeling function for a trace π, $\mathbb{L}^\pi : S^\pi \to \mathscr{T}^\mathbf{P}$ for the set of states $S^\pi \subseteq \mathbb{S}$ in the trace π.

Now, let's dive into the actual methodology of constructing Φ_n^S. Let $\mathbf{P}$ is the set of propositional symbols and Λ is the set of all logical operators.

Variables: We introduce some Boolean variables and some auxiliary multi-valued variables to reduce the problem into a satisfiability problem;

- $label_{i,\lambda} : \{0,1\}$, for $1 \leq i \leq n$ and $\lambda \in \mathbf{P} \cup \Lambda$;
- $l_{i,j} : \{0,1\}$, for $2 \leq i \leq n$, $1 \leq j \leq i-1$;
- $r_{i,j} : \{0,1\}$, for $2 \leq i \leq n$, $1 \leq j \leq i-1$;
- $val_{i,\tau}^\pi : \mathscr{T}$, for $1 \leq i \leq n, 0 \leq \tau \leq N$ and $\pi \in \pi^+ \cup \pi^-$

The variables $label_{i,\lambda}$ are Boolean variables, having domain $\{0,1\}$. They are defined for each node in the syntax DAG and denote the labeling of the corresponding node. If $label_{i,\lambda} = 1$ then the label of the ith node in the syntax DAG is λ, be it a propositional symbol or any logical operator. The variables $l_{i,j}$ (and $r_{i,j}$) denotes that the node with identifier j is the left (and right) child of the node with identifier i. The multi-valued variables $val_{i,\tau}^\pi$ are defined for each node i in the syntax DAG and each position τ of the finite example traces $\pi \in S$. The variables $val_{i,\tau}^\pi$ are introduced to capture the valuation of the subformula ϕ_i, rooted at the node i over the sub-trace π_τ, that is the portion of π starting from the position τ upto the last position.

Clauses: Two sets of clauses are constructed.

1. The Structural Clauses: The following clauses enforce some structural constraints on the syntax DAG;
   (a) Unique Labeling:
   $$\left[\bigwedge_{i=1}^{n} \bigvee_{\lambda \in P \cup \Lambda} label_{i,\lambda} = 1\right] \wedge \left[\bigwedge_{i=1}^{n} \bigwedge_{\lambda \neq \lambda' \in P \cup \Lambda} \neg(label_{i,\lambda} = 1 \wedge label_{i,\lambda'} = 1)\right]$$
   (b) Unique Left Child:
   $$\left[\bigwedge_{i=2}^{n} \bigvee_{j=1}^{i-1} l_{i,j} = 1\right] \wedge \left[\bigwedge_{i=2}^{n} \bigwedge_{1 \leq j < j'}^{i-1} \neg(l_{i,j} = 1 \wedge l_{i,j'} = 1)\right]$$
   (c) Unique Right Child:
   $$\left[\bigwedge_{i=2}^{n} \bigvee_{j=1}^{i-1} r_{i,j} = 1\right] \wedge \left[\bigwedge_{i=2}^{n} \bigwedge_{1 \leq j < j'}^{i-1} \neg(r_{i,j} = 1 \wedge r_{i,j'} = 1)\right]$$
   (d) Propositional Symbol in a Leaf Node:
   $$\bigwedge_{i=1}^{n} \left[\bigvee_{p \in P} (label_{i,p} = 1) \rightarrow \left[\bigwedge_{j=1}^{n} \neg l_{i,j} \wedge \bigwedge_{j'=1}^{n} \neg r_{i,j'}\right]\right]$$

2. The Semantic Clauses:
   (a)
   $$\left[\bigwedge_{i=1}^{n} \bigwedge_{p \in P} label_{i,p} = 1\right] \rightarrow \bigwedge_{\tau=0}^{N} val_{i,\tau}^\pi = \mathbb{L}^\pi(\tau)(p)$$
   (b)
   $$\left[\bigwedge_{i=2}^{n} \bigwedge_{j=1}^{i-1} label_{i,\neg} = 1 \wedge l_{i,j} = 1\right] \rightarrow \bigwedge_{\tau=0}^{N} val_{i,\tau}^\pi = 1 - val_{j,\tau}^\pi$$

(c)
$$\left[\bigwedge_{i=2}^{n}\bigwedge_{j,j'=1}^{i-1} label_{i,\vee}=1 \wedge l_{i,j}=1 \wedge r_{i,j'}=1\right] \rightarrow \bigwedge_{\tau=0}^{N}\left[val_{i,\tau}^{\pi}=max\{val_{j,\tau}^{\pi},val_{j',\tau}^{\pi}\}\right]$$

(d)
$$\left[\bigwedge_{i=2}^{n}\bigwedge_{j=1}^{i-1} label_{i,\Diamond}=1 \wedge l_{i,j}=1\right] \rightarrow \bigwedge_{\tau=0}^{N}\left[val_{i,\tau}^{\pi}=\max_{\tau\leq\tau'} val_{j,\tau'}^{\pi}\right]$$

(e)
$$\left[\bigwedge_{i=2}^{n}\bigwedge_{j=1}^{i-1} label_{i,\square}=1 \wedge l_{i,j}=1\right] \rightarrow \bigwedge_{\tau=0}^{N}\left[val_{i,\tau}^{\pi}=\min_{\tau\leq\tau'} val_{j,\tau'}^{\pi}\right]$$

(f)
$$\left[\bigwedge_{i=2}^{n}\bigwedge_{j=1}^{i-1} label_{i,\mathbf{X}}=1 \wedge l_{i,j}=1\right] \rightarrow \bigwedge_{\tau=0}^{N} (val_{i,\tau}^{\pi}=val_{j,\tau+1}^{\pi})$$

(g)
$$\left[\bigwedge_{i=2}^{n}\bigwedge_{j=1}^{i-1} label_{i,Soon}=1 \wedge l_{i,j}=1\right] \rightarrow \bigwedge_{\tau=0}^{N}\left[val_{i,\tau}^{\pi}=\max_{1\leq k\leq n_\eta} val_{j,\tau+k}^{\pi} \otimes \eta(k)\right]$$

(h)
$$\left[\bigwedge_{i=2}^{n}\bigwedge_{j,j'=1}^{i-1} label_{i,\mathbf{U}}=1 \wedge l_{i,j}=1 \wedge r_{i,j'}=1\right] \rightarrow$$

$$\bigwedge_{\tau=0}^{N}\left[val_{i,\tau}^{\pi}=\max_{\tau\leq\tau'\leq N} min\{val_{j,\tau'}^{\pi}, \min_{\tau\leq t\leq\tau'} val_{j',\tau}^{\pi}\}\right]$$

(i)
$$\left[\bigwedge_{i=2}^{n}\bigwedge_{j=1}^{i-1} label_{i,\mathbf{Gr}}=1 \wedge l_{i,j}=1\right] \rightarrow \bigwedge_{\tau=0}^{N}\left[val_{i,\tau}^{\pi}=\max_{1\leq k\leq n_\eta} val_{j,\tau+k}^{\pi} \otimes \eta_{\mathbf{Gr}}(k)\right]$$

(j) For $I_N = \{1,2,...,N\}$,

$$\left[\bigwedge_{i=2}^{n}\bigwedge_{j=1}^{i-1} label_{i,A\square}=1 \wedge l_{i,j}=1\right] \rightarrow \bigwedge_{\tau=0}^{N}\left[val_{i,\tau}^{\pi}=\max_{j\in I_N} \max_{H\subseteq I_N, |H|=N-j} \min_{h\in H} val_{i,\tau+h}^{\pi} \otimes \eta(j)\right]$$

The structural clauses specify the structural features of the syntax DAG. Clause 1 (a) ensures unique labeling for each node. Clauses 1 (b) and 1 (c) imposes the constraint of having unique left and unique right child. Clause 1(d) states that the leaf nodes of DAG is labeled with a propositional symbol from **P**. All these structural clauses are combined together via conjunction to obtain a formula Φ_n^{str}.

The semantic clauses are responsible to ensure that the constructed DAG is consistent with the positive and negative example traces. The clauses are obtained directly from the semantics of the corresponding operators in FTL_f. Clauses (g),(i) and (f) are modified in terms of multiplication of the weight function, where normal multiplication is replaced with Lukasiewicz t-norm to ensure closure of the valuations in the set $\mathbb{L}$. The semantic clauses as expressed here are shorthand representations of much longer clauses consisting of multi-valued literals. The formula Φ_π^{sem} is obtained from the conjunction of all the semantic clauses.

Along with the structural constraints and semantic constraints two additional constraints are imposed in order to ensure that the constructed formula is satisfied by the example traces in $\pi^+ \cup \pi^-$ upto a certain degree. We have chosen the satisfaction degree to be 0.5. So the complete set of constraints is conjoined together in the form:

$$\Phi_n^S = \Phi_n^{str} \wedge \bigwedge_{\pi \in S} \Phi_\pi^{sem} \wedge \bigwedge_{\pi \in \pi^+} (val_{n,0}^\pi \geq 0.5) \wedge \bigwedge_{\pi \in \pi^-} (val_{n,0}^\pi \leq 0.5)$$

4.2 Weight Assignment and Solution

Let, the set of propositional symbols and logical operators has a cardinality c, i.e. $|\mathbf{P} \cup \Lambda| = c$. Also, S be the set of examples and each example trace has a length N. Then the constructed formula Φ_n^S contains $O(n^2 + nN|S|)$ new variables. The size of Φ_n^S is polynomial in n.

After construction the formula Φ_n^S is converted into CNF in accordance with [23]. During the transformation every multi-valued literal, of the form $x = i$ or $x \neq i$, is treated as an atomic proposition.

After the transformation, the clauses of Φ_n^S are assigned with weights. The clauses Φ_n^{str} and $\bigwedge_{\pi \in S} \Phi_\pi^{sem}$ are assigned *infinite* weight, i.e., they are posed as *hard constraints* or *hard clauses*. The last two clauses are given unity weight, i.e. $weight = 1$; so that the calculated loss directly accounts for the number of unsatisfied example traces. Hence, these clauses are treated as *soft clauses* or *soft constraints*. Now that, all the soft constraints are assigned with same weight this multi-valued assignment problem takes the form of a partial MaxSAT (PMaxSAT) problem [3]. This problem can be solved using any off-the-shelf MVSAT solvers [2,3,7,8,17].

Termination and Correctness: Given a sample $S = \{\pi^+, \pi^-\}$ and specified loss threshold **k**, Algorithm 1 is bound to terminate after at most T steps, which bounds the maximum size of the derived FTL_f formula. At termination, it either returns an FTL_f formula of size $\leq T$ such that the formula gives a loss of $\leq \mathbf{k}$ over the sample S; or it returns 'Unsatisfiable', which states that it is not possible to construct any FTL_f formula with size $\leq T$ which makes the loss $\leq \mathbf{k}$ over the sample S.

Lemma 1. *For a given sample $S = \{\pi^+, \pi^-\}$, some number $n \in \mathbb{N}\setminus\{0\}$ and a given set $\mathcal{T} = \{0, \frac{1}{m}, \frac{2}{m}, ..., 1\}$ for defining the truth values of FTL_f formulas; let Φ_n^S be the formula as defined before. Then,*

1. *the hard constraints are satisfiable,*
2. *if we find an assignment, say ν, that satisfies the hard constraints and maximizes the number of satisfied soft constraints, then we get an FTL_f, Φ_{FTL_f}, of size n from ν, such that for any other FTL_f formula Φ'_{FTL_f}, $L^S_{\Phi_{FTL_f}} \leq L^S_{\Phi'_{FTL_f}}$.*

Proof: The two hard constraints, that are the structural constraint Φ_n^{str} and semantical constraint Φ_n^{sem}, are satisfiable. The constraint Φ_n^{str} encodes the structure of a syntax DAG corresponding to an FTL_f formula of size n, and it is satisfiable. On the other hand the constraint Φ_n^{sem} simply represents the semantical valuation of the formula on the traces of S. Hence, an appropriate assignment of the multi-valued variable can be found to satisfy the constraint. Only alteration from the semantics as stated in Sect. 2.1 is the use of Lukasiewicz t-norm instead of the multiplication. This won't be problematic, because the alteration of the multiplication operator has just the effect as if the weighting function η is altered. So, with proper choice of η this effect can be minimized. Also, this multiplication results in penalization of the valuation for 'almost' operators; so the use of Lukaseiwicz t-norm just alters the amount of penalization, keeping the process essentially same.

As for the second part, the multi-valued MAXSAT solver gives an assignment ν, that satisfies the hard clauses and maximizes the sum of weights of the satisfied soft clauses, i.e., maximizes the number of satisfied soft clauses. Hence the assignment minimizes the loss.

Theorem 1. *Given a sample $S = \{\pi^+, \pi_-\}$, a threshold $\mathbf{k} \in \mathbb{R}$ and a finite set $\mathcal{T} = \{0, \frac{1}{m}, \frac{2}{m}, ..., 1\}$, Algorithm 1 terminates either by returning 'Unsatisfiable' or by returning an FTL_f formula Φ of size $\leq T$, such that $L^S_\Phi \leq \mathbf{k}$ and it is of the minimal size.*

The guarantee of termination has been already discussed. The rest can be obtained from Lemma 1. The obtained solution is of minimal possible size, because Algorithm 1 proceeds with gradual increment of the size of syntax DAG. The algorithm terminates at the attainment of the minimal size formula that satisfies the hard constraints and minimizes the loss L^S_Φ keeping is lesser than some predefined threshold $\mathbf{k}$.

5 Conclusion

This work focuses on the problem of synthesizing fuzzy temporal rules from a set of given example traces of finite length. Fuzzy time linear temporal logic over finite traces (FTL_f) is chosen to express the synthesized rules. FTL_f more expressive than linear temporal logic (LTL) rules, because FTL_f enables evaluation of statements that includes fuzzy temporal modalities, like 'almost always', 'soon after', 'gradually' etc. These fuzzy modalities provide more natural explanations that is much more understandable. The problem is solved by reducing it to a multi-valued maximum satisfiablity problem.

In this work the multi-valued variable domain is searched exhaustively which results in a computational complexity that is double exponential in the size of formula. There are some recently proposed approaches to efficiently scanning the search space and make the all over algorithm more efficient. Incorporation of such methodologies is the future scope. Implementation of the system for real life examples and to experiment with the variation of the running time with the sample size, admissible size of the FTL_f formula and incorporation of some other fuzzy temporal operators are subjects of future research.

This work is significant from the perspective of behaviour analysis of complex systems, process mining, explainable AI, neuro-symbolic reinforcement learning, specially imitation learning.

References

1. Angluin, D.: Learning regular sets from queries and counterexamples. Inf. Comput. **75**(2), 87–106 (1987)
2. Argelich, J., Domingo, X., Li, C.M., Manya, F., Planes, J.: Towards solving many-valued maxsat. In: 36th International Symposium on Multiple-Valued Logic (ISMVL 2006), pp. 26–26. IEEE (2006)
3. Argelich, J., Li, C.M., Manyà, F.: Exploiting many-valued variables in maxsat. In: 2017 IEEE 47th International Symposium on Multiple-Valued Logic (ISMVL), pp. 155–160. IEEE (2017)
4. Burgess, J.P., Gurevich, Y.: The decision problem for linear temporal logic. Notre Dame J. Formal Logic **26**(2), 115–128 (1985)
5. Camacho, A., McIlraith, S.A.: Learning interpretable models expressed in linear temporal logic. In: Proceedings of the International Conference on Automated Planning and Scheduling, vol. 29, pp. 621–630 (2019)
6. De Giacomo, G., Vardi, M.Y., et al.: Linear temporal logic and linear dynamic logic on finite traces. In: IJCAI, vol. 13, pp. 854–860 (2013)
7. El Halaby, M., Abdalla, A.: Fuzzy maximum satisfiability. In: Proceedings of the 10th International Conference on Informatics and Systems, pp. 50–55 (2016)
8. Feldman, A., Pietersma, J., van Gemund, A.: A multi-valued sat-based algorithm for faster model-based diagnosis. In: Proceedings of DX-06, pp. 93–100 (2006)
9. Frigeri, A., Pasquale, L., Spoletini, P.: Fuzzy time in linear temporal logic. ACM Trans. Comput. Logic (TOCL) **15**(4), 1–22 (2014)
10. Gaglione, J.-R., Neider, D., Roy, R., Topcu, U., Xu, Z.: Learning linear temporal properties from noisy data: a MaxSAT-based approach. In: Hou, Z., Ganesh, V. (eds.) ATVA 2021. LNCS, vol. 12971, pp. 74–90. Springer, Cham (2021). https://doi.org/10.1007/978-3-030-88885-5_6

11. Gaglione, J.R., Neider, D., Roy, R., Topcu, U., Xu, Z.: Maxsat-based temporal logic inference from noisy data. Innov. Syst. Softw. Eng. **18**(3), 427–442 (2022)
12. Ghiorzi, E., Colledanchise, M., Piquet, G., Bernagozzi, S., Tacchella, A., Natale, L.: Learning linear temporal properties for autonomous robotic systems. IEEE Rob. Autom. Lett. **8**(5), 2930–2937 (2023)
13. Ielo, A., Law, M., Fionda, V., Ricca, F., De Giacomo, G., Russo, A.: Towards ilp-based ltl f passive learning. In: International Conference on Inductive Logic Programming, pp. 30–45. Springer, Heidelberg (2023). https://doi.org/10.1007/978-3-031-49299-0_3
14. Komatsu, K., Horita, H.: Generating ltl formulas for process mining by example of trace. J. Data Sci. Intell. Syst. (2024)
15. Lamine, K.B., Kabanza, F.: Using fuzzy temporal logic for monitoring behavior-based mobile robots. In: Proceedings of IASTED International Conference on Robotics and Applications, pp. 116–121 (2000)
16. Li, Y., Wei, J.: Possibilistic fuzzy linear temporal logic and its model checking. IEEE Trans. Fuzzy Syst. **29**(7), 1899–1913 (2020)
17. Liu, C., Kuehlmann, A., Moskewicz, M.W.: Cama: a multi-valued satisfiability solver. In: ICCAD-2003, International Conference on Computer Aided Design (IEEE Cat. No. 03CH37486), pp. 326–333. IEEE (2003)
18. Neider, D., Gavran, I.: Learning linear temporal properties. In: 2018 Formal Methods in Computer Aided Design (FMCAD), pp. 1–10. IEEE (2018)
19. Raha, R., Roy, R., Fijalkow, N., Neider, D., Pérez, G.A.: Synthesizing efficiently monitorable formulas in metric temporal logic. In: International Conference on Verification, Model Checking, and Abstract Interpretation, pp. 264–288. Springer, Heidelberg (2023). https://doi.org/10.1007/978-3-031-50521-8_13
20. Riener, H.: Exact synthesis of ltl properties from traces. In: 2019 Forum for Specification and Design Languages (FDL), pp. 1–6. IEEE (2019)
21. Roy, R., Fisman, D., Neider, D.: Learning interpretable models in the property specification language. arXiv preprint arXiv:2002.03668 (2020)
22. Roy, R., Gaglione, J.R., Baharisangari, N., Neider, D., Xu, Z., Topcu, U.: Learning interpretable temporal properties from positive examples only. In: Proceedings of the AAAI Conference on Artificial Intelligence, vol. 37, pp. 6507–6515 (2023)
23. Tseitin, G.S.: On the complexity of derivation in propositional calculus. In: Automation of Reasoning: 2: Classical Papers on Computational Logic 1967–1970, pp. 466–483 (1983)

A Mīmāṃsā Inspired Framework Towards Temporal Reasoning in Large Language Models

Bama Srinivasan[✉] and Mohan Raj Vijayan

Department of Information Science and Technology, CEG, Anna University,
Chennai, India
bama@auist.net

Abstract. Mīmāṃsā, one of the systems of Indian Philosophy, deals with the interpretation of the Vedas. The *Brāhmaṇas*, a division of the Vedas, include precise instructions (*Vidhi*) for the execution of rituals. Interpreting these immediately can be confusing. In order to fully understand this, the interpretive processes from Mīmāṃsā are utilized. The procedures encompass various components, such as linguistic proficiency, grammatical comprehension, the individual's capacity to execute the ritual, and logical attributes. In addition, Mīmāṃsā incorporates a mention of diverse sequencing systems known as *krama*, which precisely outlines the specified sequence in which rituals should be performed. This paper takes inspiration from these sequential techniques and proposes a framework for temporal reasoning from a Logical perspective. This approach is incorporated into the existing MIRA (Mīmāṃsā Inspired Representation of Actions) work, resulting in activity sequencing. Subsequently, it is utilized in the Large Language Models to autonomously produce a sequence of instructions in real-life situations.

Keywords: Mīmāṃsā Inspired Representation of Actions (MIRA) · Imperative sequencing · Temporal Reasoning · Large Language Models

1 Introduction

Take a hypothetical situation in which the robot has to be instructed on how to prepare Tea. In this case, the process must be deconstructed into explicit and feasible stages, which is a laborious process. As an illustration, the process of "boiling water" needed for tea preparation should be divided into the following steps: "Fill the kettle with water", "Place the kettle on the stove", "Switch on the stove", and "Wait until water boils". In addition to these specifics, the offered jobs may be lacking in completeness and clarity. Moreover, the tasks must be executed exclusively using accessible objects and specific actions. Modeling such action representation and reasoning in a formal manner is a challenging endeavour.

With the recent substantial progress in generative AI capabilities, research has concentrated on the challenge of deducing actions through reasoning. A successful strategy is to establish a clearly defined structure of tasks that specify the object, action, and other relevant details to guarantee the efficient completion of the tasks. Moreover, in addition to the framework, it is necessary to develop protocols to generate the sequence of actions.

Towards this direction, this paper provides a framework for structured representation of instructions with sequencing mechanisms. This methodology is inspired by Mīmāṃsā, a system of Indian philosophy which deals with the interpretation of *Vedas*. This interpretation includes different sequencing methods when the injunctions that appear are ambiguous.

The present work expands upon the existing research on MIRA [1] and the classification of injunctions [2], which offers a systematic representation of instructions based on Mīmāṃsā. It formalizes the sequencing methods derived from Mīmāṃsā, enabling the sequencing of instructions under various scenarios. In order to automatically structure and sequence the instruction generation in practical scenarios, the Generative AI model GPT4o [3] is utilized.

Thus the objective of this paper is two-fold: (i) To offer a succinct representation for generating an instruction sequence from a set of instructions (inspired by Mīmāṃsā tenets) and (ii) To present a practical scenario utilizing a Generative AI model to organize the instructions and produce the sequence.

This paper is structured as follows. In Sect. 2, an overview is given on the several sequencing techniques as classified by Mīmāṃsā. Section 3 provides a comprehensive explanation of the action representation formalism inspired by Mīmāṃsā, which serves as the foundation for this work. Section 4 provides a comprehensive explanation of the organization of several sequencing methods. In Sect. 5, a practical implementation of this structure using a Large Language Model is presented, followed by the related work in Sect. 6. The conclusion is provided in Sect. 7 of the paper.

2 Sequencing Methods from Mīmāṃsā

Mīmāṃsā specifies different injunctions, that include principal injunctions (*Utpattividhi*), injunctions enjoining auxiliaries (*viniyogavidhi*), injunction of qualification (*Adhikāravidhi*) and procedural injunction (*prayogavidhi*). In the type of procedural injunctions, there is a requirement of performing the defined set of rituals without any interruption. In order to do so, the sequence or ordering of the rituals must be predetermined. To arrive at the sequence, six methods are provided at different prioirity levels in descending order. These are Direct Assertion (*Śrutikrama*), Purpose or Meaning (*Arthakrama*), Reading (*Pāṭhakrama*), Position (*Sthānakrama*), Principal activities (*Mukhyakrama*) and Iterative procedure (*Pravrittikrama*). Direct assertion is given the most importance among these techniques, while iterative procedures are given the lowest [4,5]. An outline of each of these are provided below.

1. Direct Assertion (*Śrutikrama* [1]): In this type, explicit reference of sequence is specified. For example, in the statement *"make vedi after making veda"*, the sequence is indicated through the word "after".
2. Sequence based on Purpose (Arthakrama [2]): In this type, the purpose is used to indicate the sequence. The injunctions specified may not be in sequence during the execution of multiple activities. In such instances, it is necessary to consider the purpose in order to reorder the activities. For instance, the injunction *"Perform Agnihotra"* followed by another injunction, *"Cook Rice Porridge"* is mentioned in the text of Mīmāṃsā. Subsequently, the injunction *"Use Rice Porridge to perform Agnihotra"* is also specified in the same text. These three injunctions suggest that Rice Porridge should be prepared first, and then agnihotra should be performed using it. The purpose of preparing rice porridge in this context is to utilize it in Agnihotra [5].
3. Sequence based on Reading (*Pāṭhakrama* [3]): This type suggests the direct way of performing activities as prescribed in the text.
4. Sequence based on Position (*Sthānakrama* [4]): When multiple activities need to be performed simultaneously, this method suggests that the one that comes to mind first should be prioritized. Consider the example of the *Soma Yaga*. In one part of the text, it is stated that the *Soma Yaga* should be conducted over five days, with three specific rituals—*Daiksha*, *savanīya*, and *Anubadhya*—to be performed on the fourth day, the middle of the fifth day, and the end of the fifth day, respectively. However, another part of the text mentions that all three rituals should be conducted together, creating ambiguity. In such cases, the position of the *savanīya Yaga* is considered. Since it is closely related to the main sacrifice of *Soma Yaga*, it is inferred that all three rituals should be performed together in the middle of the fifth day.
5. Sequence based on principal activities (*Mukhyakrama* [5]): This method can be applied when multiple primary sacrifices need to be performed along with their associated subsidiary rituals. For instance, in the *Darśapūrṇamāsa* ritual, which is the main sacrifice, there is a specified order for pouring ghee. The remaining ghee should be poured on the materials offered to the deity (*havis*) in the same sequence, as it is a subsidiary part of the main ritual.

---

[1] *Śrutikrama*: Order by Direct Statement - 'means a text intimating order. It is of two kinds viz. initimating mere order and intimating things particularized by that order'. This takes the highest precendence among all other types [6].

[2] *Arthakrama*: Order by sense - The sequence is determined according to the purpose. This is weaker than *śrutikrama*, but stronger than *Pāṭhakrama* [6].

[3] *Pāṭhakrama*: Order by Text - 'means the order of sentences which intimate certain things. There are two types: *Mantra* text and *Brāhmana* text. The later is more stronger than the former and the order remains so, since *mantra* is intimately connected with the performance'. This type is stronger than *Sthānakrama* [6].

[4] *Sthānakrama*: 'Order by position - Position means presentation or presenting oneself'. This type is stronger than *Mukhyakrama* [6].

[5] *Mukhyakrama*: 'Order by Principal - The order of subsidiaries based on the respective principals'. This type is stronger than *Pravṛttikrama*. [6].

6. Iterative Procedure (*Pravṛttikrama* [6]): In this method, a specific sequence of actions must be repeated several times. For instance, in the ritual of *Prājāpatya*, 17 sacrifices are to be offered to the deity *Prājāpati*, each with 17 material offerings. The actions should be carried out by performing the first action 17 times, followed by the second action 17 times, and so on.

This paper focuses on formalizing the sequencing processes using a few methods, namely Direct Assertion, Sequence Based on Purpose, and Iterative Procedure. The fundamental concept behind these approaches is integrated with the formalism of MIRA. The following section presents a concise overview of MIRA.

3 Overview of MIRA

A formalism inspired by Mīmāṃsā has been worked out based on imperative logic which combines imperatives and propositions. Here, the focus is towards the imperatives, rather than the propositions [1]. The formalism operates with the specification of instruction at the syntactic level and the execution of these instructions at the semantic level.

At the syntactic level, the Language of imperatives is specified as $\mathcal{L}_i = \langle I, R, P, B \rangle$, with the formation of imperatives as $\mathcal{F}_i$ as shown in Eq. 1. Here, I denotes imperatives $\{i_1, i_2, \ldots, i_n\}$, R denotes reasons $\{r_1, r_2, \ldots, r_n\}$, P denotes purpose in terms of goals $\{p_1, p_2, \ldots, p_n\}$ and B indicate the binary connectives $\{\wedge, \vee, \rightarrow_r, \rightarrow_i, \rightarrow_p\}$.

$$\mathcal{F}_i = \{i | i \rightarrow_p p | (i \rightarrow_p p_1) \wedge (j \rightarrow_p p_2) | (i \rightarrow_p \theta) \oplus (j \rightarrow_p \theta) | (\varphi \rightarrow_i \psi) | (\tau \rightarrow_r \varphi)\} \quad (1)$$

where:

- $i \in I$: i is an imperative, which belongs to the set I. I includes both positive unconditional imperatives (I^v) and negative unconditional imperatives (I^n).
- $i \rightarrow_p p$: This represents a conditional imperative where i implies a purpose p (i.e., i leads to or is intended to achieve the goal or subgoal p).
- $(i \rightarrow_p p_1) \wedge (j \rightarrow_p p_2)$: This is a conjunction of two conditional imperatives. It states that imperative i implies subgoal p_1 and imperative j implies subgoal p_2. The imperatives i and j are different, as are the subgoals p_1 and p_2.
- $(i \rightarrow_p \theta) \oplus (j \rightarrow_p \theta)$: This represents an exclusive disjunction (either-or) between two imperatives. It states that either imperative i or j (but not both) leads to the same goal θ, where θ is a propositional formula from the set P.

---

[6] *Pravṛttikrama*: Order by Procedure - When a series of actions is to be performed a specific number of times within a procedure, the first action should be repeated a set number of times, followed by the second action for its specified number of times, and so forth [6].

- $\varphi \to_i \psi$: This is a conditional imperative where φ implies ψ based on temporal action. φ and ψ are formulas within the set $\mathcal{F}_i$.
- $\tau \to_r \varphi$: This represents a conditional imperative where τ (a reason from the set R, a proposition formula) implies φ, which is a formula in $\mathcal{F}_i$. Here, $\to_r$ indicates that φ is implied by the reason τ.

Different deduction rules such as conjunction introduction, conjunction elimination, disjunction elimination, conditional introduction and conditional elimination rules are prescribed, which helps in the reasoning process. While disjunction introduction is possible, it is not explicitly provided as a part of this formalism inaccordance with Mīmāṃsā [1,7,8].

At the semantic level, the evaluation takes place in terms of *Satisfaction* (S), *Violation*(V) and *Lack of intention to reach the goal* (N). The main idea behind this evaluation rests on the presence of intention to reach the goal on the agent's part.

In this formalism, I from $\mathcal{L}_i$ is given by:

$$I = (I^v \cup I^n) \tag{2}$$
$$I^v = (i_1{}^+, i_2{}^+, i_3{}^+, ..., i_n{}^+) \tag{3}$$
$$I^n = (i_n{}', i_2{}', i_3{}', ..., i_n{}') \tag{4}$$

where, I is the set of imperatives that consists of positive imperatives (*Vidhi*) - I^v and negative imperatives (*Niṣedha*) - I^n.

Each instruction here includes an action part and the object part. For example, consider the instructions: *'Take pen'* and *'Pour water into vessel'*. Here, actions are *'take, pour'* and objects are *'pen,water,vessel'*. This aspect of defining actions and objects as a part of instructions is extended in this paper, which helps in formalising the techniques of sequencing as prescribed by Mīmāṃsā. This is described in the next section.

4 Sequencing Mechanisms

In this section, the sequencing methods of Direct Assertion, Sequence based on purpose and Iterative methods are presented. From Eq. 2, Let $I = \{i_1, i_2, i_3, ..., i_n\}$, where $\{i_1, i_2, ..., i_n\}$ are positive imperatives[7].

Since each instruction is associated with action (a) and object (o), the set of Actions(A) and set of Objects(O) from Instruction I can be represented as:

$$A = a_1, a_2, a_3, ..., a_k \tag{5}$$

$$O = o_1, o_2, o_3, ..., o_n \tag{6}$$

---

[7] The same formalism holds for negative imperatives also.

Equations (5) and (6) specify the Actions and Objects from the Instruction. Here, the action can range from $\{1, 2, \ldots, k\}$ and objects from $\{1, 2, \ldots, n\}$, where $n \neq k$. This indicates that the number of actions need not be equal to the number of objects, as in the case of iterative actions (*Pravṛttikrama*). For example, consider the statement: *"take 5 bread pieces, place those on the table"*. The actions in this case are *take* and *place*. Five pieces of bread are connected to each of these actions. In this case, a specific number of objects can be associated with the same action.

This representation in terms of actions and objects helps in formalising the sequencing mechanisms, which is described below.

4.1 Direct Assertion (*Śrutikrama*)

As indicated in MIRA, Direct Assertion (Explicit rendering of sequencing of actions) is indicated as conditional imperative enjoining temporal actions (cia). For example, it follows the pattern *"Perform instruction 1, then instruction 2"* in a sequence. This is formalised as $\varphi \rightarrow_i \psi$ as shown in Eq. 1, where $\varphi, \psi \in \mathcal{F}_i$.

Following Eq. 5, if Instruction at the initial step (in a two-step sequence) includes $\{a_t, o_t\}$ and instruction at the subsequent step includes $\{a_{(t+1)}, o_{(t+1)}\}$, where t indicates the time-step, then Direct Assertion is given by:

$$(a_t \ o_t \rightarrow_i a_{(t+1)} \ o_{(t+1)}) \tag{7}$$

Example: Given the instruction as *"Take a pen, then write with it in paper"*, here $a_t, a_{(t+1)}$ are *"Take"* and *"Write"*, respectively; and o_t, $o_{(t+1)}$ are *"pen"* and *"paper"*.

4.2 Sequence Based on Purpose (*Arthakrama*)

Each instruction that is laid out has a meaning or purpose. This aspect is taken into consideration for this method of sequencing. According to MIRA (from Eq. 1), this imperative can be expressed as:

$$(\tau \rightarrow_r (i \rightarrow_p p)) \tag{8}$$

which indicates that the instruction (i) holds a condition (τ) for the action to take place with the intention of purpose (p).

The condition for an imperative is in the form of a reason, which leads towards a purpose. Take, for example, the statement: *"If it is raining, take an umbrella, so that you don't get wet"*. Here, *"if it is raining"* is the reason (a proposition belonging to R), *"take an umbrella"* is an imperative (belonging to set I), and *"you don't get wet"* is the purpose (belonging to set P). It is in this sense that we have considered imperatives in this paper. The basis for this approach can be referred in the formalism of MIRA [1].

Let the series of instructions be given as $\{i_1, i_2, i_3, \ldots, i_n\}$. Following Eq. 8, each of these instructions can be expressed as:

$$(r_1 \rightarrow_r (i_1 \rightarrow_p p_1)), (r_2 \rightarrow_r (i_2 \rightarrow_p p_2)), (r_3 \rightarrow_r (i_2 \rightarrow_p p_3)), \ldots$$
$$(r_n \rightarrow_r (i_n \rightarrow_p p_n)) \tag{9}$$

Applying the semantics of MIRA to Eq. 8, if r_1 is *True* and the agent has the intention of achieving the goal or purpose (p_1 is *True*), assuming the agent performs the task as per the given instrucion i_1, then Eq. 8 evaluates to S. This also indicates that the purpose p_1 is attained, which forms another condition (r_2) for future instruction (i_2) to take place. Subsequently, when the second imperative i_2 is evaluated to S, p_2 is attained, which forms the condition of r_3.

Thus in a series of instructions, if imperative (φ_k) as in Eq. 8 evaluates to S, then the corresponding purpose p_k is attained. In such cases:

$$r_{k+1} = p_k \tag{10}$$

This aspect can be used for sequencing, even if the given instructions are not in order. Computationally, the series of instructions can be included as a knowledge base in the structure of Eq. 9. Assume a query in the form appears as i_5, i_3, i_4 with the initial condition of r_3 and p_3 as *True* value. On checking with the knowledge base, the evaluation of i_5 leads to N, since r_5 is not present. Only the instruction i_3 activates in such cases, thereby evaluating to S. This causes p_3 to be attained, and on applying Eq. 10, r_4 results, which in turn activates i_4.

Thus through this method, given instructions i_5, i_4, i_6 sequences to i_4, i_5, i_6. This notion has already been applied in the work of task analysis for special education [9]. But the formal representation is presented only in this paper.

4.3 Sequential Completion and Step-by-Step Parallel or Iterative Procedure Method (*Pravrittikrama*)

Under specific circumstances, it becomes necessary to perform a certain activity multiple times. Consider, as an illustration, the domestic chore of cleaning 5 different containers. Commonly, there are four stages involved. These procedures involve rinsing each *container with water, putting a cleaning chemical on it, brushing the surface, and then rinsing it again with water*. Regarding the execution of this task, two approaches might be utilized. Under one approach, the whole series of four activities is executed for a single vessel, and subsequently the identical duties are performed for 2, 3, 4, and 5 vessels. An alternative strategy would be to divide and carry out the individual stages of action as five separate tasks (for each vessel) in the first step, five independent tasks in the second step, and so on for the remaining phases. In this paper, the former method is called the **Sequential Completion Method** and the latter is the **step-by-Step parallel method or Iterative Procedure**. The formalisation of these two methods are given below.

To represent this case, the repeated number of performance of actions can be considered. This repeated actions are to be performed on N objects[8]. Thus Eq. 5 becomes:

---

[8] This value of N is different from the semantic evaluation of N in MIRA, which is indicated as the absence of intention to reach the goal.

$$A = a_1, a_2, a_3, ..., a_n \qquad (11)$$
$$O_1 = o_{11}, o_{12}, o_{13}, ..., o_{1N}$$
$$O_2 = o_{21}, o_{22}, o_{23}, ..., o_{2N}$$
$$O_3 = o_{31}, o_{32}, o_{33}, ..., o_{3N}$$
$$\vdots$$
$$O_n = o_{n1}, o_{n2}, o_{n3}, ..., o_{nN}$$

Here, each action a_k, where $1 \leq k \leq n$ is associated with the corresponding object O_k.

Assuming the series of instructions are given as $(i_1, i_2, i_3, ..., i_n)$ with the corresponding action and objects as in Eq. 11, then Sequential Completion Method and Step-by-Step parallel methods can be expressed as shown in Eq. 12 and 13, respectively.

$$(a_1 o_{11} \rightarrow_i a_2 o_{21} \rightarrow a_3 o_{31} \rightarrow_i ... \rightarrow_i a_n o_{n1}) \qquad (12)$$
$$(a_1 o_{12} \rightarrow_i a_2 o_{22} \rightarrow a_3 o_{32} \rightarrow_i ... \rightarrow_i a_n o_{n2})$$
$$(a_1 o_{13} \rightarrow_i a_2 o_{23} \rightarrow a_3 o_{33} \rightarrow_i ... \rightarrow_i a_n o_{n3})$$
$$\vdots$$
$$(a_1 o_{1N} \rightarrow_i a_2 o_{2N} \rightarrow a_3 o_{3N} \rightarrow_i ... \rightarrow_i a_n o_{nN})$$

According to Eq. 12, the series of actions are to be performed completely first time and then repeated N number of times.

$$(a_1 o_{11} \rightarrow_i a_1 o_{12} \rightarrow a_1 o_{13} \rightarrow_i ... \rightarrow_i a_1 o_{1N}) \qquad (13)$$
$$(a_2 o_{21} \rightarrow_i a_2 o_{22} \rightarrow a_2 o_{23} \rightarrow_i ... \rightarrow_i a_2 o_{2N})$$
$$(a_3 o_{31} \rightarrow_i a_3 o_{32} \rightarrow a_3 o_{33} \rightarrow_i ... \rightarrow_i a_3 o_{3N})$$
$$\vdots$$
$$(a_n o_{n1} \rightarrow_i a_n o_{n2} \rightarrow a_n o_{n3} \rightarrow_i ... \rightarrow_i a_n o_{nN})$$

In Eq. 13, the initial action is iterated N times, followed by subsequent actions up to n^{th}, each involving associated objects.

The practical applications of these sequencing methods can be implemented utilizing Large Language Models, as detailed in the next section.

5 Application in Large Language Models (LLM)

The MIRA formalism's structure can be applied in Large Language Models (LLMs) to generate and reason about instructions. To achieve this, we have conducted experiments with LLMs to create instruction sequences using customized step-by-step prompts. This approach is implemented in two ways: At the first level, an AI agent is created with action protocols, and in the second one, the sequencing method employed. In this paper, the first level is described in Sect. 5.1 and the step-by-step parallel method in Sect. 5.2. Similar approaches can be applied for other sequencing methods also.

5.1 MIRA AI Agent and Protocol for Temporal Reasoning

To create an AI agent in LLM, the agent's role and other details have to be specified. These details include agent description, protocol for defining the structure of instructions, namely 'MIRA protocol', task description, valid actions, available objects, initial condition and the end goal. These are described below with a running example of preparation of bread sandwiches.

- **MIRA AI agent:** Initially, the LLM is asked to take up the role of AI agent that assists in temporal reasoning. The description of what it has to do such as following certain protocols, constructing instructions with available objects, classifying those according to MIRA, and generating sequences for temporal reasoning are provided. For example, the following description for the AI agent according to MIRA is given below.

"You are an AI agent for the purpose of temporal reasoning. Your name is MIRA AI agent. You have to follow the protocols listed in **MIRA PROTOCOL**'. I'll give you a list of available Objects (OBJ) and Valid Actions. Pickout correct objects from the list of Available Objects. Each action duration (DUR) is also given in valid action. The starting condition (START) and the end goal (FINAL GOAL) for the sequence is also given. With these you have to generate a sequence of instructions for the given task description. For each instruction also provide the type (TYPE) of instruction from MIRA PROTOCOL"

Consequently the corresponding response generated from the environment of GPT4o is shown below

---

Sure! To help you with the task, please provide the following details:
1. *MIRA Protocol: Any specific guidelines or rules that MIRA AI Agent must follow.*
2. *Available Objects (OBJ): A list of objects that can be used in the task.*
3. *Valid Actions: A list of actions along with their durations (DUR).*
4. *Starting Condition: The initial condition or setup for the task.*
5. *Task Description: A description of the task that MIRA AI agent needs to accomplish.*
   With these details, I can generate the sequence of instructions, calculate the total time for the actions, and determine the overall time for the task.

- **MIRA Protocol:** Next, the details of the structure of instructions according to MIRA formalism is given according to Eq. 1 mentioning the type of instruction (*vidhi*) in protocol. In addition, an example is given against each type, which helps the AI agent to understand better. For the same example, the following protocol is provided to MIRA agent.

- *You have to generate one action for each instruction : TYPE: SIMPLE INSTRUCTION*
- *Instructions can include one object or multiple objects (e.g. take pen or take pens and papers) TYPE: SIMPLE INSTRUCTIONS*
- *Each instruction can include a goal - here the instruction should have the intention of goal perspective (e.g., take pen if you have the intention of writing, take pen if you want to write) TYPE: INSTRUCTION WITH PURPOSE*
- *Each instruction can include multiple objects which are exclusive (e.g. take pen or pencil) - Instructions: TYPE: EXCLUSIVE INSTRUCTION*
- *If there are options between actions, include or within that instructions, to represent exclusive or (e.g. Go by walk or take a car to reach destination) TYPE: EXCLUSIVE INSTRUCTION*
- *If two actions are provided in a sequence, use "then" and do not use "and" (e.g. Take pen then write) TYPE: INSTRUCTION WITH SEQUENCE*

- **Available Objects** and **Valid Actions** as per Eq. 5 are provided in addition to **Task Description**. For the example of preparing sandwich, these are given as:

Available Objects = {Slices of bread White bread, Whole wheat bread, Multi-grain bread, Condiments, Mayonnaise, Mustard, Butter, Ketchup, Spreads, Vegetables, Meat, Cheese, Sandwich maker, Grill, Water, Teapot}

Valid Actions = { Take OBJ: DUR 1 min, Heat OBJ: DUR 3 min, Pickup OBJ: 30 s, Add OBJ: 1 min}

Task Description - Prepare five bread sandwiches

- **Start and End Goal:** Start is the initial condition and the goal is the desired outcome of the entire task. For the example, these are given below.

START White bread bread is available

GOAL Sandwich with cheese is ready to serve

Given these inputs, the sequence generated from LLM is shown in Fig. 1. Here, the instructions are provided in a 4-step sequence with the type of instruction according to MIRA. But the classification of instruction in Step-2 is given as 'Instruction with purpose' which is incorrect, since the statement *"add cheese to 5 slices of white bread"* does not indicate any purpose. This reflects the current

> **Task: Prepare Five Bread Sandwiches with Cheese**
>
> **Sequence of Instructions:**
>
> 1. **Instruction 1:** Take 10 slices of White bread.
>   - **Type:** SIMPLE INSTRUCTION
>   - **Duration:** 1 minute
> 2. **Instruction 2:** Add cheese to 5 slices of White bread.
>   - **Type:** INSTRUCTION WITH PURPOSE
>   - **Duration:** 1 minute
> 3. **Instruction 3:** Optionally, heat the sandwiches using a Sandwich maker or Grill.
>   - **Type:** EXCLUSIVE INSTRUCTION
>   - **Duration:** 3 minutes (optional)
> 4. **Instruction 4:** Pickup the sandwiches then serve them.
>   - **Type:** INSTRUCTION WITH SEQUENCE
>   - **Duration:** 30 seconds
>
> **Total Time Calculation:**
>
> - **Without Heating:** 2 minutes 30 seconds
> - **With Heating:** 5 minutes 30 seconds
>
> This sequence includes all the necessary instructions for preparing and serving the sandwiches, using the appropriate instruction types as per the MIRA protocol.

Fig. 1. Response from GPT4o

LLM incapability in understanding the language constructs. However, with more examples and fine-tuning of LLM, this issue can be solved.

With the above structure of MIRA AI agent, the sequencing capabilities according to Mīmāṃsā can be taught to LLM. This method is described in the next section.

5.2 Sequencing According to Step-by-Step Parallel Method

In this case, the protocol described in Sect. 5.1 can be extended with the following constructs for step-by-step parallel method

---

MIRA Protocol
Check the number of tasks in task description. Perform each action the number of times of the task (e.g. Suppose there are five sandwiches to be prepared, the instruction should be take 1st set of two slices of bread, take 2nd set of two slices of bread, take 3rd set of two slices of bread..., then add condiments to 1st set, add condiments to 2nd set...)

The result for this input is given in Appendix 1, where the first action is repeated 5 times, followed by the second action 5 times and so on. This represents the method if step-by-step parallel method through LLM.

6 Related Work

There has been a wide range of research with respect to actions from the view of logic and language perspectives. Different types of logic has been used for task planning and scheduling. There has been work on temporal logic [10] for planning [11]. From language perspective, Natural Language Processing has been used for planning [12] and navigation systems [13]. Oflate, with the advent of Generative AI with the significant progress in LLM, more work can be seen in understanding human connotations [14,15]. Among those, a number of work has been oriented towards reasoning out actions [16]. Towards this angle, it has been shown that LLMs can handle sequential task generation that is required for a feasible understanding of the temporal relationships between actions and objects [15,16]. Specific real time applications such as Auction has also been considered for generating strategic planning using LLM [17]. There has been work on using LLM over temporal data [18].

Recent developments in Generative AI such as nl2spec and TIMEARENA frameworks have highlighted the use of interactive prompting to refine unstructured language into temporal reasoning [19]. These frameworks are particularly effective in multitasking and parallel processing environments, where the correct sequence and timing of tasks are paramount. This approach addresses existing challenges in temporal reasoning for LLMs, particularly in real-world situations like cooking tasks, household tasks, and laboratory tasks in which the accurate sequencing of actions is critical in timing and crucial for success in modern AI systems. In this work, multiple tasks can be fed to LLM to automate the task sequence. However, it lacks the details of different instruction types and sequencing methods. This aspect is addressed in the proposed work, which is unique in the following ways:

- The work presented in this paper establishes the conceptual foundation for representing instructions using Simple instruction, Instruction with Purpose, Instruction with Sequence, and Exclusive instructions.
- An independent customized protocol using the representation scheme can be developed to enable the LLM to comprehend and make optimal decisions when presented with the necessary actions and objects.
- The suggested work introduces a new item sequencing mechanism, namely the iterative method or step-by-step parallel method, which is not found in any other previously published studies.

7 Conclusion

In this paper, the task sequencing methods have been explored from Mīmāṃsā. Among the six methods of sequencing, the structure of sequencing formally is provided for the three methods namely Direct Assertion, Sequence based on Purpose and Iterative Procedure. These are then shown to be useful in practical scenarios using the Generative AI model.

The integration of Mīmāṃsā principles into LLMs has demonstrated potential in enhancing the accuracy and coherence of temporal reasoning. However, further refinement is needed to achieve precise reasoning capabilities. This is evident in Fig. 1, where the agent incorrectly classifies an instruction. Our immediate future work will focus on this area. Additionally, formalizing other sequencing methods from Mīmāṃsā, such as position-based sequencing, represents another area for future research.

Appendix 1: Output Response of Step-By-Step Parallel Method

The output response is shown in Fig. 2, where each action is repeated 'N' times. Here, 'N = 5'.

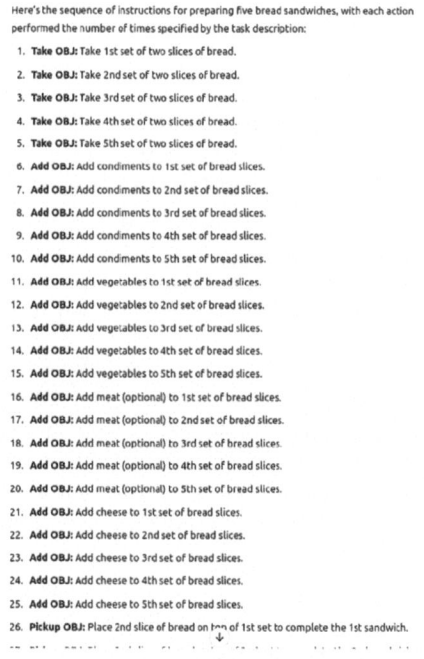

Fig. 2. Part of the output response from GPT4o

References

1. Srinivasan, B., Parthasarathi, R.: A formalism to specify unambiguous instructions inspired by Mīmāsā in computational settings. Log. Univers. **16**, 27–55 (2022)

2. Srinivasan, B., Parthasarathi, R.: Classification of natural language instructions inspired from Mīmāsā. In: Kulkarni, A. (ed.) Sanskrit and Computational Linguistics, Select Papers Presented in the Sanskrit and the IT World section of the 16th World Sanskrit Conference,. DK Publishers, India (2015)
3. OpenAI (2023) GPT-4 Technical report. arXiv, abs/2303.08774
4. Madhavananda: Mīmāsā-Paribhāmā: of Krishna Yajvan. Advaita Ashrama, India (1986)
5. Ranganathan, S.V.: Mīmāsā Shastram: Online Introductory Course (2022). https://velukkudidiscourses.com/product/mimamsa-shastram/. Accessed 13 Nov 2024
6. Gajendragadkar, A.B., Karmarkar, R.D.: The Arthasamgraha of Laugākṣi Bhāskara. Motilal Banarsidass Publishers Private Limited, India (1934)
7. Basu, B.D., Sandal, P.M.L.: The Mīmāsā sutras of Jaimini. In: The Sacred Books of the Hindus, Volume XXVII: Part I. Dr. Sudhindre Nath Basu M.B, the Panini Office, Bhuvaneswari Asrama, Bahadurganj (1923)
8. Pandurangi, K.T.: Purvamimamsa from an interdisciplinary point of view. In: History of Science, Philosophy, and Culture in Indian Civilization. Volume II Part 6, Motilal Banarsidass, Delhi (2006)
9. Bama, S., Ranjani, P.: An intelligent task analysis approach for special education based on MIRA. J. Appl. Logic **11**(1), 137–145 (2013)
10. Huth, M., Ryan, M.D.: Logic in Computer Science: Modelling and Reasoning About Systems, 2nd edn. Cambridge University Press, Cambridge (2004)
11. Allen, J.F.: Planning as temporal reasoning. In: Proceedings of the Second International Conference on Principles of Knowledge Representation and Reasoning (KR 1991), pp. 3–14. Morgan Kaufmann Publishers Inc., San Francisco (1991)
12. Hua, X., Hu, Z., Wang, L.: Argument generation with retrieval, planning, and realization. In: Proceedings of the 57th Annual Meeting of the Association for Computational Linguistics, pp. 2661–2672. Association for Computational Linguistics, Florence (2019)
13. Chen, D.L., Mooney, R.J.: Learning to interpret natural language navigation instructions from observations. In: Proceedings of the Twenty-Fifth AAAI Conference on Artificial Intelligence (AAAI 2011), pp. 859–865. AAAI Press (2011)
14. Chang, Y., et al.: A survey on evaluation of large language models. ACM Trans. Intell. Syst. Technol. **15**(3), 1–45 (2024)
15. Ouyang, L., et al.: Training language models to follow instructions with human feedback. In: Proceedings of the 36th International Conference on Neural Information Processing Systems (NIPS 2022), Curran Associates Inc., Red Hook (2024)
16. Xiong, S., Payani, A., Kompella, R., Fekri, F. Large language models can learn temporal reasoning. In: Proceedings of the 62nd Annual Meeting of the Association for Computational Linguistics, Bangkok, Thailand, vol. 1, pp. 10452–10470 (2024)
17. Chen, J., Yuan, S., Ye, R., Majumder, B.P., Richardson, K.: Put your money where your mouth is: evaluating strategic planning and execution of LLM agents in an auction arena. arXiv preprint arXiv:2310.05746 (2023)
18. Dhingra, B., et al.: Time-aware language models as temporal knowledge bases. Trans. Assoc. Comput. Linguist. **10**, 257–273 (2022)
19. Zhang, Y., et al.: TimeArena: shaping efficient multitasking language agents in a time-aware simulation. In: Proceedings of the 62nd Annual Meeting of the Association for Computational Linguistics, Bangkok, vol. 1, pp. 3894–3916 (2024)

Measurement-Theoretic Foundations of Logic of Inexact Knowledge

Satoru Suzuki(✉)

Faculty of Arts and Sciences, Komazawa University,
1-23-1, Komazawa, Setagaya-ku, Tokyo 154-8525, Japan
bxs05253@nifty.com

Abstract. In this paper, we propose a new version of *complete* logic—Logic of Inexact Knowledge (LIK)—that has the following *eight* features out of which the *seven* can reflect Williamson (1994)'s arguments but out of which the *only one* is essentially *different* from the features based on Williamson's *knowledge first epistemology*: 1. This model based on *additively-semiordered qualitative conditional probability* that is a qualitatively-probabilistic counterpart of a *JND* which is a psychophysical counterpart of a *margin for error* can reflect the essence of *inexact knowledge*. 2. We can formalize a *margin for error principle* in LIK. 3. The *width* of a margin for error (JND) depends on the *cognitive capacities*. 4. In LIK, the *direct indiscriminability relation* is a *non-transitive* relation. 5. The reason why we introduce qualitative not absolute but *conditional* probability is to make it possible to express the *direct indiscriminability between the two events on the condition that either of the two occurs*. 6. LIK has so *rich expressive power* as to fully formalize such inferences as (6)–(8). 7. The *KK principle* is not valid in LIK. 8. *Essential Difference*: In Williamson's LC inexact knowledge is *primitive* based on his *knowledge first epistemology*, whereas in LIK inexact knowledge is *defined* in terms of the direct indiscriminability relation based on our *direct indiscriminability relation first epistemology*.

Keywords: additively-semiordered qualitative conditional probability relation · bounded rationality · direct indiscriminability relation first epistemology · epistemic logic · indiscriminability · inexact knowledge · just noticeable difference (JND) · KK principle · knowledge first epistemology · margin for error principle · measurement theory · representation theorem · psychophysics · Timothy Williamson

1 Motivation

Williamson [21, p. 217] makes the following observation about the notion of *inexact knowledge*:

> One sees roughly but not exactly how many books a room contains, for example: it is certainly more than two hundred and less than twenty thousand, but one does not know the exact number. Yet there need be no

relevant vagueness in the number. The *inexactness* was in the *knowledge*, not in the *object* about which it was acquired.

He [21, p. 227] states that each case of inexact knowledge is governed by a *margin for error principle* saying that 'φ' is true in all cases similar to cases in which 'It is known that φ' is true. He [21, p. 226] argues that the width of a margin for error depends on the *cognitive capacities*. He [21, pp. 270–275] proposes a logic for a margin for error that he calls the *logic of clarity* LC. He [21, pp. 237–238] classifies indiscriminability into two types: *direct indiscriminability* and *indirect indiscriminability*, and then defines the latter by the former: Two things are indirectly indiscriminable in a certain respect just in case they are directly indiscriminable in that respect from exactly the same things. He observes that direct indiscriminability is a *non-transitive* relation, while 'indirect indiscriminability is by definition a *transitive* relation' ([21, p. 238]) and that 'indirect discrimination is not a genuinely cognitive form of discrimination at all' ([21, pp. 240]). He [21, pp. 238–242] argues that, with the *KK principle*, the following consistent sentences (6)–(8) and Closure in which *direct indiscriminability* occurs *would* result in absurdity:

(6) I know that the height on day 0 is not the same as the height on day 5,000....

(7) I know that if the height on day i and the height on day j differ by less than a millimetre, then I do not know that the height on day i is not the same as the height on j....

(8) I know that the height on day i and the height on day $i+1$ differ by less than a millimetre....

Once the KK principle is dropped, (6)–(8) and Closure form a consistent set.

However, because Williamson's LC has no direct indiscriminability relation symbol in the object language, it cannot fully express the inferences about (6)–(8). According to Halpern's proposal that possible worlds are individuated by objective facts and the subject's state, Dutant [7] tries to give the condition under which the KK principle holds in inexact knowledge. Bonnay and Égré [3] propose centered semantics for inexact knowledge that has reason to reject (7) and preserve the KK principle. In terms of not static but dynamic epistemic logic, Cohen [5] considers inexact knowledge. Neither Dutant [7], Bonnay and Égré [3], nor Cohen [5] has a direct indiscriminability relation symbol in the respective object language. In this paper, we would like to propose a version of the logic of inexact knowledge the language of which has a direct indiscriminability relation symbol as a primitive symbol and has an inexact knowledge expression as a defined symbol in terms of this direct indiscriminability relation symbol and so can fully express such inferences as (6)–(8). In this paper, *measurement theory* (cf. Sect. 2) makes it possible to provide this language with its semantics. Like this paper, van Rooij [16] and Cobreros et al. [4] introduce a direct indiscriminability relation symbol into the respective object language. The essential *difference* between our direct indiscriminability relation and theirs is as follows:

Our aim of introducing a direct indiscriminability relation is to define inexact knowledge by this relation. Because a knowledge operator operates on a sentence the semantic value of which is a proposition (event), in order to define knowledge by a direct indiscriminability relation in a direct way, this relation should a relation on the set of propositions (events) or relation on the set of possible worlds (elementary events). In fact, our direct indiscriminability relation is a relation on the set of events. On the other hand, the direct indiscriminability relation in van Rooij [16] and Cobreros et al. [4] is not originally intended to define knowledge by this relation. This relation is a relation on the set not of possible worlds but of *individuals*. So it is unclear how to define knowledge by this relation in a direct way. We discuss Williamson's arguments above in a broader context. The standard models of many sciences are based on *global rationality* that requires an *optimising behavior*. But according to Simon [18], cognitive and information-processing constrains on the capabilities of agents, together with the complexity of their environment, render an optimising behavior an *unattainable ideal*. Simon dismisses the idea that agents should exhibit global rationality and suggests that they in fact exhibit *bounded rationality* that allows a *satisficing behavior*. If an agent has only a *limited* ability of discrimination, he may be considered to be only *boundedly rational*. The *margin for error principle* can be regarded as an *instance of bounded rationality*. From a *psychophysical* point of view, we consider a margin for error. A psychophysicist Fechner [8] explains this limited ability by the concept of a *threshold of discrimination*, that is, *just noticeable difference (JND)*. Given a measure function f that an examiner could assign to an examinee for an object a, its JND δ is the *lowest intensity increment* such that $f(a) + \delta$ is recognized to be higher than $f(a)$ by the examinee. JND can be considered to be a psychophysical counterpart of a *margin for error*. We can consider a JND from a *qualitatively-probabilistic* point of view. Domotor [6] introduces the concept of *additively-semiordered qualitative conditional probability relation* that represents a conditional probability measure accompanied with a JND. An additively-semiordered qualitative conditional probability relation is a qualitatively-conditional-probabilistic version of a semiorder that originates with Luce [13] and that represents a function accompanied with a JND. The *aim* of this talk is to propose a new version of *complete* logic—Logic of Inexact Knowledge (LIK)—that has the following *eight* features out of which the *seven* can reflect Williamson's arguments but out of which the *only one* is essentially *different* from the features based on Williamson's *knowledge first epistemology*:

1. This model based on *additively-semiordered qualitative conditional probability* that is a qualitatively-probabilistic counterpart of a *JND* which is a psychophysical counterpart of a *margin for error* can reflect the essence of *inexact knowledge*.
2. We can formalize a *margin for error principle* in LIK.
3. The *width* of a margin for error (JND) depends on the *cognitive capacities*.
4. In LIK, the *direct indiscriminability relation* is a *non-transitive* relation.
5. The reason why we introduce qualitative not absolute but *conditional* probability is to make it possible to express the *direct indiscriminability between the two events on the condition that either of the two occurs*.

6. LIK has so *rich expressive power* as to fully formalize such inferences as (6)–(8).
7. The *KK principle* is not valid in LIK.
8. *Essential Difference*: In Williamson's LC inexact knowledge is *primitive* based on his *knowledge first epistemology*, whereas in LIK inexact knowledge is *defined* in terms of the direct indiscriminability relation based on our *direct indiscriminability relation first epistemology*.

The structure of this paper is as follows. In Sect. 2, we give a brief introduction of measurement theory. In Sect. 3, we state Domotor [6]'s representation conjecture of an additively-semiordered qualitative conditional probability relation. In Sect. 4, we propose LIK. In Subsect. 4.1, we define the language $\mathscr{L}_{\mathsf{LIK}}$ of LIK. In Subsect. 4.2, we define a model $\mathfrak{M}$ of LIK, provide LIK with a truth definition and a validity definition, and consider inexact knowledge semantically in terms of LIK. In Subsect. 4.3, we provide LIK with its proof system. In Subsect. 4.4, we prove the soundness and completeness of LIK. In Sect. 5, we finish with brief concluding remarks.

2 Brief Introduction to Measurement Theory

Measurement plays an essential role not only in science but also in our daily lives. Measurement theory investigates not concrete methods for measurement but conditions for *measurability*. The mathematical foundation of measurement had not been studied before Hölder [10] developed his axiomatization for the measurement of mass. Krantz et al. [12], Suppes et al. [19] and Luce et al. [14] are seen as milestones in the history of measurement theory. We would like to give brief explanations of the following important concepts of measurement theory:

Homomorphism Suppose a relational structure $\mathfrak{U} := (A, R_1, R_2, \ldots, R_p, \circ_1, \circ_2, \ldots, \circ_q)$ and another $\mathfrak{V} := (B, R'_1, R'_2, \ldots, R'_p, \circ'_1, \circ'_2, \ldots, \circ'_q)$, where A and B are sets, $R_1, R_2, \ldots, R_p$ are relations on A, $R'_1, R'_2, \ldots, R'_p$ are relations on B, $\circ_1, \circ_2, \ldots, \circ_q$ are operations on A, and $\circ'_1, \circ'_2, \ldots, \circ'_q$ are operations on B. f is called a *homomorphism* from $\mathfrak{U}$ into $\mathfrak{V}$ if, for any $a_1, a_2, \ldots, a_{r_i} \in A$, $R_i(a_1, a_2, \ldots, a_{r_i})$ iff $R'_i(f(a_1), f(a_2), \ldots, f(a_{r_i}))$, $i = 1, 2, \ldots, p$, and for any $a, b \in A$, $f(a \circ_i b) = f(a) \circ'_i f(b)$, $i = 1, 2, \ldots, q$.

Scale A *scale* is a triple $(\mathfrak{U}, \mathfrak{V}, f)$ where $\mathfrak{U}$ is an observed relational structure that is qualitative, $\mathfrak{V}$ is a numerical relational structure (e.g., $(\mathbb{R}, \geq)$) that is quantitative, and f is a homomorphism from $\mathfrak{U}$ into $\mathfrak{V}$.

Admissible Transformation Suppose that f is one homomorphism from a relational structure $\mathfrak{U}$ into another $\mathfrak{V}$, and suppose that A is the set underlying $\mathfrak{U}$ and B is the set underlying $\mathfrak{V}$. Suppose that Φ is a function that maps the range of f, that is, the set $f(A) = \{f(a) : a \in A\}$ into B. Then the composition $\Phi \circ f$ is a function from A into B. If $\Phi \circ f$ is a homomorphism from $\mathfrak{U}$ into $\mathfrak{V}$, we call Φ an *admissible transformation* of f.

Representation and Uniqueness Theorems There are two main problems in measurement theory[1]:

1. the *representation problem*: Given a *numerical* relational structure $\mathfrak{V}$, find conditions on an *observed* relational structure $\mathfrak{U}$ (necessary and) sufficient for the *existence* of a homomorphism f from $\mathfrak{U}$ to $\mathfrak{V}$ that preserves all the relations and operations in $\mathfrak{U}$.
2. the *uniqueness problem*: Find the type of transformation of the homomorphism f under which all the relations and operations in $\mathfrak{U}$ are preserved.

The theorem stating conditions on $\mathfrak{U}$ are (necessary and) sufficient for the existence of f is called a *representation theorem* that can furnish a solution to the representation problem. On the other hand, the theorem stating the type of transformation up to which f above is unique is called a *uniqueness theorem*, which can furnish a solution to the uniqueness problem.

3 Hypothesis About Representation of Additively-Semiordered Qualitative Conditional Probability Relation

Domotor [6] tries to prove the following proposition (Theorem 12 in [6, p. 97]) in which δ is interpreted to mean a *normalized JND*. The proof of Theorem 12 is based on Theorem 10 in [6, p. 85]. However, we agree to Ibeling et al. [11, pp. 30–32]'s argument that the proof of Theorem 10 "very briefly appeals to an *unstated* result in geometry of webs [2]" ([11, p. 31]). As far as we know, the verification of the proof has been carried out only in the case of $|\Omega| = 2$ (Ibeling et al. [11, pp. 42–43]). In order to regard the following proposition a theorem (fact), the verification of the proof in the case of $|\Omega| = n \geq 1$ must be carried out. So in this paper, for the sake of argument, we would like to regard the following proposition as a *hypothesis*. Indeed because Corollary 2 depends directly on this hypothesis, its proof is conditional. But since all the other technical results, including Theorem 2 (Completeness), in this paper than Corollary 2 do not directly depend on this hypothesis, their proofs are *not conditional*.

Hypothesis 1 (Representation). *Suppose that Ω is a nonempty finite set of elementary events, and that $\mathscr{F}$ is the Boolean algebra of subsets of Ω, and that*

---
[1] Roberts [15] gives a comprehensive survey of measurement theory.

$>_\omega$ is a quaternary *relation relative to* $\omega \in \Omega$ *on* $\mathscr{F}$, *and that* $(A,B) \sim_\omega (C,D)$ *is defined as* $(A,B) \not>_\omega (C,D)$ *and* $(C,D) \not>_\omega (A,B)$, *and that* $\mathscr{F}_0 := \{A : A \in \mathscr{F} \text{ and } (A,\Omega) \not\sim_\omega (\emptyset,\Omega)\}$. *Then there exist both a finitely additive conditional probability measure* $P_\omega : \mathscr{F} \times \mathscr{F}_0 \to \mathbb{R}$ *relative to* $\omega \in \Omega$ *and* $\delta \in \mathbb{R}$ *satisfying, for any* $A,C \in \mathscr{F}$ *and any* $B,D \in \mathscr{F}_0$,

$$(A,B) >_\omega (C,D) \text{ iff } P_\omega(A,B) \geq P_\omega(C,D) + \delta,$$

where $0 < \delta \leq 1$ *and* $P_\omega(A,B)$ *is a probability relative to* w *of* A *on the condition of* B *iff the following conditions are met:*

- **Nontriviality**: $(\Omega,\Omega) >_\omega (\emptyset,\Omega)$.
- **Irreflexivity**: Not $((A,B) >_\omega (A,B))$, for any $A \in \mathscr{F}$ and any $B \in \mathscr{F}_0$.
- **Absorption**: $(A,B) \sim_\omega (A \cap B, B)$ for any $A \in \mathscr{F}$ and any $B \in \mathscr{F}_0$.
- **Dominance**: For any $A,B,C \in \mathscr{F}$ and any $D,E \in \mathscr{F}_0$, if $A \subseteq B$, then if $(C,D) >_\omega (B,E)$, then $(C,D) >_\omega (A,E)$.
- **Finite Cancellation (Addition)**: For any $A_1,\ldots,A_n, C_1,\ldots,C_n, E_1,\ldots, E_n, G_1,\ldots,G_n \in \mathscr{F}$ and any $B_1,\ldots,B_n, D_1,\ldots,D_n, F_1,\ldots,F_n, H_1,\ldots,H_n \in \mathscr{F}_0$, if, for any $i < n$, $((A_i,B_i) >_\omega (C_i,D_i)$ and $(E_i,F_i) \not>_\omega (G_i,H_i))$, then if $(A_n,B_n) >_\omega (C_n,D_n)$, then $(E_n,F_n) >_\omega (G_n,H_n)$, given that

$$\bigcup_{1 \leq i_1 < \cdots < i_k \leq n} (((A_{i_1} \cap B_{i_1}) \cup (G_{i_1} \cap H_{i_1})) \cap \cdots \cap ((A_{i_k} \cap B_{i_k}) \cup (G_{i_k} \cap H_{i_k})))$$
$$= \bigcup_{1 \leq i_1 < \cdots < i_k \leq n} (((C_{i_1} \cap D_{i_1}) \cup (E_{i_1} \cap F_{i_1})) \cap \cdots \cap ((C_{i_k} \cap D_{i_k}) \cup (E_{i_k} \cap F_{i_k})))$$

holds for any k with $1 \leq k \leq n$.

- **Finite Cancellation (Multiplication)**: For any $A_1,\ldots,A_n, B_1,\ldots,B_n \in \mathscr{F}_0$, if, for any k with $1 \leq k \leq n-1$, $(B_{k+1}, \bigcap_{0 \leq i \leq k} B_i) \not>_\omega (\bigcap_{0 \leq i \leq k} B_i, B_0)$, then if, for any l with $0 < l \leq n$, $(A_l, \bigcap_{0 \leq i < l} A_i) >_\omega (B_{\beta(l)}, \bigcap_{0 \leq i < l} B_i)$, then $(\bigcap_{0 \leq i \leq n} A_i, A_0) >_\omega (\bigcap_{0 \leq i \leq n} B_{\beta(i)}, B_0)$, where β is a permutation on $\{1,\ldots,n\}$.
- **Factivity**: $(\{\omega\},\Omega) >_\omega (\emptyset,\Omega)$.

The relation $>_\omega$ *satisfying the above conditions is called an* additively-semiordered qualitative conditional probability relation.

Remark 1 (Finite Cancellations (Addition) and (Multiplication)). Finite Cancellation (Addition) *and* Finite Cancellation (Multiplication) *are the qualitative versions of the addition and multiplication laws of conditional probability, respectively.*

Remark 2 (Factivity). Factivity *means that if* ω *is an actual elementary event, then the event* $\{\omega\}$ *is believed to be possible. As Gärdenfors [9, p. 180] argues,* Factivity *is the semantic counterpart of* $K\varphi \to \varphi$ (T).

4 Logic of Inexact Knowledge LIK

4.1 Language

Definition 1 (Language).

- Let $\mathscr{S}$ denote a set of sentential variables and a qualitative conditional probability relation symbol $>$ a quaternary sentential operator.
- The language $\mathscr{L}_{\mathsf{LIK}}$ of LIK is given by the following BNF grammar:

$$\varphi ::= s \mid \top \mid \neg\varphi \mid \varphi \wedge \psi \mid (\varphi, \psi) > (\chi, \tau)$$

such that $s \in \mathscr{S}$.
- $\bot, \vee, \rightarrow$ and $\leftrightarrow$ are introduced by the standard definitions.
- $\varphi \approx \psi$ (φ is directly indiscriminable from ψ) $:= \neg((\varphi, \varphi \vee \psi) > (\psi, \varphi \vee \psi)) \wedge \neg((\psi, \varphi \vee \psi) > (\varphi, \varphi \vee \psi))$.
- $\varphi \approxeq \psi$ (φ is indirectly indiscriminable from ψ) $:= (\varphi \approx \chi) \leftrightarrow (\psi \approx \chi)$.
- $K\varphi$ (It is known that φ) $:= \varphi \approx \top$.
- The set of all well-formed formulae of $\mathscr{L}_{\mathsf{LIK}}$ is denoted by $\Phi_{\mathscr{L}_{\mathsf{LIK}}}$.

Remark 3 (Not Absolute But Conditional). *The reason why we introduce a qualitative not absolute but conditional probability relation symbol is to make it possible to express direct indiscriminability between φ and ψ on the condition that $\varphi \vee \psi$.*

4.2 Semantics

We define a structured model $\mathfrak{M}$ of LIK as follows:

Definition 2 (Model). $\mathfrak{M}$ *is a triple* (Ω, ρ, V) *in which*

- Ω *is a nonempty finite set of elementary events.*
- ρ *is an additively-semiordered qualitative conditional probability space assignment that assigns to each $\omega \in \Omega$ an additively-semiordered qualitative conditional probability space* $(\Omega, \mathscr{F}, >_\omega)$ *in which* $>_\omega$ *is an additively-semiordered qualitative conditional probability relation relative to $\omega \in \Omega$ on $\mathscr{F}$ that satisfies all of* **Nontriviality, Irreflexivity, Absorption, Dominance, Finite Cancellation (Addition), Finite Cancellation (Multiplication),** *and* **Factivity** *of Hypothesis 1, and in which $\mathscr{F}$ contains higher-order qualitative conditional probability events (e.g.,* $\{\omega : (\{\omega_1 : (A_1, B_1) >_{\omega_1} (C_1, D_1)\}, \{\omega_2 : (A_2, B_2) >_{\omega_2} (C_2, D_2)\}) >_\omega (\{\omega_3 : (A_3, B_3) >_{\omega_3} (C_3, D_3)\}, \{\omega_4 : (A_4, B_4) >_{\omega_4} (C_4, D_4)\})\}$, *where* $A_1, \ldots, A_4, C_1, \ldots, C_4 \in \mathscr{F}$ *and* $B_1, \ldots, B_4, D_1, \ldots, D_4 \in \mathscr{F}_0$).
- V *is a truth assignment to each $s \in \mathscr{S}$ for each $\omega \in \Omega$.*

We provide LIK with the following truth definition at $\omega \in \Omega$ in $\mathfrak{M}$, define the truth in $\mathfrak{M}$, and then define validity as follows:

Definition 3 (Truth and Validity).

- The notion of $\varphi \in \Phi_{\mathscr{L}_{\mathsf{LIK}}}$ being true at $\omega \in \Omega$ in $\mathfrak{M}$, in symbols $(\mathfrak{M}, \omega) \vDash_{\mathsf{LIK}} \varphi$, is inductively defined as follows:
  - $(\mathfrak{M}, \omega) \vDash_{\mathsf{LIK}} s$ iff $V(\omega)(s) =$ true.
  - $(\mathfrak{M}, \omega) \vDash_{\mathsf{LIK}} \top$.
  - $(\mathfrak{M}, \omega) \vDash_{\mathsf{LIK}} \neg \varphi$ iff $(\mathfrak{M}, \omega) \nvDash_{\mathsf{LIK}} \varphi$.
  - $(\mathfrak{M}, \omega) \vDash_{\mathsf{LIK}} \varphi \wedge \psi$ iff $(\mathfrak{M}, \omega) \vDash_{\mathsf{LIK}} \varphi$ and $(\mathfrak{M}, \omega) \vDash_{\mathsf{LIK}} \psi$.
  - $(\mathfrak{M}, \omega) \vDash_{\mathsf{LIK}} (\varphi, \psi) > (\chi, \tau)$ iff $([\![\varphi]\!]^{\mathfrak{M}}, [\![\psi]\!]^{\mathfrak{M}}) >_\omega ([\![\chi]\!]^{\mathfrak{M}}, [\![\tau]\!]^{\mathfrak{M}})$, where $[\![\varphi]\!]^{\mathfrak{M}} \coloneqq \{\omega' \in \Omega : (\mathfrak{M}, \omega') \vDash_{\mathsf{LIK}} \varphi\}$ and $[\![\varphi]\!]^{\mathfrak{M}}, [\![\chi]\!]^{\mathfrak{M}} \in \mathscr{F}$ and $[\![\psi]\!]^{\mathfrak{M}}, [\![\tau]\!]^{\mathfrak{M}} \in \mathscr{F}_0$.
- If $(\mathfrak{M}, \omega) \vDash_{\mathsf{LIK}} \varphi$ for any $\omega \in \Omega$, we write $\mathfrak{M} \vDash_{\mathsf{LIK}} \varphi$ and say that φ is true in $\mathfrak{M}$.
- If φ is true in any model $\mathfrak{M}$, we write $\vDash_{\mathsf{LIK}} \varphi$ and say that φ is valid.

The next corollary follows from Definitions 1 and 3.

Corollary 1 (Truth Conditions).

- $(\mathfrak{M}, \omega) \vDash_{\mathsf{LIK}} \varphi \approx \psi$ iff $([\![\varphi]\!]^{\mathfrak{M}}, [\![\varphi]\!]^{\mathfrak{M}} \cup [\![\psi]\!]^{\mathfrak{M}}) \sim_\omega ([\![\psi]\!]^{\mathfrak{M}}, [\![\varphi]\!]^{\mathfrak{M}} \cup [\![\psi]\!]^{\mathfrak{M}})$.
- $(\mathfrak{M}, \omega) \vDash_{\mathsf{LIK}} \varphi \cong \psi$ iff $\Big(([\![\varphi]\!]^{\mathfrak{M}}, [\![\varphi]\!]^{\mathfrak{M}} \cup [\![\chi]\!]^{\mathfrak{M}}) \sim_\omega ([\![\chi]\!]^{\mathfrak{M}}, [\![\varphi]\!]^{\mathfrak{M}} \cup [\![\chi]\!]^{\mathfrak{M}})$ iff $([\![\psi]\!]^{\mathfrak{M}}, [\![\psi]\!]^{\mathfrak{M}} \cup [\![\chi]\!]^{\mathfrak{M}}) \sim_\omega ([\![\chi]\!]^{\mathfrak{M}}, [\![\psi]\!]^{\mathfrak{M}} \cup [\![\chi]\!]^{\mathfrak{M}})\Big)$.
- $(\mathfrak{M}, \omega) \vDash_{\mathsf{LIK}} K\varphi$ iff $([\![\varphi]\!]^{\mathfrak{M}}, \Omega) \sim_\omega (\Omega, \Omega)$.

Then the next corollary follows from Hypothesis 1 and Corollary 1.

Corollary 2 (Truth Conditions by Probability Measure). There exist both $P_\omega : \mathscr{F} \times \mathscr{F}_0 \to \mathbb{R}$ and $\delta \in \mathbb{R}$ with $0 < \delta \leq 1$ satisfying

- $(\mathfrak{M}, \omega) \vDash_{\mathsf{LIK}} (\varphi, \psi) > (\chi, \tau)$ iff $P_\omega([\![\varphi]\!]^{\mathfrak{M}}, [\![\psi]\!]^{\mathfrak{M}}) \geq P_\omega([\![\chi]\!]^{\mathfrak{M}}, [\![\tau]\!]^{\mathfrak{M}}) + \delta$.
- $(\mathfrak{M}, \omega) \vDash_{\mathsf{LIK}} \varphi \approx \psi$ iff $P_\omega([\![\psi]\!]^{\mathfrak{M}}, [\![\varphi]\!]^{\mathfrak{M}} \cup [\![\psi]\!]^{\mathfrak{M}}) - \delta < P_\omega([\![\varphi]\!]^{\mathfrak{M}}, [\![\varphi]\!]^{\mathfrak{M}} \cup [\![\psi]\!]^{\mathfrak{M}}) < P_\omega([\![\psi]\!]^{\mathfrak{M}}, [\![\varphi]\!]^{\mathfrak{M}} \cup [\![\psi]\!]^{\mathfrak{M}}) + \delta$.
- $(\mathfrak{M}, \omega) \vDash_{\mathsf{LIK}} \varphi \cong \psi$ iff $P_\omega([\![\varphi]\!]^{\mathfrak{M}}, [\![\varphi]\!]^{\mathfrak{M}} \cup [\![\psi]\!]^{\mathfrak{M}}) = P_\omega([\![\psi]\!]^{\mathfrak{M}}, [\![\varphi]\!]^{\mathfrak{M}} \cup [\![\psi]\!]^{\mathfrak{M}})$.
- $(\mathfrak{M}, \omega) \vDash_{\mathsf{LIK}} K\varphi$ iff $1 - \delta < P_\omega([\![\varphi]\!]^{\mathfrak{M}}, W \cup [\![\varphi]\!]^{\mathfrak{M}}) \leq 1$.

Remark 4 (Not Absolute But Conditional Revisited)

- According to this corollary, in such a sorites series as (6)–(8) of heights of a tree on day i, the truth condition of $\varphi_i \approx \varphi_{i+1}$ is given not by $P_\omega([\![\varphi_{i+1}]\!]^{\mathfrak{M}}) - \delta < P_\omega([\![\varphi_i]\!]^{\mathfrak{M}}) < P_\omega([\![\varphi_{i+1}]\!]^{\mathfrak{M}}) + \delta$ but by $P_\omega([\![\varphi_{i+1}]\!]^{\mathfrak{M}}, [\![\varphi_i]\!]^{\mathfrak{M}} \cup [\![\varphi_{i+1}]\!]^{\mathfrak{M}}) - \delta < P_\omega([\![\varphi_i]\!]^{\mathfrak{M}}, [\![\varphi_i]\!]^{\mathfrak{M}} \cup [\![\varphi_{i+1}]\!]^{\mathfrak{M}}) < P_\omega([\![\varphi_{i+1}]\!]^{\mathfrak{M}}, [\![\varphi_i]\!]^{\mathfrak{M}} \cup [\![\varphi_{i+1}]\!]^{\mathfrak{M}}) + \delta$.
- The former truth condition in terms of absolute probability measures cannot be interpreted as direct indiscriminability between φ_i and φ_{i+1} on the condition that φ_i or φ_{i+1} holds, whereas the latter truth condition in terms of conditional probability measures can be interpreted as this indiscriminability.

LIK has the following desirable properties on which we have argued in Sect. 1:

Proposition 1 (Invalidity of Transitivity of Direct Indiscriminability).
$\not\models_{\mathsf{LIK}} ((\varphi \approx \psi) \wedge (\psi \approx \chi)) \to (\varphi \approx \chi)$.

Proposition 2 (Validity of Transitivity of Indirect Indiscriminability).
$\models_{\mathsf{LIK}} ((\varphi \cong \psi) \wedge (\psi \cong \chi)) \to (\varphi \cong \chi)$.

Moreover, we can formulate a *margin for error principle* in LIK:

Formulation 1 (Margin for Error Principle). $(\mathfrak{M}, \omega') \models_{\mathsf{LIK}} \varphi$, for any $\omega' \in \Omega$ such that $(\mathfrak{M}, \omega) \models_{\mathsf{LIK}} K\varphi$ and $(\{\omega'\}, \{\omega, \omega'\}) \sim_\omega (\{\omega\}, \{\omega, \omega'\})$.

Williamson [21, p. 270] gives the truth definition in the *Fixed Margin Model* $\mathfrak{M}'$ of LC satisfying:

$(\mathfrak{M}', w) \models_{\mathsf{LC}} K\varphi$ iff for any $w' \in W$, if $d(w, w') \leq \alpha$, then $w' \in [\![\varphi]\!]^{\mathfrak{M}'}$,

where d is a metric on the set W of possible worlds and α is given. He [21, p. 272] also gives the truth definition in the *Variable Margin Model* $\mathfrak{M}''$ of LC satisfying:

$(\mathfrak{M}'', w) \models_{\mathsf{LC}} K\varphi$ iff there exists $\delta > \alpha$ such that, for any $w' \in W$, if $d(w, w') \leq \delta$, then $w' \in [\![\varphi]\!]^{\mathfrak{M}''}$.

On the other hand, we give the following truth condition of $K\varphi$:

There exist $P_\omega : \mathscr{F} \times \mathscr{F}_0 \to \mathbb{R}$ and $\delta \in \mathbb{R}$ with $0 < \delta \leq 1$ satisfying
$(\mathfrak{M}, \omega) \models_{\mathsf{LIK}} K\varphi$ iff $1 - \delta < P_\omega([\![\varphi]\!]^{\mathfrak{M}}, W \cup [\![\varphi]\!]^{\mathfrak{M}}) \leq 1$.

In Williamuson's formulations, a metric d on W is considered to be a measure of *similarity*. Because d is a *mere metric*, it is not clear how it can contribute to specifying *direct indiscriminability*. Because d is a mere metric, it is too weak even as a measure of similarity. On the other hand, in our formulation, direct indiscriminability is defined by an additively-semiordered qualitative conditional probability relation $\succ_\omega$, and knowledge is defined by this direct indiscrimminability. Because Williamson's LC has no direct indiscriminability relation symbol in the object language, it cannot fully express the inferences about (6)–(8). On the other hand, when a suitable model $\mathfrak{U}$ that makes (6)–(8) true is given, we have the following proposition:

Proposition 3 (Knowledge on Height of Tree).

- $\mathfrak{U} \models_{\mathsf{LIK}} (K\neg(\varphi_0 \approx \varphi_{5,000}) \wedge \bigwedge_{0 \leq i,j \leq 5,000} K((\varphi_i \approx \varphi_j) \to \neg K\neg(\varphi_i \approx \varphi_j)) \wedge \bigwedge_{0 \leq i < 5,000} K(\varphi_i \approx \varphi_{i+1}) \wedge (K\psi \to KK\psi)) \to \bot$.
- $\mathfrak{U} \not\models_{\mathsf{LIK}} (K\neg(\varphi_0 \approx \varphi_{5,000}) \wedge \bigwedge_{0 \leq i,j \leq 5,000} K((\varphi_i \approx \varphi_j) \to \neg K\neg(\varphi_i \approx \varphi_j)) \wedge \bigwedge_{0 \leq i < 5,000} K(\varphi_i \approx \varphi_{i+1})) \to \bot$.

Proof. The proof is based on the fact that $(\omega, \mathfrak{U}) \models_{\mathsf{LIK}} K(\varphi_i \approx \varphi_{i+1})$ iff $(\{\omega : ([\![\varphi_i]\!]^{\mathfrak{M}}, [\![\varphi_i]\!]^{\mathfrak{M}} \cup [\![\varphi_{i+1}]\!]^{\mathfrak{M}}) \sim_\omega (\{\omega : ([\![\varphi_{i+1}]\!]^{\mathfrak{M}}, [\![\varphi_i]\!]^{\mathfrak{M}} \cup [\![\varphi_{i+1}]\!]^{\mathfrak{M}})\}, \Omega) \sim_\omega (\Omega, \Omega)$ and the fact that $(\omega, \mathfrak{U}) \models_{\mathsf{LIK}} KK\psi$ iff $(\{\omega : ([\![\psi]\!]^{\mathfrak{M}}, \Omega) \sim_\omega (\Omega, \Omega)\}, \Omega) \sim_\omega (\Omega, \Omega)$. ∎

We can verify the invalidity of KK principle in LIK:

Proposition 4 (Invalidity of KK Principle).
$\not\models_{\mathsf{LIK}} K\varphi \to KK\varphi$.

4.3 Axiomatization

The proof system of LIK consists of the following:

Definition 4 (Proof System).

- All tautologies of classical propositional logic.
- $K: K(\varphi \to \psi) \to (K\varphi \to K\psi)$.
- $T: K\varphi \to \varphi$.
- Replacement of Known Equivalents on $>$: $(K(\varphi_1 \leftrightarrow \varphi_2) \wedge K(\psi_1 \leftrightarrow \psi_2) \wedge K(\chi_1 \leftrightarrow \chi_2) \wedge K(\tau_1 \leftrightarrow \tau_2)) \to (((\varphi_1, \chi_1) > (\psi_1, \tau_1)) \leftrightarrow ((\varphi_2, \chi_2) > (\psi_2, \tau_2)))$.
- Axiomatic Counterpart of **Nontriviality**: $(\top, \top) > (\bot, \top)$.
- Axiomatic Counterpart of **Irreflexivity**: $(\psi \leftrightarrow \bot) \to \neg((\varphi, \psi) > (\varphi, \psi))$.
- Axiomatic Counterpart of **Absorption**: $(\psi \leftrightarrow \bot) \to ((\varphi, \psi) \approx (\varphi \wedge \psi, \psi))$.
- Axiomatic Counterpart of **Finite Cancellation (Addition)**:

$$\bigwedge_{1 \le i \le n} \left(((\psi_i, \top) \not\approx (\bot, \top)) \wedge ((\tau_i, \top) \not\approx (\bot, \top)) \wedge ((\eta_i, \top) \not\approx (\bot, \top)) \wedge ((\nu_i, \top) \not\approx (\bot, \top)) \right)$$

$$\to \left(\bigwedge_{1 \le k \le n} \left(\bigvee_{1 \le i_1 < \cdots < i_k \le n} (((\varphi_{i_1} \wedge \psi_{i_1}) \vee (\mu_{i_1} \wedge \nu_{i_1})) \wedge \cdots \wedge ((\varphi_{i_k} \wedge \psi_{i_k}) \vee (\mu_{i_k} \wedge \nu_{i_k}))) \right. \right.$$

$$\left. \leftrightarrow \bigvee_{1 \le i_1 < \cdots < i_k \le n} (((\chi_{i_1} \wedge \tau_{i_1}) \vee (\zeta_{i_1} \wedge \eta_{i_1})) \wedge \cdots \wedge ((\chi_{i_k} \wedge \tau_{i_k}) \vee (\zeta_{i_k} \wedge \eta_{i_k}))) \right)$$

$$\to \left(\bigwedge_{i < n} (((\varphi_i, \psi_i) > (\chi_i, \tau_i)) \wedge \neg ((\zeta_i, \eta_i) > (\mu_i, \nu_i))) \right.$$

$$\left. \left. \to (((\varphi_n, \psi_n) > (\chi_n, \tau_n)) \to ((\zeta_n, \eta_n) > (\mu_n, \nu_n))) \right) \right).$$

- Axiomatic Counterpart of **Finite Cancellation (Multiplication)**:

$$\bigwedge_{1 \le i \le n} \left(((\varphi_i, \top) \not\approx (\bot, \top)) \wedge ((\psi_i, \top) \not\approx (\bot, \top)) \right)$$

$$\to \left(\bigwedge_{1 \le k \le n-1} \neg \left((\psi_{k+1}, \bigwedge_{0 \le i \le k} \psi_i) > (\bigwedge_{0 \le i \le k} \psi_i, \psi_0) \right) \right)$$

$$\to \left(\bigwedge_{0 < l \le n} ((\varphi_l, \bigwedge_{0 \le i < l} \varphi_i) > (\psi_{\beta(l)}, \bigwedge_{0 \le i < l} \psi_i)) \to ((\bigwedge_{0 \le i \le n} \varphi_i, \varphi_0) > (\bigwedge_{0 \le i \le n} \psi_{\beta(i)}, \psi_0)) \right),$$

where β is a permutation on $\{1, \ldots, n\}$.
- Axiomatic Counterpart of **Dominance**:

$$\frac{(\tau, \top) \not\approx (\bot, \top) \quad (\zeta, \top) \not\approx (\bot, \top) \quad \varphi \to \psi}{((\chi, \tau) > (\psi, \zeta)) \to ((\chi, \tau) > (\varphi, \zeta))}.$$

- Modus Ponens.
- Epistemic Necessitation.
- A proof of $\varphi \in \Phi_{\mathscr{L}_{\mathsf{LIK}}}$ is a finite sequence of $\mathscr{L}_{\mathsf{LIK}}$-formulae having φ as the last formula such that either each formula is an instance of an axiom or it can be obtained from formulae that appear earlier in the sequence by applying an inference rule.

– If there is a proof of φ, we write $\vdash_{\mathsf{LIK}} \varphi$.

Remark 5 (Infinite Schema). *Both the axiomatic counterparts of* **Finite Cancellation (Addition)** *and* **Finite Cancellation (Multiplication)** *are infinite schemata of axioms.*

Remark 6 (KT). *The proof system of epistemic-logical part of* LIK *is* **KT**. *The proof system of Williamson's* LC *on a variable margin model is* **KT**, *whereas* LC *on a fixed margin model is* **KTB**.

4.4 Metalogic

4.4.1 Soundness
We prove the soundness of LIK:

Theorem 1 (Soundness of LIK). *For any $\psi \in \Phi_{\mathscr{L}_{\mathsf{LIK}}}$, if $\vdash_{\mathsf{LIK}} \psi$, then $\vDash_{\mathsf{LIK}} \psi$.*

Proof. It is trivial to verify the conditions except the axiomatic counterparts of **Finite Cancellation (Addition)** and **Finite Cancellation (Multiplication)**. The verification of the former is much the same as Segerberg [17, pp. 344–346]'s. The verification of the latter can be conducted in much the same way as that of the former. ∎

4.4.2 Completeness of LIK.
We would like to *outline* the proof of the completeness of LIK referencing Gärdenfors [9] as follows. Let $\varphi \in \Phi_{\mathscr{L}_{\mathsf{LIK}}}$ any such formula that $\nvdash_{\mathsf{LIK}} \varphi$. We would like to construct a structural model $\mathfrak{M}^\varphi := (\Omega^\varphi, \rho^\varphi, V^\varphi)$ where $\rho^\varphi := (\Omega^\varphi, \mathscr{F}^\varphi, >_\omega^\varphi)$ in which $\mathscr{F}^\varphi$ is a *finite* Boolean algebra such that $\mathfrak{M}^\varphi \nvDash_{\mathsf{LIK}} \varphi$. First, we define equivalence class $[\varphi]$:

Definition 5 (Equivalence Class). *For any $\varphi \in \Phi_{\mathscr{L}_{\mathsf{LIK}}}$, the equivalence class $[\varphi] := \{\psi \in \Phi_{\mathscr{L}_{\mathsf{LIK}}} : \vdash_{\mathsf{LIK}} \varphi \leftrightarrow \psi\}$.*

The following lemma follows directly from Definition 5:

Lemma 1 (Provability and Equivalence Class). $\vdash_{\mathsf{LIK}} \varphi$ *iff* $[\varphi] = [\top]$.

We define Boolean operators $\overline{\cdots}$ and $\cap$ on the set of equivalence classes:

Definition 6 ($\overline{\cdots}$ and $\cap$). $\overline{\cdots}$ *and $\cap$ on the set of equivalence classes are defined in such a way that $\overline{[\varphi]} := [\neg\varphi]$ and $[\varphi] \cap [\psi] := [\varphi \wedge \psi]$ for any $\varphi, \psi \in \Phi_{\mathscr{L}_{\mathsf{LIK}}}$.*

By Definition 6, we can construct $\mathscr{F}^\varphi$ as a *finite* Boolean algebra[2] of $[\psi]$'s with $[\top]$ and $[\bot]$ as unit and zero elements respectively, for any $\psi \in \Phi_{\mathscr{L}_{\mathsf{LIK}}}$, and can specify Ω^φ as the set of elementary events ω's of this $\mathscr{F}^\varphi$. Then we define $([\varphi], [\psi]) >_\omega ([\chi], [\tau])$ in terms of $[(\varphi, \psi) > (\chi, \tau)]$:

---
[2] Because of space limitations, we cannot demonstrate how to construct $\mathscr{F}^\varphi$ as a finite Boolean algebra. About this construction, refer to Gärdenfors [9, p. 177].

Definition 7 ($>_\omega$ on Set of Equivalence Classes). *For any $\omega \in \Omega^\varphi$, for any $\varphi, \psi, \chi, \tau \in \Phi_{\mathscr{L}_{\mathsf{LIK}}}$, $([\varphi], [\psi]) >_\omega ([\chi], [\tau])$ iff $\omega \in [(\varphi, \psi) > (\chi, \tau)]$.*

Then we prove the next lemma:

Lemma 2 (Satisfaction of Conditions for Representation). *For any $\omega \in \Omega^\varphi$, $[\psi]$ satisfies **Nontriviality, Irreflexivity, Absorption, Dominance, Finite Cancellation (Addition), Finite Cancellation (Multiplication),** and **Factivity** of Hypothesis 1.*

Proof. It is trivial to verify that **Nontriviality, Irreflexivity, Absorption, Dominance,** and **Factivity** are satisfied given their axiomatic counterparts. For the verification of **Finite Cancellation (Addition)**, we make the following four assumptions:

1. Assume that $[\varphi_{i_1}], \ldots, [\varphi_{i_k}], [\mu_{i_1}], \ldots, [\mu_{i_k}], [\chi_{i_1}], \ldots, [\chi_{i_k}], [\zeta_{i_1}], \ldots, [\zeta_{i_k}], \in \mathscr{F}^\varphi$ and that $[\psi_{i_1}], \ldots, [\psi_{i_k}], [\nu_{i_1}], \ldots, [\nu_{i_k}], [\tau_{i_1}], \ldots, [\tau_{i_k}], [\eta_{i_1}], \ldots, [\eta_{i_k}], \in \mathscr{F}^\varphi_0$.
2. Assume that the equation part of **Finite Cancellation (Addition)** holds for these elements:

$$\bigcup_{1 \leq i_1 < \cdots < i_k \leq n} ((([\varphi_{i_1}] \cap [\psi_{i_1}]) \cup ([\mu_{i_1}] \cap [\nu_{i_1}])) \cap \cdots \cap $$
$$ ((([\varphi_{i_k}] \cap [\psi_{i_k}]) \cup ([\mu_{i_k}] \cap [\nu_{i_k}]))) $$
$$ = \bigcup_{1 \leq i_1 < \cdots < i_k \leq n} ((([\chi_{i_1}] \cap [\tau_{i_1}]) \cup ([\zeta_{i_1}] \cap [\eta_{i_1}])) \cap \cdots \cap $$
$$ ((([\chi_{i_k}] \cap [\tau_{i_k}]) \cup ([\zeta_{i_k}] \cap [\eta_{i_k}])))),$$

for any k with $1 \leq k \leq n$
3. Assume that $([\varphi_i], [\psi_i]) >_\omega ([\chi_i], [\tau_i])$ and $([\zeta_i], [\eta_i]) \not>_\omega ([\mu_i], [\nu_i])$ for any $i < n$.
4. Assume that $([\varphi_n], [\psi_n]) >_\omega ([\chi_n], [\tau_n])$.

Then by Assumption 2 and Definition 6,

$$\omega \in \Big[\bigvee_{1 \leq i_1 < \cdots < i_k \leq n} (((\varphi_{i_1} \wedge \psi_{i_1}) \vee (\mu_{i_1} \wedge \nu_{i_1})) \wedge \cdots \wedge ((\varphi_{i_k} \wedge \psi_{i_k}) \vee (\mu_{i_k} \wedge \nu_{i_k}))) $$
$$ \leftrightarrow \bigvee_{1 \leq i_1 < \cdots < i_k \leq n} (((\chi_{i_1} \wedge \tau_{i_1}) \vee (\zeta_{i_1} \wedge \eta_{i_1})) \wedge \cdots \wedge ((\chi_{i_k} \wedge \tau_{i_k}) \vee (\zeta_{i_k} \wedge \eta_{i_k}))) \Big]$$

for any k with $1 \leq k \leq n$. This holds for any $\omega \in \Omega^\varphi$. So

$$\Big[\bigvee_{1 \leq i_1 < \cdots < i_k \leq n} (((\varphi_{i_1} \wedge \psi_{i_1}) \vee (\mu_{i_1} \wedge \nu_{i_1})) \wedge \cdots \wedge ((\varphi_{i_k} \wedge \psi_{i_k}) \vee (\mu_{i_k} \wedge \nu_{i_k}))) $$
$$ \leftrightarrow \bigvee_{1 \leq i_1 < \cdots < i_k \leq n} (((\chi_{i_1} \wedge \tau_{i_1}) \vee (\zeta_{i_1} \wedge \eta_{i_1})) \wedge \cdots \wedge ((\chi_{i_k} \wedge \tau_{i_k}) \vee (\zeta_{i_k} \wedge \eta_{i_k}))) \Big] = [\top],$$

for any k with $1 \leq k \leq n$. From the axiomatic counterpart of **Finite Cancellation (Addition)**, it follows that

$$\Big[\bigwedge_{i < n} (((\varphi_i, \psi_i) > (\chi_i, \tau_i)) \wedge \neg ((\zeta_i, \eta_i) > (\mu_i, \nu_i))) $$
$$ \rightarrow (((\varphi_n, \psi_n) > (\chi_n, \tau_n)) \rightarrow ((\zeta_n, \eta_n) > (\mu_n, \nu_n))) \Big] = [\top].$$

So from the above assumptions and Definition 5, we have $\omega \in [(\zeta_n, \eta_n) > (\mu_n, \nu_n)]$. So from Definition 7, $([\zeta_n], [\eta_n]) >_\omega ([\mu_n], [\nu_n])$. The verification of **Finite Cancellation (Multiplication)** can be conducted in much the same way as that of **Finite Cancellation (Addition)**. ∎

Then we proves the following lemma from Lemma 2 as follows:

Lemma 3 (Equality Between Syntactic Value of ψ and Semantic Value of ψ). *For any $\psi \in \Phi_{\mathscr{L}_{\mathsf{LIK}}}$, $[\psi] = [\![\psi]\!]^{\mathfrak{M}^\varphi}$.*

Proof. Let V^φ be *any* truth assignment such that, for any sentential variable $s \in \mathscr{S}$, $[s] = \{\omega \in \Omega^\varphi : V^\varphi(\omega)(s) = true\}$[3] This equivalence class belongs to $\mathscr{F}^\varphi$. We show by induction on the length of ψ that $[\psi] = [\![\psi]\!]^{\mathfrak{M}^\varphi}$ as follows:

1. When ψ is a sentential variable, it holds by definition.
2. When ψ is a ⊤, trivially $[\top] = [\![\top]\!]^{\mathfrak{M}^\varphi}$.
3. When ψ is $\neg \chi$, $[\neg \chi] = \overline{[\chi]}$ by Definition 6. $\overline{[\chi]} = \overline{[\![\chi]\!]^{\mathfrak{M}^\varphi}}$ by the induction hypotheses. $\overline{[\![\chi]\!]^{\mathfrak{M}^\varphi}} = [\![\neg \chi]\!]^{\mathfrak{M}^\varphi}$ by Definition 3.
4. When ψ is $\chi \wedge \tau$, $[\chi \wedge \tau] = [\chi] \cap [\tau]$ by Definition 6. $[\chi] \cap [\tau] = [\![\chi]\!]^{\mathfrak{M}^\varphi} \cap [\![\tau]\!]^{\mathfrak{M}^\varphi}$ by the induction hypotheses. $[\![\chi]\!]^{\mathfrak{M}^\varphi} \cap [\![\tau]\!]^{\mathfrak{M}^\varphi} = [\![\chi \wedge \tau]\!]^{\mathfrak{M}^\varphi}$ by Definition 3.
5. When ψ is $(\chi, \tau) > (\mu, \nu)$, $[(\chi, \tau) > (\mu, \nu)] = \{\omega \in \Omega^\varphi : \omega \in [(\chi, \tau) > (\mu, \nu)]\}$ by the property of equivalence classes. $\{\omega \in \Omega^\varphi : \omega \in [(\chi, \tau) > (\mu, \nu)]\} = \{\omega \in \Omega^\varphi : ([\chi], [\tau]) >_\omega ([\mu], [\nu])\}$ by Definition 7. $\{\omega \in \Omega^\varphi : ([\chi], [\tau]) >_\omega ([\mu], [\nu])\} = \{\omega \in \Omega^\varphi : ([\![\chi]\!]^{\mathfrak{M}^\varphi}, [\![\tau]\!]^{\mathfrak{M}^\varphi}) >_\omega ([\![\mu]\!]^{\mathfrak{M}^\varphi}, [\![\nu]\!]^{\mathfrak{M}^\varphi})\}$ by the induction hypotheses. By Lemma 2, we can apply Definition 3 to $>_\omega$. So $\{\omega \in \Omega^\varphi : ([\![\chi]\!]^{\mathfrak{M}^\varphi}, [\![\tau]\!]^{\mathfrak{M}^\varphi}) >_\omega ([\![\mu]\!]^{\mathfrak{M}^\varphi}, [\![\nu]\!]^{\mathfrak{M}^\varphi})\} = [\![(\chi, \tau) > (\mu, \nu)]\!]^{\mathfrak{M}^\varphi}$. ∎

Finally, the completeness of LIK follows from Lemma 1 and 3:

Theorem 2 (Completeness). *For any $\psi \in \Phi_{\mathscr{L}_{\mathsf{LIK}}}$, if $\vDash_{\mathsf{LIK}} \psi$, then $\vdash_{\mathsf{LIK}} \psi$.*

Proof. Suppose that $\nvdash_{\mathsf{LIK}} \psi$. Then, if $\nvdash_{\mathsf{LIK}} \psi$, then $[\psi] \neq [\top]$ by Lemma 1. So $[\![\psi]\!]^{\mathfrak{M}^\varphi} \neq \Omega^\varphi$ by Lemma 3. So $\mathfrak{M}^\varphi \nvDash_{\mathsf{LIK}} \psi$. Therefore, there is a model $\mathfrak{M}$ such that $\mathfrak{M} \nvDash_{\mathsf{LIK}} \psi$. ∎

5 Concluding Remarks

We have proposed a new version of a complete logic (LIK) the model of which is suitable as a model of inexact knowledge in the sense that the feature 1 of LIK in Sect. 1 is explained by the comparison in Subsect. 4.2 between our and Williamson's models, the feature 2 is realized in Formulation 1, the feature 3 in Hypothesis 1, the feature 4 in Proposition 1, the feature 5 is elaborated in Remarks 3 and 4, the feature 6 is realized in Proposition 3, the feature 7 in Proposition 4, and the feature 8 in Definition 1 and Corollary 1. In a future

---

[3] Generally speaking, in probability theory, the status of an elementary event ω is not specified. This non-specificity guarantees that we can equate the set of ω's with the equivalence class $[s]$ that is a syntactic object.

study, we would like to carry out the verification of the proof of Theorem 10 in [6, p. 85] on which Hypothesis 1 is based in the case of $|\Omega| = n \geq 1$ by appeal to geometry of webs (nets) of [2] and [1]. We [20] now study about a version of the logic for Phenomenal Sorites Paradox the model of which is based on the basis of an additively-semiordered qualitative conditional probability relation on which the model of LIK is based.

Acknowledgements. The author would like to thank the three reviewers of ICLA 2025 for their very helpful comments.

References

1. Aczél, J.: Quasigroups, nets, and nomograms. Adv. Math. **1**, 383–450 (1965)
2. Aczél, J., et al.: Nomogramme, gewebe und quasigruppen. Math. (Cluj) **2**, 5–24 (1960)
3. Bonnay, D., Égré, P.: Inexact knowledge with introspection. J. Philos. Log. **38**, 179–227 (2009)
4. Cobreros, P., et al.: Tolerant, classical, strict. J. Philos. Log. **41**, 347–385 (2012)
5. Cohen, M.: Inexact knowledge and dynamic introspection. Synthese, 5509–5531 (2021). https://doi.org/10.1007/s11229-021-03033-7
6. Domotor, Z.: Probabilistic relational structures and their applications. Technical report No. 144, Institute for Mathematical Studies in the Social Sciences, Stanford University (1969)
7. Dutant, J.: Inexact knowledge, margin for error and positive introspection. In: Proceedings of Tark XI, pp. 118–124 (2007)
8. Fechner, G.T.: Elemente der Psychophysik. Breitkopf and Hartel, Leipzig (1860)
9. Gärdenfors, P.: Qualitative probability as an intensional logic. J. Philos. Log. **4**, 171–185 (1975)
10. Hölder, O.: Die axiome der Quantität und die Lehre vom mass. Berichte über die Verhandlungen der Königlich Sächsischen Gesellschaft der Wissenschaften zu Leipzig. Math. Phys. Klasse **53**, 1–64 (1901)
11. Ibeling, D., et al.: Probing the quantitative-qualitative divide in probabilistic reasoning. Ann. Pure Appl. Logic **175**, 103339 (2024)
12. Krantz, D.H., et al.: Foundations of Measurement, vol. 1. Academic Press, New York (1971)
13. Luce, R.D.: Semiorders and a theory of utility discrimination. Econometrica **24**, 178–191 (1956)
14. Luce, R.D., et al.: Foundations of Measurement, vol. 3. Academic Press, San Diego (1990)
15. Roberts, F.S.: Measurement Theory. Addison-Wesley, Reading (1979)
16. van Rooij, R.: Vagueness and linguistics. In: Ronzitti, G. (ed.) Vagueness: A Guide, pp. 123–170. Springer, Heidelberg (2011)
17. Segerberg, K.: Qualitative probability in a modal setting. In: Fenstad, J.E. (ed.) Proceedings of the Second Scandinavian Logic Symposiu, pp. 341–352. North-Holland (1971)
18. Simon, H.A.: Models of Bounded Rationality, vol. 1. The MIT Press, Cambridge (1982)

19. Suppes, P., et al.: Foundations of Measurement, vol. 2. Academic Press, San Diego (1989)
20. Suzuki, S.: Measurement-theoretic foundations of logic of probabilistic indiscriminability (2024). manuscript
21. Williamson, T.: Vagueness. Routledge, London (1994)

Craig Interpolation for Awareness Logics

Kosuke Udatsu[1](✉) and Katsuhiko Sano[2]

[1] Graduate School of Humanities and Human Sciences, Hokkaido University, Sapporo, Japan
udatsu.logic.work@gmail.com
[2] Faculty of Humanities and Human Sciences, Hokkaido University, Sapporo, Japan
v-sano@let.hokudai.ac.jp

Abstract. In this paper, we establish the Craig interpolation theorems of awareness logics by introducing semi-analytic sequent calculi for them. For a semantic clause for an awareness operator, we choose an idea of propositional awareness, which is introduced by Fagin and Halpern and means that an agent is aware of a formula if and only if he is aware of all the atomic propositions contained in it. Although our sequent calculi are not cut-free (due to the fact that our epistemic logic is based on **S5**), we are able to restrict all applications of the rule of cut to semi-analytic ones. This enables us to employ the Maehara method in order to compute Craig interpolants.

1 Introduction

Awareness logic [6] is introduced by Fagin and Halpern to solve the problem of logical omniscience in epistemic logic. Let K_i be an epistemic modal operator, whose reading is "agent i knows that". The traditional epistemic logic admits the necessitation law (Nec): if φ is a theorem then $K_i\varphi$ is also a theorem, and K axiom: $K_i(\varphi \to \psi) \to (K_i\varphi \to K_i\psi)$. The necessitation law implies that each agent knows all tautologies. The axiom K implies that each agent knows all consequences obtained from his knowledge. These are called the problem of logical omniscience. However, we cannot assume that all agents are idealized, e.g. due to lack of computational resources.

Awareness logic [6] divides the notion of knowledge into *implicit knowledge* and *explicit knowledge* by introducing the notion of *awareness*. Implicit knowledge of an agent is the same as knowledge in ordinary epistemic logic, independent of the agent's awareness. Explicit knowledge of an agent is implicit knowledge of which the agent is aware. Then we may avoid the problem of logical omniscience in terms of explicit knowledge. So the question boils down to how the notion of awareness is defined. Awareness is defined in terms of a set of formulas: if a formula is in this set of awareness, then each agent is aware of the formula, otherwise, it is not aware of it. There are different ways to define this set of awareness for an agent. This paper employs the notion of propositional or primitive awareness [4–6,8], which means that, for an agent i and a

---

We would like to thank the anonymous referees for their valuable comments. The work of the first author was supported by Hokkaido University EXEX Doctoral Fellowship. The work of the second author was partially supported by JSPS KAKENHI Grant-in-Aid for Scientific Research (B) Grant Number JP 22H00597.

state w, we define a set $\mathscr{A}(i,w)$ of formulas of which i is aware at state w as consisting of *propositional variables* alone. Let us explain why the necessitation rule fails for explicit knowledge defined as $K_i\varphi := \Box_i\varphi \wedge A_i\varphi$, where $\Box_i$ represents implicit knowledge for which we allow the necessitation rule. Suppose that φ is a (very long) tautology in terms of atomic propositions, say, p, q, and r. We explain why $\Box_i\varphi \wedge A_i\varphi$ cannot be obtained. Because we can use the necessitation rule for implicit knowledge, we get $\Box_i\varphi$. But we cannot obtain $A_i\varphi$ in general, because agent i may not be aware of all the atomic propositions p, q, and r in φ (he may be aware of p alone). So we can say that awareness restricts the necessity rule for explicit knowledge. In their seminal work, Fagin and Halpern [6] stated that their Hilbert system for multi-agent epistemic logic (based on **KD45**) with propositional awareness is both sound and complete with respect to the intended Kripke semantics (see also [8]). To the best of the authors' knowledge, no studies of alternative proof theory exist beyond those pertaining to Hilbert systems.

This paper proposes Gentzen-style sequent calculi for minimal multi-agent epistemic logic with propositional awareness and its extensions as a technical tool to obtain the Craig interpolation theorem. In sequent calculi, a basic unit is the notion of *sequent* $\Gamma \Rightarrow \Delta$ whose reading is "if all of Γ hold then some of Δ holds". All of our sequent calculi are based on a sequent calculus for modal logic **S5**, which is a well-known system where the cut rule is not eliminable [14, p.116]:

$$\frac{\Gamma \Rightarrow \Delta, \varphi \quad \varphi, \Pi \Rightarrow \Sigma}{\Gamma, \Pi \Rightarrow \Delta, \Sigma} \ (Cut).$$

However, as Takano [21,22] established, we can restrict all applications of (Cut) in the sequent calculus for **S5** into analytic ones, i.e., an application where a cut formula φ in (Cut) is a subformula of the conclusion $\Gamma, \Pi \Rightarrow \Delta, \Sigma$ of (Cut). This restriction still allows us to compute a Craig interpolant of a derivable sequent $\varphi \Rightarrow \psi$ in **S5** (see [15]), where χ is a Craig interpolant of $\varphi \Rightarrow \psi$ if both $\varphi \Rightarrow \chi$ and $\chi \Rightarrow \psi$ are derivable and the set of all propositional variables in χ are common to both φ and ψ.

This paper shows that this proof-theoretic aspect still holds, by carefully incorporating the notion of propositional awareness into the sequent calculus for the multi-agent version of **S5**, although we need to generalize the notion of subformula to deal with the existing axioms for propositional awareness given in [6]. As a result, we newly prove that Hilbert system for multi-agent epistemic logic with propositional awareness in [6] enjoys the Craig interpolation theorem.

The structure of this paper is organized in the following manner. Section 2 reviews the syntax of multi-agent epistemic logic with *propositional awareness* and semantics based on Kripke models, and then considers two restrictions on Kripke models. Moreover, we review Hilbert systems of awareness logics (cf. [6,8]). Section 3 proposes sequent calculi for three awareness logics, which are equipollent with Hilbert systems (Theorem 1). While our sequent calculi are not cut-free as expected, we prove that all sequent calculi enjoy the semi-analytic cut property, i.e., all applications of the cut rule can be replaced by applications of the cut whose cut formulas are taken from the extended subformulas of the conclusions of the rule (Corollary 2). The notion of semi-analytic cut has been used as a variant of the analytic cut rule in a study of tableau method [7] (cf. [20,22]). Our argument for this semi-analytic cut property is semantic (cf. [10,12,16,20,22]): we prove that sequent calculi with semi-analytic cuts are semantically complete for the corresponding class of finite models (Theorem 2). As corollaries, the finite model property and decidability are established for three awareness logics (Corollary 1). Section 4 establishes the Craig interpolation theorem for the three awareness logics (Theorem 3). Finally, Sect. 5 concludes the paper with further directions for research.

2 Syntax, Semantics and Hilbert Systems

Let Atom and Agt be a fixed countably infinite set of propositional variables and a fixed finite set of agents, respectively (we assume they are mutually disjoint). The language $\mathcal{L}$ is defined by the following:

$$\mathcal{L} \ni \varphi ::= \bot \mid p \mid \neg \varphi \mid \varphi \to \varphi \mid \Box_i \varphi \mid A_i \varphi,$$

where $p \in$ Atom and $i \in$ Agt. Boolean connectives $\vee, \wedge, \leftrightarrow$, and the truth constant $\top$ are defined in the usual manner. A formula of the form $\Box_i \varphi$ is read as "*agent i implicitly knows that φ*". The explicit knowledge operator K_i is defined as $K_i \varphi := A_i \varphi \wedge \Box_i \varphi$. The set At($\varphi$) of all propositional variables in φ is defined as follows:

$$\begin{aligned}
\mathsf{At}(\bot) &:= \emptyset \\
\mathsf{At}(p) &:= \{p\} \\
\mathsf{At}(\bullet \varphi) &= \mathsf{At}(\varphi) \quad (\bullet \in \{\neg, \Box_i, A_i, \mid i \in \mathsf{Agt}\}) \\
\mathsf{At}(\varphi \to \psi) &= \mathsf{At}(\varphi) \cup \mathsf{At}(\psi)
\end{aligned}$$

Given a set Δ of formulas, we define $\mathsf{At}(\Delta) := \bigcup_{\varphi \in \Delta} \mathsf{At}(\varphi)$.

An *epistemic awareness model* (or simply *model*) M is a tuple $(W, R, \mathscr{A}, V)$ where W is a non-empty set of possible worlds, $R : \mathsf{Agt} \to \wp(W \times W)$ is a function assigning to each agent $i \in \mathsf{Agt}$ an equivalence relation R_i on W which is called an *epistemic indistinguishability relation*, $\mathscr{A} : \mathsf{Agt} \times W \to \wp(\mathsf{Atom})$ is a function assigning a set of propositional variables to each pair of an agent and a state, and $V : \mathsf{Atom} \to \wp(W)$ is a *valuation function*.

We say that a function $\mathscr{A} : \mathsf{Agt} \times W \to \wp(\mathsf{Atom})$ is *locally rigid* if, for all $i \in \mathsf{Agt}$, wR_iv implies $\mathscr{A}(i,w) = \mathscr{A}(i,v)$ holds for all $w, v \in W$. We say that a function $\mathscr{A} : \mathsf{Agt} \times W \to \wp(\mathsf{Atom})$ is *globally rigid* if, for all $i \in \mathsf{Agt}$, wR_jv implies $\mathscr{A}(i,w) = \mathscr{A}(i,v)$ holds for all $w, v \in W$ and *all* $j \in \mathsf{Agt}$. We define that a model M is locally or globally rigid if a function $\mathscr{A} : \mathsf{Agt} \times W \to \wp(\mathsf{Atom})$ in M is locally or globally rigid, respectively.

Given a model $M = (W, R, \mathscr{A}, V)$, $w \in W$ and a formula φ of $\mathcal{L}$, the satisfaction relation $M, w \models \varphi$ is defined as follows:

$$\begin{aligned}
M, w &\not\models \bot ; \\
M, w &\models p &&\iff w \in V(p); \\
M, w &\models \neg \varphi &&\iff M, w \not\models \varphi; \\
M, w &\models \varphi \to \psi &&\iff M, w \not\models \varphi \text{ or } M, w \models \psi; \\
M, w &\models A_i \varphi &&\iff \mathsf{At}(\varphi) \subseteq \mathscr{A}(i, w); \\
M, w &\models \Box_i \varphi &&\iff \text{for all } v \in W, wR_iv \text{ implies } M, v \models \varphi.
\end{aligned}$$

When $M, w \models \varphi$ for all states w in M, we say that φ is *valid* in M. We denote by **AL** the set of all formulas that are valid in every model and by **LRAL** the set of all formulas that are valid in every locally rigid model. We denote by **GRAL** the set of all formulas that are valid in every globally rigid model.

Table 1 provides three Hilbert systems. Our axiomatization H(**AL**) is based on [8] though a system in [8] is a single-agent version. For H(**AL**), we add axioms for awareness to axioms and rules for multi-agent epistemic logic **S5** (4 is derivable from T and 5 but we include it for simplicity). Axioms for awareness are for assuring that "awareness being generated by primitive propositions" [8, p.331]. Halpern [9, p.508] also states that these axioms can be captured by the following one axiom: $A_i\varphi \leftrightarrow \bigwedge_{p \in \mathrm{At}(\varphi)} A_i p$. To obtain each of H(**LRAL**) and H(**GRAL**), we add the two axioms specified in Table 1 to H(**AL**). Axioms $(\Box_i \mathtt{A}_i)$ and $(\Box_i \neg \mathtt{A}_i)$ of H(**LRAL**) mean "an agent must know what formulas he is aware of"(we replace "might" in [6] to "must"). Our axioms $(\Box_i \mathtt{A}_i)$ and $(\Box_i \neg \mathtt{A}_i)$ of H(**LRAL**) are named as "KA" and "NKA", respectively in [9].

In the previous literature on the Hilbert systems for awareness logic, the reader can find our axioms (AA) and $(\mathtt{A}_i \Box_j)$ in [4,6], and **S5** axioms in [8,18], although the axiomatization in [8] is for a unimodal language. We also note that the axiomatization in [4] is based on the multi-agent version of modal logic **K** and the axiomatization [6] is based on the multi-agent version of modal logic **KD45**. As for the axiom (AA) of Table 1, we note that [18] has an axiom of the form $A_i A_i \varphi \leftrightarrow A_i \varphi$.

Remark 1. Axioms similar to $(\Box_j \mathtt{A}_i)$ and $(\Box_j \neg \mathtt{A}_i)$ are studied in [11,23]. The semantics in [23] requires that each agent's awareness is the same for all states in a model, and this requirement is written in terms of the global modality instead of $\Box_j$. On the other hand, [11]'s awareness operators are indexed with a pair of agents such as $A_j^i \varphi$, whose reading is "From i's viewpoint, j is aware of φ" [11, Section 3.1], and they used a variant of the common knowledge operator in place of $\Box_j$ in axioms $(\Box_j \mathtt{A}_i)$ and $(\Box_j \neg \mathtt{A}_i)$.

Similar to $(\Box_i \mathtt{A}_i)$ and $(\Box_i \neg \mathtt{A}_i)$, if we read axioms $(\Box_j \mathtt{A}_i)$ and $(\Box_j \neg \mathtt{A}_i)$ of H(**GRAL**) literally, then they mean "an agent j must know what formulas a possibly different agent i is aware of". In H(**LRAL**), the awareness is generated from primitive propositions and self-reflective in the sense that $A_j \varphi \to A_j A_j \varphi$ is a theorem and each agent knows which formulas he knows himself. Thus we may regard each agent j's awareness as *private* in this sense, i.e., the other agents than j need not know what formulas j is aware of. On the other hand, axioms $(\Box_j \mathtt{A}_i)$ and $(\Box_j \neg \mathtt{A}_i)$ of H(**GRAL**) make each agent's awareness *public* but still keep each agent's (implicit) knowledge private. Let us make sense of these new axioms from a different viewpoint. Following the example in [6, p.57], we may model agent j as a process within a distributed system. We assume that each process is executing an algorithm to determine its knowledge. However, unlike [6, p.57], we introduce the concept of on-topic information (cf. [1, 19]). Each process j maintains a set of *on-topic* information pieces within its local state w. A piece of information not in the set is *off-topic* for j at the local state w. We postulate that communication channels exist between all agents, allowing any agent i to implicitly access the on-topic/off-topic information of agent j. This is because both $A_j \varphi \to \Box_i A_j \varphi$ and $\neg A_j \varphi \to \Box_i \neg A_j \varphi$ hold, and so we can regard that $A_j \varphi$ holds in w iff all atomic propositions in φ are on-topic for j in w. These postulates allow a group of agents to cooperate and achieve common goals.

The notion of theorem in these systems is defined as usual. The following result is essentially due to [8, Theorem 5.2] and [4, Theorem 24], where we note that the former [8] only deals with the unimodal case and the latter [4] is based on the multimodal **K**. Note that a Hilbert system in [6] for propositional awareness is based on a multimodal **KD45** [6] and a base system of [4, Theorem 24] for propositional awareness is a multimodal **K**. For the second item, see also [18, Theorem 3.1]. We can also easily extend the completeness argument in [8, Theorem 5.2] to H(**GRAL**), but we prove the soundness and completeness of H(**GRAL**) indirectly via those properties for a sequent calculus for **GRAL** in Sect. 3 (Corollary 3).

Proposition 1 ([8]). *Let φ be a formula of $\mathcal{L}$. (i) $\varphi \in$ **AL** iff φ is a theorem of H(**AL**); (ii) $\varphi \in$ **LRAL** iff φ is a theorem of H(**LRAL**).*

Now it is easy to see that $\textbf{AL} \subsetneq \textbf{LRAL} \subsetneq \textbf{GRAL}$.

Table 1. Hilbert Systems of Awareness Logics

	Hilbert System H(**AL**)
(Taut)	all instances of propositional tautologies
(K)	$\Box_i(\varphi \to \psi) \to (\Box_i\varphi \to \Box_i\psi)$
(T)	$\Box_i\varphi \to \varphi$
(4)	$\Box_i\varphi \to \Box_i\Box_i\varphi$
(5)	$\neg\Box_i\varphi \to \Box_i\neg\Box_i\varphi$
(A⊥)	$A_i\bot$
(A→)	$A_i(\varphi \to \psi) \leftrightarrow (A_i\varphi \wedge A_i\psi)$
(A¬)	$A_i\neg\varphi \leftrightarrow A_i\varphi$
(AA)	$A_iA_j\varphi \leftrightarrow A_i\varphi$
(A□)	$A_i\Box_j\varphi \leftrightarrow A_i\varphi$
(MP)	From φ and $\varphi \to \psi$, we may infer ψ
(Nec)	From φ, we may infer $\Box_i\varphi$
Additional Axioms for Hilbert System H(**LRAL**)	
($\Box_i\mathbf{A}_i$)	$A_i\varphi \to \Box_i A_i\varphi$
($\Box_i\neg\mathbf{A}_i$)	$\neg A_i\varphi \to \Box_i\neg A_i\varphi$
Additional Axioms for Hilbert System H(**GRAL**)	
($\Box_j\mathbf{A}_i$)	$A_i\varphi \to \Box_j A_i\varphi$
($\Box_j\neg\mathbf{A}_i$)	$\neg A_i\varphi \to \Box_j\neg A_i\varphi$

3 Semi-analytic Sequent Calculi

A sequent is a pair $\Gamma \Rightarrow \Delta$ of finite multisets of formulas and $\Gamma \Rightarrow \Delta$ is read as "if all formulas of Γ hold then some formula in Δ holds". Let Δ be a finite (possibly empty) multiset. For an agent $i \in \mathsf{Agt}$ and a finite list $(i_\delta)_{\delta \in \Delta} \subseteq \mathsf{Agt}$, we define

$\Box_i \Delta := \{\Box_i \delta \mid \delta \in \Delta\}, \qquad A_i \Delta := \{A_i \delta \mid \delta \in \Delta\}, \qquad A\Delta := \{A_{i_\delta} \delta \mid \delta \in \Delta\}.$

When $\Delta = \{\varphi_1, \varphi_2, \varphi_3\}$ and $(i_\delta)_{\delta \in \Delta} = (i, j, k)$, then $A\Delta = \{A_i\varphi_1, A_j\varphi_2, A_k\varphi_3\}$. Table 2 provides inference rules of sequent calculi G(**AL**), G(**LRAL**) and G(**GRAL**). To obtain the sequent calculus G(**LRAL**) and G(**GRAL**), we replace the modal rules for G(**AL**) with those for G(**LRAL**) and G(**GRAL**) in Table 2, respectively.

We comment on the design of our sequent calculi. For G(**AL**), axioms and rules without awareness rules of Table 2 form a sequent calculus of a multi-agent epistemic logic **S5** where sequent rule ($\Rightarrow \Box_i$) correspond to axioms (K), (4) and (5) and rule (Nec) and sequent rule ($\Box_i \Rightarrow$) is for axiom T in Table 1. Moreover, the awareness rules of Table 2 corresponds to the awareness axioms in H(**AL**) from Table 1. For G(**LRAL**), rule ($\Rightarrow \Box_i^{\mathbf{LR}}$), we add contexts $A_i\Pi$ and $A_i\Sigma$ to rule ($\Rightarrow \Box_i$) to capture axioms ($\Box_i A_i$) and ($\Box_i \neg A_i$). Similarly, for G(**GRAL**), rule ($\Rightarrow \Box_i^{\mathbf{GR}}$), we add contexts $A\Pi$ and $A\Sigma$ to rule ($\Rightarrow \Box_i$) to capture axioms ($\Box_i A_{i_\delta}$) and ($\Box_i \neg A_{i_\delta}$) of Table 1 where $A\Delta := \{A_{i_\delta}\delta \mid \delta \in \Delta\}$ where $(i_\delta)_{\delta \in \Delta} \subseteq \mathsf{Agt}$ is a finite list of possibly different agents. Note that, if we were to restrict initial sequences (id) to atomic propositions, then other inference rules would not derive sequents of the form $A_i p \Rightarrow A_i p$.[1]

We denote the semi-analytic variants $\mathsf{G}^a(\mathbf{AL})$ and $\mathsf{G}^a(\mathbf{GRAL})$ of $\mathsf{G}(\mathbf{AL})$ and $\mathsf{G}(\mathbf{GRAL})$ by restricting all applications of (Cut) to the following semi-analytic application:

$$\frac{\Gamma \Rightarrow \Delta, \varphi \quad \varphi, \Pi \Rightarrow \Sigma}{\Gamma, \Pi \Rightarrow \Sigma, \Delta} \ (Cut)^a$$

where $\varphi \in \mathsf{CL}(\Gamma, \Pi, \Delta, \Sigma)$, $\mathsf{CL}(\Theta) := \bigcup_{\theta \in \Theta} \mathsf{CL}(\theta)$, and the closure $\mathsf{CL}(\varphi)$ of a given formula φ is defined as follows:

$$\begin{aligned}
\mathsf{CL}(p) &:= \{p\} \\
\mathsf{CL}(A_i p) &:= \{A_i p, p\} \\
\mathsf{CL}(\bot) &:= \{\bot\} \\
\mathsf{CL}(A_i \bot) &:= \{A_i \bot, \bot\} \\
\mathsf{CL}(\neg\varphi) &:= \mathsf{CL}(\varphi) \cup \{\neg\varphi\} \\
\mathsf{CL}(A_i \neg\varphi) &:= \mathsf{CL}(A_i\varphi) \cup \mathsf{CL}(\neg\varphi) \cup \{A_i \neg\varphi\} \\
\mathsf{CL}(\varphi \to \psi) &:= \mathsf{CL}(\varphi) \cup \mathsf{CL}(\psi) \cup \{\varphi \to \psi\} \\
\mathsf{CL}(A_i(\varphi \to \psi)) &:= \mathsf{CL}(A_i\varphi) \cup \mathsf{CL}(A_i\psi) \cup \mathsf{CL}(\varphi \to \psi) \cup \{A_i(\varphi \to \psi)\} \\
\mathsf{CL}(\Box_i\varphi) &:= \mathsf{CL}(\varphi) \cup \{\Box_i\varphi\} \\
\mathsf{CL}(A_i\Box_j\varphi) &:= \mathsf{CL}(A_i\varphi) \cup \mathsf{CL}(\Box_j\varphi) \cup \{A_i\Box_j\varphi\} \\
\mathsf{CL}(A_i\varphi) &:= \mathsf{CL}(\varphi) \cup \{A_i\varphi\} \\
\mathsf{CL}(A_i A_j\varphi) &:= \mathsf{CL}(A_i\varphi) \cup \mathsf{CL}(A_j\varphi) \cup \{A_i A_j\varphi\}.
\end{aligned}$$

It is noted that $\mathsf{CL}(\Theta)$ is closed under taking subformulas and that if Θ is finite then $\mathsf{CL}(\Theta)$ is finite.

Let $\Lambda \in \{\mathbf{AL}, \mathbf{LRAL}, \mathbf{GRAL}\}$. We say that a sequent $\Gamma \Rightarrow \Delta$ is *derivable* in $\mathsf{G}(\Lambda)$ (or $\mathsf{G}^a(\Lambda)$) if there is a finite tree generated by inference rules of $\mathsf{G}(\Lambda)$ (or $\mathsf{G}^a(\Lambda)$) from initial sequents of $\mathsf{G}(\Lambda)$ (or $\mathsf{G}^a(\Lambda)$, respectively) and the root of the tree is $\Gamma \Rightarrow \Delta$. We write $\mathsf{G}(\Lambda) \vdash \Gamma \Rightarrow \Delta$ and $\mathsf{G}^a(\Lambda) \vdash \Gamma \Rightarrow \Delta$ to mean that $\Gamma \Rightarrow \Delta$ is derivable in $\mathsf{G}(\Lambda)$ and $\mathsf{G}^a(\Lambda)$, respectively.

---

[1] We owe this point to one of the reviewers.

We say that $\Gamma \Rightarrow \Delta$ is valid in a model M (written: $M \models \Gamma \Rightarrow \Delta$) if the translation $\bigwedge \Gamma \Rightarrow \bigvee \Delta$ is valid in M, where $\bigwedge \Gamma$ is the conjunction of all formulas in Γ and $\bigvee \Delta$ is the disjunction of all formulas in Δ (we stipulate $\bigwedge \emptyset := \bot$ and $\bigvee \emptyset := \top$). We write $(M, w) \models \Gamma \Rightarrow \Delta$ to mean $(M, w) \models \bigwedge \Gamma \to \bigvee \Delta$.

Proposition 2. (i) *If $\Gamma \Rightarrow \Delta$ is derivable in $\mathtt{G}(\mathbf{AL})$ (or $\mathtt{G}^a(\mathbf{AL})$), then it is valid in all models M.* (ii) *If $\Gamma \Rightarrow \Delta$ is derivable in $\mathtt{G}(\mathbf{LRAL})$ (or $\mathtt{G}^a(\mathbf{LRAL})$), then it is valid in all locally rigid models M.* (iii) *If $\Gamma \Rightarrow \Delta$ is derivable in $\mathtt{G}(\mathbf{GRAL})$ (or $\mathtt{G}^a(\mathbf{GRAL})$), then it is valid in all globally rigid models M.*

Table 2. Sequent Calculi of Awareness Logics

Initial Sequents:

$$\dfrac{}{\varphi \Rightarrow \varphi}\ (\mathtt{id}) \qquad \dfrac{}{\bot \Rightarrow}\ (\bot \Rightarrow)$$

Structural Rules:

$$\dfrac{\Gamma \Rightarrow \Delta}{\Gamma \Rightarrow \Delta, \varphi}\ (\Rightarrow w) \qquad \dfrac{\Gamma \Rightarrow \Delta}{\varphi, \Gamma \Rightarrow \Delta}\ (w \Rightarrow) \qquad \dfrac{\Gamma \Rightarrow \Delta, \varphi, \varphi}{\Gamma \Rightarrow \Delta, \varphi}\ (\Rightarrow c) \qquad \dfrac{\varphi, \varphi, \Gamma \Rightarrow \Delta}{\varphi, \Gamma \Rightarrow \Delta}\ (c \Rightarrow)$$

$$\dfrac{\Gamma \Rightarrow \Delta, \varphi \quad \varphi, \Pi \Rightarrow \Sigma}{\Gamma, \Pi \Rightarrow \Sigma, \Delta}\ (Cut)$$

Logical Rules:

$$\dfrac{\varphi, \Gamma \Rightarrow \Delta}{\Gamma \Rightarrow \Delta, \neg\varphi}\ (\Rightarrow \neg) \qquad \dfrac{\Gamma \Rightarrow \Delta, \varphi}{\neg\varphi, \Gamma \Rightarrow \Delta}\ (\neg \Rightarrow) \qquad \dfrac{\varphi, \Gamma \Rightarrow \Delta, \psi}{\Gamma \Rightarrow \Delta, \varphi \to \psi}\ (\Rightarrow \to) \qquad \dfrac{\Gamma_1 \Rightarrow \Delta_1, \varphi \quad \psi, \Gamma_2 \Rightarrow \Delta_2}{\varphi \to \psi, \Gamma_1, \Gamma_2 \Rightarrow \Delta_1, \Delta_2}\ (\to \Rightarrow)$$

Modal Rules for $\mathtt{G}(\mathbf{AL})$:

$$\dfrac{\Box_i \Gamma \Rightarrow \Box_i \Delta, \varphi}{\Box_i \Gamma \Rightarrow \Box_i \Delta, \Box_i \varphi}\ (\Rightarrow \Box_i) \qquad \dfrac{\varphi, \Gamma \Rightarrow \Delta}{\Box_i \varphi, \Gamma \Rightarrow \Delta}\ (\Box_i \Rightarrow)$$

Awareness Rules:

$$\dfrac{\bot, \Gamma \Rightarrow \Delta}{\Gamma \Rightarrow \Delta, A_i \bot}\ (\Rightarrow A_i \bot) \qquad \dfrac{\Gamma \Rightarrow \Delta, \bot}{A_i \bot, \Gamma \Rightarrow \Delta}\ (A_i \bot \Rightarrow)$$

$$\dfrac{\Gamma \Rightarrow \Delta, A_i \varphi}{\Gamma \Rightarrow \Delta, A_i \neg \varphi}\ (\Rightarrow A_i \neg) \qquad \dfrac{A_i \varphi, \Gamma \Rightarrow \Delta}{A_i \neg \varphi, \Gamma \Rightarrow \Delta}\ (A_i \neg \Rightarrow)$$

$$\dfrac{\Gamma \Rightarrow \Delta, A_i \varphi \quad \Gamma \Rightarrow \Delta, A_i \psi}{\Gamma \Rightarrow \Delta, A_i(\varphi \to \psi)}\ (\Rightarrow A_i \to) \qquad \dfrac{A_i \varphi, \Gamma \Rightarrow \Delta}{A_i(\varphi \to \psi), \Gamma \Rightarrow \Delta}\ (A_i \to_1 \Rightarrow) \qquad \dfrac{A_i \psi, \Gamma \Rightarrow \Delta}{A_i(\varphi \to \psi), \Gamma \Rightarrow \Delta}\ (A_i \to_2 \Rightarrow)$$

$$\dfrac{\Gamma \Rightarrow \Delta, A_i \varphi}{\Gamma \Rightarrow \Delta, A_i \Box_j \varphi}\ (\Rightarrow A_i \Box_j) \qquad \dfrac{A_i \varphi, \Gamma \Rightarrow \Delta}{A_i \Box_j \varphi, \Gamma \Rightarrow \Delta}\ (A_i \Box_j \Rightarrow)$$

$$\dfrac{\Gamma \Rightarrow \Delta, A_i \varphi}{\Gamma \Rightarrow \Delta, A_i A_j \varphi}\ (\Rightarrow A_i A_j) \qquad \dfrac{A_i \varphi, \Gamma \Rightarrow \Delta}{A_i A_j \varphi, \Gamma \Rightarrow \Delta}\ (A_i A_j \Rightarrow)$$

Modal Rules for $\mathtt{G}(\mathbf{LRAL})$:

$$\dfrac{A_i \Pi, \Box_i \Gamma \Rightarrow \Box_i \Delta, A_i \Sigma, \varphi}{A_i \Pi, \Box_i \Gamma \Rightarrow \Box_i \Delta, A_i \Sigma, \Box_i \varphi}\ (\Rightarrow \Box_i^{\mathbf{LR}}) \qquad \dfrac{\varphi, \Gamma \Rightarrow \Delta}{\Box_i \varphi, \Gamma \Rightarrow \Delta}\ (\Box_i \Rightarrow)$$

Modal Rules for $\mathtt{G}(\mathbf{GRAL})$:

$$\dfrac{A \Pi, \Box_i \Gamma \Rightarrow \Box_i \Delta, A \Sigma, \varphi}{A \Pi, \Box_i \Gamma \Rightarrow \Box_i \Delta, A \Sigma, \Box_i \varphi}\ (\Rightarrow \Box_i^{\mathbf{GR}}) \qquad \dfrac{\varphi, \Gamma \Rightarrow \Delta}{\Box_i \varphi, \Gamma \Rightarrow \Delta}\ (\Box_i \Rightarrow)$$

Theorem 1. *For all* $\Lambda \in \{\mathbf{AL}, \mathbf{LRAL}, \mathbf{GRAL}\}$, *a formula* φ *is a theorem of* $\mathsf{H}(\Lambda)$ *iff* $\mathsf{G}(\Lambda) \vdash \Rightarrow \varphi$.

Proof. We focus on the case of $\Lambda = \mathbf{GRAL}$. For the left to right direction, it suffices to show that all axioms of $\mathsf{H}(\Lambda)$ are derivable in $\mathsf{G}(\Lambda)$ and all rules of $\mathsf{H}(\Lambda)$ are admissible in $\mathsf{G}(\Lambda)$. We show axioms $(\Box_j A_i)$ and $(\Box_j \neg A_i)$ of Table 1 are derivable in $\mathsf{G}(\mathbf{GRAL})$.

$$\dfrac{\dfrac{\dfrac{A_i\varphi \Rightarrow A_i\varphi}{A_i\varphi \Rightarrow \Box_j A_i\varphi}\,(\Rightarrow \Box_j^{\mathbf{GR}})}{\Rightarrow A_i\varphi \to \Box_j A_i\varphi}\,(\Rightarrow \to)}{}\,(\mathrm{id}) \qquad , \qquad \dfrac{\dfrac{\dfrac{\dfrac{\dfrac{A_i\varphi \Rightarrow A_i\varphi}{\Rightarrow \neg A_i\varphi, A_i\varphi}\,(\Rightarrow \neg)}{\Rightarrow \Box_j\neg A_i\varphi, A_i\varphi}\,(\Rightarrow \Box_j^{\mathbf{GR}})}{\neg A_i\varphi \Rightarrow \Box_j\neg A_i\varphi}\,(\neg \Rightarrow)}{\Rightarrow \neg A_i\varphi \to \Box_j\neg A_i\varphi}\,(\Rightarrow \to)}{}\,(\mathrm{id}).$$

Moreover (MP) can be simulated in $\mathsf{G}(\Lambda)$ as follows:

$$\dfrac{\Rightarrow \varphi \to \psi \qquad \dfrac{\Rightarrow \varphi \quad \psi \Rightarrow \psi}{\varphi \to \psi \Rightarrow \psi}\,(\to\Rightarrow)}{\Rightarrow \psi}\,(Cut).$$

For the right to left direction, we show a generalized statement: if $\mathsf{G}(\Lambda) \vdash \Gamma \Rightarrow \Delta$ then $\mathsf{H}(\Lambda) \vdash \bigwedge \Gamma \to \bigvee \Delta$ by induction on a derivation $\Gamma \Rightarrow \Delta$ in $\mathsf{G}(\Lambda)$. When the last applied rule is an awareness rule of Table 2, it suffices to use the corresponding awareness axiom in Table 1. When the last applied rule is $(\Rightarrow \Box_i^{\mathbf{GR}})$, we use both axioms $(\Box_j A_i)$ and $(\Box_j \neg A_i)$ of Table 1 in addition to axioms (K), (T), (4) and rule (Nec) in Table 1. □

It is remarked that a formula $p \to \Box_i \neg \Box_i \neg p$ (axiom B) is derivable in $\mathsf{G}(\Lambda)$ ($\Lambda \in \{\mathbf{AL}, \mathbf{LRAL}, \mathbf{GRAL}\}$) with the help of (Cut) as follows:

$$\dfrac{\dfrac{\dfrac{\dfrac{\Box_i\neg p \Rightarrow \Box_i\neg p}{\Rightarrow \neg\Box_i\neg p, \Box_i\neg p}\,(\Rightarrow \neg)}{\Rightarrow \Box_i\neg\Box_i\neg p, \Box_i\neg p}\,(\Rightarrow \Box_i) \qquad \dfrac{\dfrac{p \Rightarrow p}{\neg p, p \Rightarrow}\,(\neg \Rightarrow)}{\Box_i\neg p, p \Rightarrow}\,(\Box_i \Rightarrow)}{p \Rightarrow \Box_i\neg\Box_i\neg p}\,(Cut)}{\Rightarrow p \to \Box_i\neg\Box_i\neg p}\,(\Rightarrow \to).$$

As noted in [14], however, we cannot eliminate the above application of (Cut) as in a sequent calculus of modal logic **S5**, though $\Box_i \neg p$ is a subformula of the conclusion $p \Rightarrow \Box_i \neg \Box_i \neg p$ of (Cut). For the sequent calculus of modal logic **S5**, Takano [21] showed that all applications of (Cut) can be restricted to analytic ones, i.e., the cut formula is a subformula of the conclusion of (Cut) and later he also established the same fact in terms of a semantic argument [22]. We follow a semantic approach to establish that all of our sequent calculi enjoy this semi-analytic cut property below, though we need to slightly extend the notion of subformula in terms of the closure $\mathsf{CL}(\varphi)$.

In what follows, we let $\Lambda \in \{\mathbf{AL}, \mathbf{LRAL}, \mathbf{GRAL}\}$. We are going to prove that $\mathsf{G}^a(\Lambda)$ is semantically complete for the corresponding class of all finite models to Λ. Let us say that a set Ξ of formulas is *closed* if $\mathsf{CL}(\Xi) \subseteq \Xi$. In what follows, we fix Ξ as a closed finite set of formulas. The following notion is employed also in [12, 16], though we do not need to extend the notion of subformula in [12, 16].

Definition 1. *Let $\Gamma \cup \Delta \subseteq \Xi$. Then a pair (Γ, Δ) is a Ξ-partial valuation in $\mathsf{G}^a(\Lambda)$ if the following hold:* (i) $\mathsf{G}^a(\Lambda) \not\vdash \Gamma \Rightarrow \Delta$ *and* (ii) $\mathsf{CL}(\Gamma, \Delta) = \Gamma \cup \Delta$.

Lemma 1. *Let $\Gamma, \Delta \subseteq \Xi$. If $\mathsf{G}^a(\Lambda) \not\vdash \Gamma \Rightarrow \Delta$ then there exists a Ξ-partial valuation (Γ^+, Δ^+) in $\mathsf{G}^a(\Lambda)$ such that $\Gamma \subseteq \Gamma^+$, $\Delta \subseteq \Delta^+$ and $\mathsf{CL}(\Gamma, \Delta) = \Gamma^+ \cup \Delta^+$.*

Lemma 2. *Let a pair (Γ, Δ) be a Ξ-partial valuation in $\mathsf{G}^a(\Lambda)$. Then, (Γ, Δ) is saturated, i.e., it satisfies all the following:*

($\rightarrow$**r**) *If $\varphi \rightarrow \psi \in \Delta$ then $\varphi \in \Gamma$ and $\psi \in \Delta$.*
($\rightarrow$**l**) *If $\varphi \rightarrow \psi \in \Gamma$ then $\varphi \in \Delta$ or $\psi \in \Gamma$.*
($\neg$**r**) *If $\neg \varphi \in \Delta$ then $\varphi \in \Gamma$.*
($\neg$**l**) *If $\neg \varphi \in \Gamma$ then $\varphi \in \Delta$.*
($\Box_i$**l**) *If $\Box_i \varphi \in \Gamma$ then $\varphi \in \Gamma$.*
($A_i \neg$**r**) *If $A_i \neg \varphi \in \Delta$ then $A_i \varphi \in \Delta$.*
($A_i \neg$**l**) *If $A_i \neg \varphi \in \Gamma$ then $A_i \varphi \in \Gamma$.*
($A_i \Box_j$**r**) *If $A_i \Box_j \varphi \in \Delta$ then $A_i \varphi \in \Delta$.*
($A_i \Box_j$**l**) *If $A_i \Box_j \varphi \in \Gamma$ then $A_i \varphi \in \Gamma$.*
($A_i A_j$**r**) *If $A_i A_j \varphi \in \Delta$ then $A_i \varphi \in \Delta$.*
($A_i A_j$**l**) *If $A_i A_j \varphi \in \Gamma$ then $A_i \varphi \in \Gamma$.*
($A_i \rightarrow$ **r**) *If $A_i(\varphi \rightarrow \psi) \in \Delta$ then $A_i \varphi \in \Delta$ or $A_i \psi, \in \Delta$.*
($A_i \rightarrow$ **l**) *If $A_i(\varphi \rightarrow \psi) \in \Gamma$ then $A_i \varphi, A_i \psi \in \Gamma$.*

Definition 2. *We define $M_\Lambda^\Xi := (W^\Xi, (R_i^\Xi)_{i \in \mathsf{Agt}}, \mathscr{A}^\Xi, V^\Xi)$ as follows:*

- $W^\Xi := \{(\Gamma, \Delta) \mid (\Gamma, \Delta)$ *is a Ξ-partial valuation in $\mathsf{G}^a(\Lambda)\}$.*
- *When $\Lambda = \mathbf{AL}$, R_i^Ξ is defined as:*

$$(\Gamma_1, \Delta_1) R_i^\Xi (\Gamma_2, \Delta_2) \iff \text{for all } \Box_i \varphi \in \Xi (\Box_i \varphi \in \Gamma_1 \text{ iff } \Box_i \varphi \in \Gamma_2).$$

When $\Lambda = \mathbf{LRAL}$, R_i^Ξ is defined as:

$$(\Gamma_1, \Delta_1) R_i^\Xi (\Gamma_2, \Delta_2) \iff \text{for all } \Box_i \varphi \in \Xi (\Box_i \varphi \in \Gamma_1 \text{ iff } \Box_i \varphi \in \Gamma_2) \text{ and}$$
$$\text{for all } A_i \psi \in \Xi (A_i \psi \in \Gamma_1 \text{ iff } A_i \psi \in \Gamma_2).$$

When $\Lambda = \mathbf{GRAL}$, R_i^Ξ is defined as:

$$(\Gamma_1, \Delta_1) R_i^\Xi (\Gamma_2, \Delta_2) \iff \text{for all } \Box_i \varphi \in \Xi (\Box_i \varphi \in \Gamma_1 \text{ iff } \Box_i \varphi \in \Gamma_2) \text{ and}$$
$$\text{for all } j \text{ such that } A_j \psi \in \Xi (A_j \psi \in \Gamma_1 \text{ iff } A_j \psi \in \Gamma_2).$$

- $\mathscr{A}^\Xi(i, (\Gamma, \Delta)) := \{p \mid A_i p \in \Gamma\}$.
- $(\Gamma, \Delta) \in V^\Xi(p)$ *iff $p \in \Gamma$.*

Lemma 3. *For all choices of $\Lambda \in \{\mathbf{AL}, \mathbf{LRAL}, \mathbf{GRAL}\}$, R_i^Ξ and $\mathscr{A}^\Xi$ of M_Λ^Ξ satisfy all the required conditions depending on Λ*

Lemma 4. *Let $\Lambda \in \{\mathbf{AL}, \mathbf{LRAL}, \mathbf{GRAL}\}$. Then the following hold for all formulas $\varphi \in \Xi$ and $(\Gamma, \Delta) \in W^\Xi$:*

(i) $\varphi \in \Gamma$ *implies* $M_\Lambda^\Xi, (\Gamma, \Delta) \models \varphi$;   (ii) $\varphi \in \Delta$ *implies* $M_\Lambda^\Xi, (\Gamma, \Delta) \not\models \varphi$.

Proof. We only consider the case of $\Lambda = \mathbf{GRAL}$. By induction on the number of logical connectives in $\varphi \in \Xi$, i.e., the length of $\varphi \in \Xi$. We establish both items when φ is a formula of the form p, the form $\Box_i \psi$ or the form $A_i \psi$.

First, let $\varphi \equiv p$. For item (i), we proceed as follows: $p \in \Gamma \iff (\Gamma, \Delta) \in V^\Xi(p)$, which implies $M_\Lambda^\Xi, (\Gamma, \Delta) \models p$, as desired. For item (ii), assume that $p \in \Delta$. Then $p \notin \Gamma$ (since (Γ, Δ) is a Ξ-partial valuation). We proceed as follows: $(\Gamma, \Delta) \notin V^\Xi(p)$, which implies $M_\Lambda^\Xi, (\Gamma, \Delta) \not\models p$, as desired.

Second, let $\varphi \equiv \Box_i \psi$. (i) Suppose $\Box_i \psi \in \Gamma$. To show that $M_\Lambda^\Xi, (\Gamma, \Delta) \models \Box_i \psi$, let us fix any (Γ', Δ') such that $(\Gamma, \Delta) R_i^\Xi (\Gamma', \Delta')$. Since $\Box_i \psi \in \Gamma$, $\Box_i \psi \in \Gamma'$ holds by definition of R_i^Ξ. By ($\Box_i$1) of Lemma 2, $\psi \in \Gamma'$. By induction hypothesis, $M_\Lambda^\Xi, (\Gamma', \Delta') \models \psi$, as desired. (ii) Assume $\Box_i \psi \in \Delta$. Given a finite set Θ of formulas, we write

$$\Theta_{\Box_i} := \{ \Box_i \varphi \mid \Box_i \varphi \in \Theta \} \text{ and } \Theta_{A_j} := \{ A_j \varphi \mid A_j \varphi \in \Theta \}.$$

Our goal is to find a Ξ-partial valuation (Γ', Δ') such that $(\Gamma, \Delta) R_i^\Xi (\Gamma', \Delta')$ and $M_\Lambda^\Xi, (\Gamma', \Delta') \not\models \psi$ hence $\psi \in \Delta'$ by induction hypothesis. First of all, we show that a sequent $\Gamma_{\Box_i}, (\Gamma_{A_j})_{j \in \mathsf{Agt}} \Rightarrow \Delta_{\Box_i}, (\Delta_{A_j})_{j \in \mathsf{Agt}}, \psi$ is not derivable in $\mathsf{G}^a(\Lambda)$. Suppose otherwise. Then we proceed as follows.

$$\frac{\dfrac{\Gamma_{\Box_i}, (\Gamma_{A_j})_{j \in \mathsf{Agt}} \Rightarrow \Delta_{\Box_i}, (\Delta_{A_j})_{j \in \mathsf{Agt}}, \psi}{\Gamma_{\Box_i}, (\Gamma_{A_j})_{j \in \mathsf{Agt}} \Rightarrow \Delta_{\Box_i}, (\Delta_{A_j})_{j \in \mathsf{Agt}}, \Box_i \psi} (\Rightarrow \Box_i^{\mathbf{GR}})}{\Gamma \Rightarrow \Delta} (\Rightarrow w), (w \Rightarrow),$$

which implies a contradiction since (Γ, Δ) is a Ξ-partial valuation in $\mathsf{G}^a(\Lambda)$. By Lemma 1, there is a Ξ-partial valuation (Γ', Δ') such that the following hold:

- $\Gamma_{\Box_i} \cup (\Gamma_{A_j})_{j \in \mathsf{Agt}} \subseteq \Gamma'$ and $\Delta_{\Box_i} \cup (\Delta_{A_j})_{j \in \mathsf{Agt}} \cup \{\psi\} \subseteq \Delta'$,
- $\Gamma' \cup \Delta' = \mathsf{CL}(((\Gamma, \Delta)_{\Box_i}, ((\Gamma, \Delta)_{A_j})_{j \in \mathsf{Agt}}, \psi)$.

It is noted that $\Gamma' \cup \Delta' \subseteq \Gamma \cup \Delta$ by $\Box_i \psi \in \Delta$ and $\mathsf{CL}((\Gamma, \Delta)_{\Box_i}, ((\Gamma, \Delta)_{A_j})_{j \in \mathsf{Agt}}, \psi) \subseteq \mathsf{CL}(\Gamma, \Delta)$. In what follows, we show that $(\Gamma, \Delta) R_i^\Xi (\Gamma', \Delta')$ since $\psi \in \Delta'$ is immediate. We need to establish (A) $\Gamma_{\Box_i} = \Gamma'_{\Box_i}$ and (B) for all $j \in \mathsf{Agt}$, $\Gamma_{A_j} = \Gamma'_{A_j}$. For (A), we proceed as follows. Since $\Gamma_{\Box_i} \subseteq \Gamma'_{\Box_i}$ is easy by definition, we focus on the converse direction. Suppose that $\Box_i \gamma \in \Gamma'$. Our goal is to show $\Box_i \gamma \in \Gamma$. Suppose otherwise, i.e., $\Box_i \gamma \notin \Gamma$. By $\Gamma' \cup \Delta' \subseteq \Gamma \cup \Delta$, we get $\Box_i \gamma \in \Delta$. By $\Delta_{\Box_i} \subseteq \Delta'$, we obtain $\Box_i \gamma \in \Delta'$. Together with $\Box_i \gamma \in \Gamma'$, we get a desired contradiction with the underivability of $\Gamma' \Rightarrow \Delta'$ in $\mathsf{G}^a(\Lambda)$. Therefore $\Box_i \gamma \in \Gamma$, as required. As for (B), let us fix any $j \in \mathsf{Agt}$. We show that $\Gamma_{A_j} = \Gamma'_{A_j}$. Since $\Gamma_{A_j} \subseteq \Gamma'_{A_j}$ is easy by construction, we focus on showing the converse direction, i.e., $\Gamma'_{A_j} \subseteq \Gamma_{A_j}$. Suppose that $A_j \gamma \in \Gamma'$. Our goal is to show that $A_j \gamma \in \Gamma$. Then we can apply a similar argument to (A) also here.

Third, we consider the case where a formula is of the form $A_i \psi$. We divide our argument depending on the form of ψ.

- Let $\psi \equiv p$, i.e., $\varphi \equiv A_i p$. For item (i), we proceed as follows: $A_i p \in \Gamma \iff p \in \mathscr{A}(i, (\Gamma, \Delta))$, which implies $M_\Lambda^\Xi, (\Gamma, \Delta) \models A_i p$, as desired. For item (ii), assume that $A_i p \in \Delta$. Then $A_i p \notin \Gamma \iff p \notin \mathscr{A}(i, (\Gamma, \Delta))$, which implies $M_\Lambda^\Xi, (\Gamma, \Delta) \not\models A_i p$, as desired.

- Let $\psi \equiv \Box_j\theta$ i.e., $\varphi \equiv A_i\Box_j\theta$. For item (i), suppose that $A_i\Box_j\theta \in \Gamma$. By $(A_i\Box_j 1)$ of Lemma 2, $A_i\theta \in \Gamma$. By induction hypothesis (the length of $A_i\theta <$ the length of $A_i\Box_j\theta$), we obtain $M_\Lambda^\Xi, (\Gamma, \Delta) \models A_i\theta$, i.e., $\mathsf{At}(\theta) \subseteq \mathscr{A}(i, (\Gamma, \Delta))$. It follows from $\mathsf{At}(\Box_j\theta) = \mathsf{At}(\theta)$ that $\mathsf{At}(\Box_j\theta) \subseteq \mathscr{A}(i, (\Gamma, \Delta))$ hence $M_\Lambda^\Xi, (\Gamma, \Delta) \models A_i\Box_j\theta$, as desired. Item (ii) can be shown similarly. □

Theorem 2. (i) *If $\Gamma \Rightarrow \Delta$ is valid in all finite models then it is derivable in $\mathsf{G}^a(\mathbf{AL})$.* (ii) *If $\Gamma \Rightarrow \Delta$ is valid in all finite locally rigid models then it is derivable in $\mathsf{G}^a(\mathbf{LRAL})$.* (iii) *If $\Gamma \Rightarrow \Delta$ is valid in all finite globally rigid models then it is derivable in $\mathsf{G}^a(\mathbf{GRAL})$.*

Proof. We show item (iii) alone. We prove the contrapositive implication. Suppose that $\Gamma \Rightarrow \Delta$ is not derivable in $\mathsf{G}^a(\mathbf{GRAL})$. Define $\Xi := \mathsf{CL}(\Gamma, \Delta)$. By Lemma 1, there is a Ξ-partial valuation (Γ^+, Δ^+) such that $\Gamma \subseteq \Gamma^+, \Delta \subseteq \Delta^+$. By Lemma 4, $M_\Lambda^\Xi, (\Gamma^+, \Delta^+) \models \gamma$ for all $\gamma \in \Gamma$ and $M_\Lambda^\Xi, (\Gamma^+, \Delta^+) \not\models \delta$ for all $\delta \in \Delta$. Since Ξ is finite, W^Ξ is also finite. By Lemma 3, M^Ξ *is* a globally rigid model. Therefore, $\Gamma \Rightarrow \Delta$ is not valid in a finite globally rigid model M^Ξ. □

By Theorem 2 and Proposition 2, we obtain the following.

Corollary 1. *All of $\mathsf{G}(\mathbf{AL})$, $\mathsf{G}(\mathbf{LRAL})$ and $\mathsf{G}(\mathbf{GRAL})$ enjoy the finite model property, hence they are decidable.*

Corollary 2. *Let $\mathbf{\Lambda} \in \{\mathbf{AL}, \mathbf{LRAL}, \mathbf{GRAL}\}$. The following equivalence holds: $\Gamma \Rightarrow \Delta$ is derivable in $\mathsf{G}(\mathbf{\Lambda})$ iff $\Gamma \Rightarrow \Delta$ is derivable in $\mathsf{G}^a(\mathbf{\Lambda})$.*

Proof. The right-to-left direction is trivial. The left-to-right direction follows from Theorem 2 and Proposition 2. To be more precise, suppose that $\Gamma \Rightarrow \Delta$ is derivable in $\mathsf{G}(\mathbf{\Lambda})$. By Proposition 2, $\Gamma \Rightarrow \Delta$ is valid in the corresponding class of all finite models to $\mathbf{\Lambda}$. By Theorem 2, $\Gamma \Rightarrow \Delta$ is derivable in $\mathsf{G}^a(\mathbf{\Lambda})$. □

By Theorem 1, Theorem 2 and Proposition 2, we obtain the following.

Corollary 3. *For all formulas φ, $\varphi \in \mathbf{GRAL}$ iff φ is a theorem of $\mathsf{H}(\mathbf{GRAL})$.*

4 Craig Interpolation

Definition 3. *We say that $((\Gamma_1 : \Delta_1); (\Gamma_2 : \Delta_2))$ is a partition of a given sequent $\Gamma \Rightarrow \Delta$ if $\Gamma = \Gamma_1, \Gamma_2$ and $\Delta = \Delta_1, \Delta_2$. A Craig interpolant in $\mathsf{G}(\mathbf{\Lambda})$ of a partition $((\Gamma_1 : \Delta_1); (\Gamma_2 : \Delta_2))$ of $\Gamma \Rightarrow \Delta$ is a formula χ such that it satisfies the following:*

(derivability) *Both $\Gamma_1 \Rightarrow \Delta_1, \chi$ and $\chi, \Gamma_2 \Rightarrow \Delta_2$ are derivable in $\mathsf{G}(\mathbf{\Lambda})$;*
(variable sharing) $\mathsf{At}(\chi) \subseteq \mathsf{At}(\Gamma_1, \Delta_1) \cap \mathsf{At}(\Gamma_2, \Delta_2)$.

Theorem 3. *Let $\mathbf{\Lambda} \in \{\mathbf{AL}, \mathbf{LRAL}, \mathbf{GRAL}\}$. If a sequent $\Gamma \Rightarrow \Delta$ is derivable in $\mathsf{G}(\mathbf{\Lambda})$, then every partition $((\Gamma_1 : \Delta_1); (\Gamma_2 : \Delta_2))$ of it has a Craig interpolant in $\mathsf{G}(\mathbf{\Lambda})$.*

Proof. Suppose that a sequent $\Gamma \Rightarrow \Delta$ is derivable in $\mathsf{G}(\Lambda)$. By Corollary 2, it is also derivable in $\mathsf{G}^a(\Lambda)$. By induction on a derivation of $\Gamma \Rightarrow \Delta$ in $\mathsf{G}^a(\Lambda)$, we prove that every partition $((\Gamma_1 : \Delta_1); (\Gamma_2 : \Delta_2))$ of the sequent has a Craig interpolant. In what follows, we focus on the case where $\Lambda := \mathbf{GRAL}$. When the last applied rule is $(Cut)^a$, we can apply almost the same argument as given in [15, pp.245–246]. In what follows, we consider the last applied rule is $(\Rightarrow \Box_i^{\mathbf{GR}})$ as follows:

$$\frac{A\Pi_1, A\Pi_2, \Box_i\Gamma_1, \Box_i\Gamma_2 \Rightarrow \Box_i\Delta_1, \Box_i\Delta_2, A\Sigma_1, A\Sigma_2, \varphi}{A\Pi_1, A\Pi_2, \Box_i\Gamma_1, \Box_i\Gamma_2 \Rightarrow \Box_i\Delta_1, \Box_i\Delta_2, A\Sigma_1, A\Sigma_2, \Box_i\varphi} \;(\Rightarrow \Box_i^{\mathbf{GR}}).$$

Our possible partitions are of the following two forms:

- $((A\Pi_1, \Box_i\Gamma_1 : A\Sigma_1, \Box_i\Delta_1); (A\Pi_2, \Box_i\Gamma_2 : A\Sigma_2, \Box_i\Delta_2, \Box_i\varphi))$,
- $((A\Pi_1, \Box_i\Gamma_1 : A\Sigma_1, \Box_i\Delta_1, \Box_i\varphi); (A\Pi_2, \Box_i\Gamma_2 : A\Sigma_2, \Box_i\Delta_2))$.

We only consider the second partition alone. By considering the corresponding partition of the premise of the rule $(\Rightarrow \Box_i^{\mathbf{GR}})$, induction hypothesis allows us to find a Craig interpolant χ such that

- $A\Pi_1, \Box_i\Gamma_1 \Rightarrow A\Sigma_1, \Box_i\Delta_1, \varphi, \chi$ is derivable in $\mathsf{G}^a(\Lambda)$.
- $\chi, A\Pi_2, \Box_i\Gamma_2 \Rightarrow A\Sigma_2, \Box_i\Delta_2$ is derivable in $\mathsf{G}^a(\Lambda)$.
- $\mathsf{At}(\chi) \subseteq \mathsf{At}(A\Pi_1, \Box_i\Gamma_1, A\Sigma_1, \Box_i\Delta_1, \varphi) \cap \mathsf{At}(A\Pi_2, \Box_i\Gamma_2, A\Sigma_2, \Box_i\Delta_2))$.

Then $\neg\Box_i\neg\chi$ becomes our desired Craig interpolant. Since the variable sharing condition is easy to check, we can establish the derivability condition as follows:

$$\frac{\dfrac{\dfrac{\dfrac{\dfrac{A\Pi_1, \Box_i\Gamma_1 \Rightarrow A\Sigma_1, \Box_i\Delta_1, \varphi, \chi}{\neg\chi, A\Pi_1, \Box_i\Gamma_1 \Rightarrow A\Sigma_1, \Box_i\Delta_1, \varphi}\;(\neg \Rightarrow)}{\Box_i\neg\chi, A\Pi_1, \Box_i\Gamma_1 \Rightarrow A\Sigma_1, \Box_i\Delta_1, \varphi}\;(\Box_i \Rightarrow)}{\Box_i\neg\chi, A\Pi_1, \Box_i\Gamma_1 \Rightarrow A\Sigma_1, \Box_i\Delta_1, \Box_i\varphi}\;(\Rightarrow \Box_i^{\mathbf{GR}})}{A\Pi_1, \Box_i\Gamma_1 \Rightarrow A\Sigma_1, \Box_i\Delta_1, \Box_i\varphi, \neg\Box_i\neg\chi}\;(\Rightarrow \neg)},$$

$$\frac{\dfrac{\dfrac{\dfrac{\chi, A\Pi_2, \Box_i\Gamma_2 \Rightarrow A\Sigma_2, \Box_i\Delta_2}{A\Pi_2, \Box_i\Gamma_2 \Rightarrow A\Sigma_2, \Box_i\Delta_2, \neg\chi}\;(\Rightarrow \neg)}{A\Pi_2, \Box_i\Gamma_2 \Rightarrow A\Sigma_2, \Box_i\Delta_2, \Box_i\neg\chi}\;(\Rightarrow \Box_i^{\mathbf{GR}})}{\neg\Box_i\neg\chi, A\Pi_2, \Box_i\Gamma_2 \Rightarrow A\Sigma_2, \Box_i\Delta_2}\;(\neg \Rightarrow)}.$$

This finishes to show the existence of an interpolant for the second partition above. □

Corollary 4. *Let $\Lambda \in \{\mathbf{AL}, \mathbf{LRAL}, \mathbf{GRAL}\}$. If $\varphi \to \psi$ is a theorem in $\mathsf{H}(\Lambda)$ then there exists a formula χ such that both $\varphi \to \chi$ and $\chi \to \psi$ are theorems in $\mathsf{H}(\Lambda)$, and $\mathsf{At}(\chi) \subseteq \mathsf{At}(\varphi) \cap \mathsf{At}(\psi)$.*

Proof. Suppose that $\varphi \to \psi$ is a theorem in $\mathsf{H}(\Lambda)$. By Theorem 1, we obtain $\mathsf{G}(\Lambda) \vdash \Rightarrow \varphi \to \psi$. We can derive $\varphi \Rightarrow \psi$ in $\mathsf{G}(\Lambda)$ as follows:

$$\frac{\Rightarrow \varphi \to \psi \quad \dfrac{\varphi \Rightarrow \varphi \quad \psi \Rightarrow \psi}{\varphi \to \psi, \varphi \Rightarrow \psi}\;(\to\Rightarrow)}{\varphi \Rightarrow \psi}\;(Cut).$$

By Theorem 3, there exists a formula χ such that both $\varphi \Rightarrow \chi$ and $\chi \Rightarrow \psi$ are derivable in $\mathsf{G}(\Lambda)$, and $\mathsf{At}(\chi) \subseteq \mathsf{At}(\varphi) \cap \mathsf{At}(\psi)$. Therefore, we obtain our goal by Theorem 1. □

5 Further Directions

There are three directions for further research. First, we may additionally require an interpolant χ of the implication $\varphi \to \psi$ to share the information of agents, i.e., we add the following agent condition: $\mathsf{Agt}(\chi) \subseteq \mathsf{Agt}(\varphi) \cap \mathsf{Agt}(\psi)$. This amounts to consider the *strong* Craig interpolation theorem studied, e.g., in [13]. Second, we can change our base multi-agent epistemic logic **S5** into weaker multi-agent epistemic logics like **K**, **KD**, **KT**, **S4**, **K45**, **KD45**, etc., to ask whether the Craig interpolation theorems still hold. For these multi-agent epistemic logics we have listed, we may verify if we can construct cut-free sequent calculi [22]. Third, we can explore a possible connection between the on-topic/off-topic interpretation of an awareness set $\mathscr{A}(i, w)$ and weak Kleene logic, since the off-topic interpretation was originally provided for the value "undefined" of weak Kleene three-valued logic in [1] (later generalized in [19]). We conjecture that there is a strong connection between awareness operators and A. N. Prior's *stability operator* S studied in modal logic over weak Kleene three-valued logic [3, p.86]. This is because the characteristic axioms given in [17, p.85] (also see [2, p.245] and [3, p.86]) are equivalent to $S\varphi \leftrightarrow \bigwedge_{p \in \mathsf{At}(\varphi)} Sp$, which has exactly the same form of the characteristic formula $A_i\varphi \leftrightarrow \bigwedge_{p \in \mathsf{At}(\varphi)} A_i p$ for propositional awareness [9, p.508].

References

1. Beall, J.: Off-topic: a new interpretation of weak-Kleene logic. Australas. J. Logic **13**(6), 136–142 (2016). https://doi.org/10.26686/ajl.v13i6.3976
2. Correia, F.: Adequacy results for some Priorean modal propositional logics. Notre Dame J. Formal Logic **40**(2), 236–249 (1999). https://doi.org/10.1305/ndjfl/1038949539
3. Correia, F.: Weak necessity on weak Kleene matrices. In: Wolter, F., Wansing, H., de Rijke, M., Zakharyaschev, M. (eds.) Advances in Modal Logic, vol. 3, pp. 73–90. World Scientific Publishing Co Pte Ltd (2002)
4. Ditmarsch, H., French, T., Velazquez-Quesada, F., Wang, Y.: Knowledge, awareness, and bisimulation. In: TARK 2013 - Proceedings of the 14th Conference on Theoretical Aspects of Rationality and Knowledge, vol. 1, pp. 61–70. The Institute of Mathematical Sciences, India (2013), Theoretical Aspects of Rationality and Knowledge 14th Conference ; Conference Date: 07-01-2013–09-01-2013
5. van Ditmarsch, H., French, T.: Becoming aware of propositional variables. In: Banerjee, M., Seth, A. (eds.) ICLA 2011. LNCS (LNAI), vol. 6521, pp. 204–218. Springer, Heidelberg (2011). https://doi.org/10.1007/978-3-642-18026-2_17
6. Fagin, R., Halpern, J.Y.: Belief, awareness, and limited reasoning. Artif. Intell. **34**(1), 39–76 (1987). https://doi.org/10.1016/0004-3702(87)90003-8
7. Fitting, M.: Proof Methods for Modal and Intuitionistic Logics. Springer, Netherlands (1983). https://doi.org/10.1007/978-94-017-2794-5
8. Halpern, J.Y.: Alternative semantics for unawareness. Games Econom. Behav. **37**(2), 321–339 (2001). https://doi.org/10.1006/game.2000.0832
9. Halpern, J.Y., Rêgo, L.C.: Reasoning about knowledge of unawareness. Games Econom. Behav. **67**(2), 503–525 (2009). https://doi.org/10.1016/j.geb.2009.02.001
10. Kowalski, T., Ono, H.: Analytic cut and interpolation for bi-intuisionistic logic. Rev. Symbolic Logic **10**(2), 259–283 (2016). https://doi.org/10.1017/S175502031600040X

11. Kubono., Y., Racharak., T., Tojo., S.: Logic of awareness in agent's reasoning. In: Proceedings of the 15th International Conference on Agents and Artificial Intelligence, vol. 1, pp. 207–216. ICAART, INSTICC, SciTePress (2023). https://doi.org/10.5220/0011630300003393
12. Maruyama, A., Tojo, S., Ono, H.: Decidability of temporal epistemic logics for multi-agent models. In: Proceedings of the ICLP'01 Workshop on Computational Logic in Multi-Agent Systems (CLIMA-01), pp. 31–40 (2001)
13. Marx, M.: Interpolation in modal logic. In: Haeberer, A.M. (ed.) AMAST 1999. LNCS, vol. 1548, pp. 154–163. Springer, Heidelberg (1998). https://doi.org/10.1007/3-540-49253-4_13
14. Ohnishi, M., Matsumoto, K.: Gentzen method in modal calculi. II. Osaka Math. J. **11**(2), 115–120 (1959)
15. Ono, H.: Proof-theoretic methods in nonclassical logic –an introduction. In: Theories of Types and Proofs, vol. 2, pp. 207–255. Mathematical Society of Japan (1998)
16. Ono, H., Sano, K.: Analytic cut and Mints' symmetric interpolation method for bi-intuitionistic tense logic. In: Fernández-Duque, D., Palmigiano, A., Pinchinat, S. (eds.) Advances in Modal Logic, AiML 2022, Rennes, 22–25 August 2022, pp. 601–623. College Publications (2022). http://www.aiml.net/volumes/volume14/34-Ono-Sano.pdf
17. Prior, A.N.: Worlds, Times and Selves. Duckworth, London (1977). edited by Kit Fine
18. Schipper, B.C.: Awareness. In: van Ditmarsch, H., Halpern, J.Y., van der Hoek, W., Kooi, B. (eds.) Handbook of Epistemic Logic, chap. 3, pp. 77–146. College Publication (2015)
19. Song, Y., Omori, H., Arenhart, J., Tojo, S.: A generalization of Beall's off-topic interpretation. Stud. Logica. **112**(4), 893–932 (2023). https://doi.org/10.1007/s11225-023-10075-0
20. Takano, M.: A modified subformula property for the modal logics K5 and K5D. Bull. Sect. Logic **30**(2), 115–122 (2001)
21. Takano, M.: Subformula property as a substitute for cut-elimination in modal propositional logics. Math. Jpn. **37**(6), 1129–1145 (1992)
22. Takano, M.: A semantical analysis of cut-free calculi for modal logics. Rep. Math. Logic **53**, 43–65 (2018)
23. Udatsu, K.: Awareness Logic of Abstraction. Master's thesis, Japan Advanced Institute of Science and Technology (2024)

Knowable as Knowing How to Inquire

Yiting Wang and Yanjing Wang[✉]

Department of Philosophy, Peking University, Beijing, China
{yitingwang,y.wang}@pku.edu.cn

Abstract. Arbitrary Public Announcement Logic (**APAL**) and its variants are proposed to formalize knowability dynamically based on public announcements as the means to update knowledge. In this paper, we introduce yet another variant **HAPAL** of **APAL**, which is based on *questions* instead of announcements, and captures knowability as *knowing how* to know by *asking questions*. Therefore, the prime modality in our language can also be viewed as a know-how operator sharing the same ∃□ bundled structure as logics of knowing how in the literature. This change in the semantics of the arbitrary announcement operator results in a highly non-trivial logic, which departs from the existing versions of **APAL**. As we will show, it is already strictly more expressive than **APAL** and epistemic logic on S5 models in the single-agent case. Moreover, it lacks compactness and Craig interpolation property. We also provide a sound and weakly complete axiomatization.

Keywords: Knowability · Knowing how · Arbitrary public announcement logic · Inquiry · Expressivity

1 Introduction

Inspired by van Benthem's discussion of knowability in [8], the logical framework of Arbitrary Public Announcement Logic (**APAL**) is introduced in [2]. It features an extra modality $\Diamond$ on top of the framework of Public Announcement Logic [7], such that $\Diamond \varphi$ says that after *some* truthful announcement,[1] φ holds. Intuitively, by using the language of **APAL**, we can formalize knowability of φ by $\Diamond K\varphi$, i.e., *knowable as knowing after an announcement*. Technical properties of **APAL** and its variants have been studied extensively in the literature ever since [2], with abundant results on axiomatizations, (un)decidability, complexity, and expressivity [1,4,10–13]. In **APAL** and most existing variants, the number of agents matters. For example, although the multi-agent **APAL** is undecidable, its *single-agent* variant is not only decidable but just as expressive as epistemic logic and public announcement logic [2].

On the other hand, as a logical framework to capture (a specific version of) *knowability*, **APAL** has some peculiar features. For example, as shown in [9], $\Diamond(K\varphi \vee K\neg\varphi)$ is valid for any **APAL** formula φ, namely, *everything is knowable*

---
[1] The announcement is restricted to formulas in the epistemic logic fragment of **APAL** to avoid vicious cycles in the semantics.

to be *whether* true or false (*wh-knowable* in [9]). This may sound surprising given the extensive discussion on Fitch's knowability paradox initiated in [3], but note that the Fitch-style knowability $\varphi \to \Diamond \mathsf{K}\varphi$ is *not* valid in **APAL**. One can also bundle $\Diamond \mathsf{K}$ together as a *single* knowability modality and study its logic [6].

From the perspective of epistemology, **APAL** provides an *externalist* approach to capture *knowability*. That is, φ is knowable if there exists, *de facto*, a truthful announcement such that after the announcement, the agent could acquire the knowledge of φ. This external perspective also induces the rather counterintuitive validity of "everything is wh-knowable" that we discussed above. However, an *internalist* reading of knowability would also make sense in various scenarios, where a question one can ask is more suitable than an externally given announcement to justify the knowability, given that the question can be answered truthfully. Moreover, for a genuine notion of internalist knowability, it is *not* enough if we merely require the existence of a question such that after getting its true answer, the agent would know, as we will illustrate in the following example.

Example 1 (Guessing who). Suppose there are four students, Alice, Bob, Claire, and David, whose available information is in the table below.

Major \ Gender	Female	Not Female
Logic	Alice	Bob
Not Logic	Claire	David

Now, as a teacher, you are told one of them registered for your course. Is the identity of the registered student *knowable* to you if you are allowed to ask *one* yes-no question about the student's available information?

Intuitively, we would say "no", as there are four possibilities, and getting the true answer to a yes-no question can only cut these four possibilities into two parts, but you cannot make sure both parts are singletons in order to identify the student, no matter what the true answer is. On the other hand, the identity of the registered student is indeed knowable to you if *two* questions are allowed: e.g., you can first ask about the gender and then the major.

In contrast, if we take the externalist view of knowability as in **APAL**, the identity of the student is, of course, knowable via simply one announcement, as someone external can just announce the characteristic features of the registered student. For example, if Alice is the one, then there *exists* an announcement "Logic ∧ Female" to let you know it is Alice.

Note that the difference between the externalist and internalist views of knowability is *not* merely about announcements vs. questions as the means to acquire information. Simply replacing the requirement of the existence of an announcement with the existence of a yes-no question would not work either if we are aiming at the internalist knowability. Again, if Alice is the one, then there indeed exists, *de facto*, a question for you to ask such that after getting its true answer, you would know it is Alice: you can simply ask whether the student is

"Logic ∧ Female" and the true answer will be yes thus you will know. However, you *cannot know in advance* that your question will give you enough info to know the identity, no matter what answer you may get.

In this paper, we will essentially propose a new semantics of $\Diamond$ in **APAL**, such that we can capture our intuitive concept of internalist knowability illustrated above. The idea comes from the *Logics of Knowing How* studied in the literature, cf. e.g., [16]. The core idea is to formalize *knowing how to achieve* φ by *there existing* a plan such that I *know that* the plan can make sure φ. The order of quantifier and modality is crucial in the so-called bundle ∃□ behind the meaning of know-how: you must know the plan (*de re* knowledge) rather than merely knowing the *existence* of the plan (*de dicto* knowledge) [15]. In other words, a *uniform* plan should make sure φ on *all* the indistinguishable epistemic possibilities.

Now, we can define internalist knowability roughly as *knowing how to know by asking questions*. We will start from the simple case where only *one* yes-no question is allowed. In this specific setting, one's internalist knowability of φ is rendered as there exists a yes-no question such that one knows that after getting its true answer, φ is known. We will use a different symbol ⧫ in our language **HAPAL** to differentiate it from the usual $\Diamond$ in **APAL**, where the □ inside the $\Diamond$ represents the logical structure of ∃□. This paper makes the first steps in studying the resulting logic, where we focus on the single-agent case over epistemic models. We will show that even in such a simple setting, our logic already behaves quite differently from **APAL** and its existing variants.

The main contributions of this paper can be summarized as follows:

- We present a variant of **APAL**, which we call **HAPAL**, to capture the internalist knowability based on know-how to inquire (H for *how*).
- We show that **HAPAL** is strictly more expressive than **EL** and **APAL** over epistemic models in the single agent case. Besides, **HAPAL** is not compact and does not have Craig interpolation property.
- We give a sound and weakly complete finitary proof system of **HAPAL** .
- In **HAPAL**, not everything is wh-knowable, given any finite bound on the number of questions that can be asked.

In the rest of the paper, we first survey the main existing variants of **APAL** in Sect. 2, and then lay out our new framework **HAPAL** in Sect. 3. In contrast with **APAL**, we shows that not everything is wh-knowable in **HAPAL**. Section 4 studies the expressivity of **HAPAL** and show that it lacks compactness and Craig interpolation property. Section 5 gives a sound and weakly complete proof system. We conclude with open questions and further directions in Sect. 6.

2 Existing Variants of APAL

Since [2], various variants of **APAL** have been proposed. We briefly survey them in this section to highlight the differences between existing work and our framework. Most variants focus on limiting the form of the announcement being quantified, e.g., **BAPAL** proposed in [11] limits announcements to boolean formulas, and [1] introduces **GAL** with an arbitrary-announcement-like modality $\langle G \rangle$ and limits announcements to group knowledge of a finite G : $\bigwedge_{i \in G} \mathsf{K}_i \varphi$. The closest variant to our work is **ACAL** in [10] where instead of the standard knowledge and announcement operators, know-whether Kw and announce-whether operator $[?\psi]$ are used. The latter could also be viewed as asking *whether* ψ, given the true answer will be provided. Correspondingly, arbitrary announcement-whether operator ♦ is used in [10] to replace the diamond in **APAL**. Note that over epistemic models **ACAL** is equally expressive as **APAL**.[2]

The languages (in the single-agent case) are summarized as follows with our language **HAPAL** at the end:

EL $\quad \varphi ::= \top \mid p \mid \neg \varphi \mid (\varphi \wedge \varphi) \mid \mathsf{K}\varphi$
APAL $\quad \varphi ::= \top \mid p \mid \neg \varphi \mid (\varphi \wedge \varphi) \mid \mathsf{K}\varphi \mid \langle \varphi \rangle \varphi \mid \Diamond \varphi$
GAL $\quad \varphi ::= \top \mid p \mid \neg \varphi \mid (\varphi \wedge \varphi) \mid \mathsf{K}\varphi \mid \langle \varphi \rangle \varphi \mid \langle G \rangle \varphi$
ACAL $\quad \varphi ::= \top \mid p \mid \neg \varphi \mid (\varphi \wedge \varphi) \mid \mathsf{Kw}\varphi \mid \langle ?\varphi \rangle \varphi \mid \blacklozenge \varphi$
HAPAL $\varphi ::= \top \mid p \mid \neg \varphi \mid (\varphi \wedge \varphi) \mid \mathsf{K}\varphi \mid \langle ?\psi \rangle \varphi \mid \Diamondblack \varphi$

The above-mentioned existing logics share some common properties, from which our framework deviates. Over single-agent S5 models, the expressivity of all these logics is equivalent to single-agent epistemic logic **EL**, since in this setting, each formula containing arbitrary announcements could be reduced to an **EL**-formula without them [2]. Moreover, all these logics except **GAL**[3] has the property that "everything is wh-knowable" mentioned in the introduction [9]. Finally, all these arbitrary-announcement-like operators share the following schemas for a *normal* S4-Diamond (denoted as $\Diamond$):

$$(\Diamond \varphi \vee \Diamond \psi) \leftrightarrow \Diamond(\varphi \vee \psi), \qquad \varphi \to \Diamond \varphi, \qquad \Diamond \Diamond \varphi \to \Diamond \varphi, \qquad \neg \Diamond \bot$$

In the rest of the paper, we will see that our operator $\Diamondblack$ behaves differently to the above-mentioned ones as shown in the table (over **single-agent S5 models**):

Operators Properties	$\Diamond$	$\langle G \rangle$	$\blacklozenge$	$\Diamondblack$
Quantifies over...	announcements	announcements	questions	questions
More expressive than **EL**	no	no	no	yes
Everything is wh-knowable	yes	no	yes	no
$S4$ axioms of Diamond	yes	yes	yes	no

As we will check later, many valid formulas of **APAL** become invalid if we replace $\Diamond$ with $\Diamondblack$, such as $\Diamond \Box \varphi \to \Box \Diamond \varphi$, and normal S4 principles ($\Diamond \varphi \vee$

---

[2] On epistemic models, $\mathsf{K}\varphi$ can be expressed by $\mathsf{Kw}\varphi \wedge \varphi$. $\langle \psi \rangle \varphi$ can be expressed by $\psi \wedge \langle ?\psi \rangle \varphi$, and $\Diamond \varphi$ is equivalent to $\blacklozenge \varphi$, as the existence of a question is equivalent to the existence of an announcement of its *true* answer.

[3] Over single-agent S5 models, $\langle G \rangle$ in **GAL** is a trivial operator as $\vDash \langle \mathsf{K}\psi \rangle \varphi \leftrightarrow \varphi$, thus $\vDash \langle G \rangle \varphi \leftrightarrow \varphi$.

$\Diamond\psi) \leftrightarrow \Diamond(\varphi \vee \psi), \varphi \to \Diamond\varphi, \Diamond\Diamond\varphi \to \Diamond\varphi$. Interestingly, there are formulas in our language that are not wh-knowable, even when multiple questions are allowed. Furthermore, unlike **APAL** and other variants, our language **HAPAL** is *strictly more expressive* than **EL** over single-agent S5 models.

Finally, although ♦ also quantifies over questions, ⧫ cannot be simply defined in **ACAL** by the combinations of ♦ and K such as K♦ or ♦K. Albeit we have not defined ⧫ technically, the intuitive idea of the undefinability is due to the ∃□-structure behind ⧫ as ⧫φ essentially expresses $\exists \psi\, \mathsf{K}\langle ?\psi\rangle \varphi$. The modality ♦ is essentially a bundle of $\exists\psi\langle ?\psi\rangle$, and we cannot insert a K operator in the *middle* of ♦. Let us now lay out the framework formally.

3 The Framework of HAPAL

3.1 Language and Semantics

Definition 1 (Language HAPAL). *Given a countable set* **Prop** *of propositional variables, we define the language* **HAPAL** *of Know-how-based Arbitrary Announcement Logic as follows:*

$$\varphi ::= \top \mid p \mid \neg\varphi \mid (\varphi \wedge \varphi) \mid \mathsf{K}\varphi \mid [?\varphi]\varphi \mid ⧫\varphi$$

where $p \in$ **Prop**.

Given $\varphi \in$ **HAPAL**, we use $PV(\varphi)$ to denote the set of variables that occurs in φ. We also use the usual abbreviations of $\bot, \vee, \hat{\mathsf{K}}$, and $\langle?\varphi\rangle$, and use $\mathsf{Kw}\varphi$ as the abbreviation of $\mathsf{K}\varphi \vee \mathsf{K}\neg\varphi$. Moreover, we use $\boxdot\varphi$ as the abbreviation for $\neg⧫\neg\varphi$, and use $⧫^n\varphi$ as abbreviation of $\underbrace{⧫\ldots⧫}_{n}\varphi$.

Then we define a complexity measure of **HAPAL** which we adapt from complexity measure in [10]. The complexity would facilitate the induction in the completeness proof.

Definition 2 (Size and ⧫-depth). *The size of a formula* φ, *in symbols* $Size(\varphi)$ *and the* ⧫-*depth of a formula* φ, *in symbols* $d_⧫(\varphi)$ *are recursively defined as follows:*

Size(φ)   **d**$_⧫$(φ)
$Size(p) = Size(\top) = 1$ $d_⧫(p) = d_⧫(\top) = 0$
$Size(\neg\varphi) = 1 + Size(\varphi)$ $d_⧫(\neg\varphi) = d_⧫(\varphi)$
$Size(\varphi \wedge \psi) = 1 + max\{Size(\varphi), Size(\psi)\}$ $d_⧫(\varphi \wedge \psi) = max\{d_⧫(\varphi), d_⧫(\psi)\}$
$Size(\mathsf{K}\varphi) = 3 + Size(\varphi)$ $d_⧫(\mathsf{K}\varphi) = d_⧫(\varphi)$
$Size([?\psi]\varphi) = Size(\psi) + 3 \cdot Size(\varphi)$ $d_⧫([?\psi]\varphi) = d_⧫(\psi) + d_⧫(\varphi)$
$Size(⧫\varphi) = 1 + Size(\varphi)$ $d_⧫(⧫\varphi) = 1 + d_⧫(\varphi)$

We define $<^{Size}_{d_⧫}$ as a lexicographic order between formulas in **HAPAL** such that $\varphi <^{Size}_{d_⧫} \psi$ iff either $d_⧫(\varphi) < d_⧫(\psi)$, or $d_⧫(\varphi) = d_⧫(\psi)$ and $Size(\varphi) < Size(\psi)$.

Note that the ⧫-depth is not the usual modal depth counting all the modalities in the language. In particular, for each **EL** formula λ, $d_⧫(\lambda) = 0$ since **EL** formulas contain no ⧫ operator.

Definition 3 (Model and Semantics). *An epistemic (or S5) model is a tuple $\langle S, \sim, V \rangle$ where S is a nonempty set of states, $\sim \subseteq S \times S$ is an equivalence relation, $V : \mathbf{Prop} \to \wp(S)$ is a valuation function. Given an epistemic model $\mathcal{M} = \langle S, \sim, V \rangle$ and a state $s \in S$, the satisfaction relation on φ and pointed model $(\mathcal{M}, s)$ is defined as follows:*

$$\begin{aligned}
\mathcal{M}, s \vDash p &\iff s \in V(p) \\
\mathcal{M}, s \vDash \neg\varphi &\iff \mathcal{M}, s \nvDash \varphi \\
\mathcal{M}, s \vDash \varphi \wedge \psi &\iff \mathcal{M}, s \vDash \varphi \text{ and } \mathcal{M}, s \vDash \psi \\
\mathcal{M}, s \vDash \mathsf{K}\varphi &\iff \text{for each } s' \sim s,\ \mathcal{M}, s' \vDash \varphi \\
\mathcal{M}, s \vDash [?\psi]\varphi &\iff \mathcal{M}_\psi, s \vDash \varphi \\
\mathcal{M}, s \vDash \Diamond\varphi &\iff \text{there exists } \psi \in \mathbf{EL} \text{ such that for each } s' \sim s,\ \mathcal{M}_\psi, s' \vDash \varphi.
\end{aligned}$$

where given that $[\![\psi]\!]_\mathcal{M} = \{s \in S \mid \mathcal{M}, s \vDash \psi\}$, the model $\mathcal{M}_\psi = \langle S, \sim'', V \rangle$ such that $\sim'' = (\sim \cap ([\![\psi]\!]_\mathcal{M} \times [\![\psi]\!]_\mathcal{M})) \cup (\sim \cap ([\![\neg\psi]\!]_\mathcal{M} \times [\![\neg\psi]\!]_\mathcal{M}))$, we simply write $s \sim^\mathcal{M}_\psi t$ as the abbreviation of $s \sim'' t$ in $\mathcal{M}_\psi$.

Clearly, $[?\psi]\varphi \leftrightarrow \langle ?\psi \rangle \varphi$ is valid as any question is always "askable". Intuitively, $\mathcal{M}, s \vDash \Diamond\varphi$ if there exists $\psi \in \mathbf{EL}$, $\mathcal{M}, s \vDash \mathsf{K}\langle ?\psi \rangle \varphi$. Namely, it says there exists an inquiry "ψ holds or not" such that the agent knows that after inquiring and getting a deterministic answer, φ would always hold. Given the semantics, we have $\mathcal{M}, s \vDash \boxtimes\varphi$ iff for each $\psi \in \mathbf{EL}$, there *exists* $s' \sim s$ such that $\mathcal{M}, s' \vDash [?\psi]\varphi$.

Let us come back to Example 1 to see how $\Diamond$ works. Let p denote "the registered student majors in logic" and q denote "the registered student is female". Let $\mathcal{M} = \langle S, \sim, V \rangle$ be an epistemic model where $S = \{00, 01, 10, 11\}$, $\sim$ is a total relation over S and $V(p) = \{10, 11\}, V(q) = \{01, 11\}$. The model represents the uncertainty of the teacher. According to the example, the states $00, 01, 10, 11$ correspond to the four students David, Claire, Bob, and Alice respectively. Knowing the identity of the registered student is equivalent to ruling out all other states and thus can be expressed by the formula $\mathsf{Kw}p \wedge \mathsf{Kw}q$ (Figs. 1, 2 and 3).

Now we can see formally whether the identity of the registered student is knowable. Let us suppose Alice is the one, thus the state 11 is the real world. First of all, we have $\mathcal{M}, 11 \vDash \Diamond \Diamond (\mathsf{Kw}p \wedge \mathsf{Kw}q)$ since $\vDash \mathsf{K}[?p]\mathsf{K}[?q](\mathsf{Kw}p \wedge \mathsf{Kw}q)$ is valid in $\mathcal{M}$ as shown by the following figures.

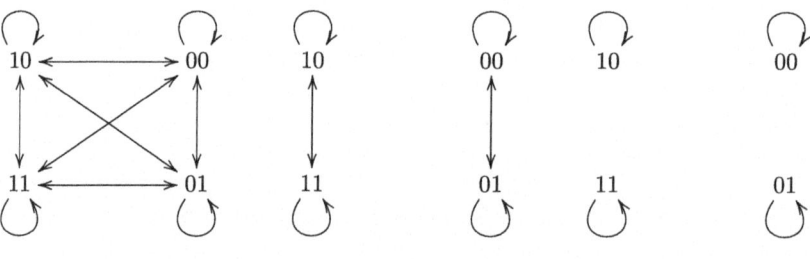

Fig. 1. $\mathcal{M}$ **Fig. 2.** $\mathcal{M}_p$ **Fig. 3.** $(\mathcal{M}_p)_q$

Now we move on to see whether $\mathcal{M}, 11 \vDash \Diamond(\mathsf{Kw}p \wedge \mathsf{Kw}q)$. For each $\psi \in \mathbf{EL}$, if there exists a state s such that $11 \sim_\psi^\mathcal{M} s$ and $s \neq 11$, then we have $\mathcal{M}_\psi, 11 \vDash \neg\mathsf{Kw}p \vee \neg\mathsf{Kw}q$. Otherwise, if there exists no state s' such $s' \sim_\psi^\mathcal{M} 11$ then there must exist two distinct states s_1 and s_2 such $s_1 \sim_\psi^\mathcal{M} s_2$ and then we will have $\mathcal{M}, s_1 \vDash \neg\mathsf{Kw}p \vee \neg\mathsf{Kw}q$. Hence, for each $\psi \in \mathbf{EL}$ there exists a state $t \sim 11$ such that $\mathcal{M}, t \vDash [?\psi](\neg\mathsf{Kw}p \vee \neg\mathsf{Kw}q)$ and then $\mathcal{M}, 11 \vDash \hat{\mathsf{K}}[?\psi](\neg\mathsf{Kw}p \vee \neg\mathsf{Kw}q)$. Therefore, we have $\mathcal{M}, 11 \vDash \boxdot(\neg\mathsf{Kw}p \vee \neg\mathsf{Kw}q)$. Namely, $\mathcal{M}, 11 \nvDash \Diamond(\mathsf{Kw}p \wedge \mathsf{Kw}q)$. In contrast, it is clear that $\mathcal{M}, 11 \vDash \Diamond(\mathsf{Kw}p \wedge \mathsf{Kw}q)$ in **APAL**, as $\mathcal{M}, 11 \vDash \langle p \wedge q \rangle(\mathsf{Kw}p \wedge \mathsf{Kw}q)$ and $\mathcal{M}, 11 \vDash \blacklozenge(\mathsf{Kw}p \wedge \mathsf{Kw}q)$ in **ACAL** similarly.

We represent some valid formulas and rules:

(i) $[?\psi]p \leftrightarrow p, [?\psi]\neg\varphi \leftrightarrow \neg[?\psi]\varphi, [?\chi](\varphi \wedge \psi) \leftrightarrow ([?\chi]\varphi \wedge [?\chi]\psi)$ (shown in [10]), $[?\psi]\mathsf{K}\varphi \leftrightarrow (\psi \to \mathsf{K}(\psi \to [?\psi]\varphi)) \wedge (\neg\psi \to \mathsf{K}(\neg\psi \to [?\psi]\varphi))$,
(ii) $\mathsf{K}[?\psi]\varphi \to [?\psi]\mathsf{K}\varphi$ (Perfect recall),
(iii) $\Diamond\alpha \leftrightarrow \mathsf{K}\alpha$ and $\Diamond\mathsf{Kw}\alpha$ where $\alpha \in \mathbf{PL}$ i.e., propositional formulas,
(iv) if $\varphi \to \psi$ is valid then $\Diamond\varphi \to \Diamond\psi$ is valid,
(v) $\neg\Diamond\bot, \Diamond\varphi \to \Diamond\Diamond\varphi, \Diamond\varphi \to \Diamond\mathsf{K}\varphi, \Diamond\varphi \to \mathsf{K}\Diamond\varphi$.

It is worth noticing that formulas in (v) are very similar to the axioms and theorems of the logic of knowing how, cf. e.g., [5].

In contrast, some formulas one would expect to be valid for **APAL** and many of its variant are actually not valid in **HAPAL**:

$$(\Diamond\varphi \vee \Diamond\psi) \leftrightarrow \Diamond(\varphi \vee \psi), \varphi \to \Diamond\varphi, \Diamond\Diamond\varphi \to \Diamond\varphi, \Diamond\boxdot\varphi \to \boxdot\Diamond\varphi$$

Now, we recall relevant notions of bisimulation which will be used later for the discussion of expressivity.

Definition 4 (Bisimulation). *Let $\mathcal{M} = \langle S, \sim, V \rangle$ and $\mathcal{M}' = \langle S', \sim', V' \rangle$ be two models. A nonempty relation $Z \subseteq S \times S'$ is a bisimulation between $\mathcal{M}$ and $\mathcal{M}'$ iff for all $s \in S$ and $s' \in S'$ with $(s, s') \in Z$ satisfying the following conditions:*

- *atoms: for all $p \in \mathbf{Prop}$, $s \in V(p)$ iff $s' \in V'(p)$.*
- *forth: for each $t \sim s$ there exists $t' \sim' s'$ such that $(t, t') \in Z$.*
- *back: for each $t' \sim' s'$ there exists $t \sim s$ such that $(t, t') \in Z$.*

If there exists a bisimulation Z between $\mathcal{M}$ and $\mathcal{M}'$ such that $(s, s') \in Z$, then $\mathcal{M}, s$ and $\mathcal{M}', s'$ are bisimilar, we use $\mathcal{M}, s \leftrightarroweq \mathcal{M}', s'$ to denote it.

Below is the restricted bisimulation w.r.t. a subset of propositional letters.

Definition 5 (Q-Bisimulation). *Let $Q \subseteq \mathbf{Prop}$. If Z between $\mathcal{M}$ and $\mathcal{M}'$ is a relation satisfying atoms for all variables $p \in Q$, forth and back, then we say Z is a Q-bisimulation, namely a bisimulation with respect to variables in Q, and we use $\rightleftharpoons_Q$ to denote two models are Q-bisimilar.*

It is routine to show that **HAPAL** is invariant under bisimulation. In contrast, in Sect. 4, we will show, somehow surprisingly, Q-bisimularity does not imply the modal equivalence of **HAPAL** formulas w.r.t. Q.

Proposition 1. *Let $\mathcal{M} = \langle S, \sim, V \rangle$ and $\mathcal{N} = \langle S', \sim', V' \rangle$ be epistemic models. If the pointed epistemic model $(\mathcal{M}, s)$ and $(\mathcal{N}, t)$ are bisimilar, then for all $\varphi \in$ **HAPAL**, $(\mathcal{M}, s) \vDash \varphi$ iff $(\mathcal{N}, t) \vDash \varphi$.*

3.2 Not Everything Is Wh-Knowable

As we have mentioned in the introduction and Sect. 2, both **APAL** and **ACAL** have the property that every proposition is wh-knowable. More formally, we have $\vDash \Diamond(K\varphi \vee K\neg\varphi)$ for each $\varphi \in$ **APAL** [10] and $\vDash \blacklozenge Kw\varphi$ for each $\varphi \in$ **ACAL** [9]. This property is guaranteed by the existence of "characteristic formula". That is given a model $\mathcal{M}, s$ and a formula φ, we have $\delta_s^\varphi = \bigwedge\{p \mid p \in PV(\varphi)$ and $\mathcal{M}, s \vDash p\} \wedge \bigwedge\{\neg p \mid p \in PV(\varphi)$ and $\mathcal{M}, s \not\vDash p\}$ and after announcing δ_s^φ (or asking whether δ_s^φ), $Kw\varphi$ always holds.

In the contrast, this property does not hold under the setting of $\Diamond$ since characteristic formula only works "locally" while $\Diamond$ works "globally". In fact, there exists $\varphi \in$ **HAPAL** and $\mathcal{M}$ such that $\mathcal{M}, s \vDash \neg \Diamond Kw\varphi$.

Proposition 2. *Let $\mathcal{M} = \langle S, \sim, V \rangle$ be an epistemic model where $S = \{00, 10, 11\}$, $\sim$ is the total relation over S, and $V(p) = \{10, 11\}, V(q) = \{11\}$ and for any other variable r, $V(r) = \emptyset$. Let δ^* be $(\neg p \wedge \neg q) \vee ((p \wedge \neg q) \wedge \hat{K}(p \wedge q))$, we have $\mathcal{M}, 00 \vDash \neg \Diamond Kw\delta^*$.*

Proof. First we have $\mathcal{M}, 00 \vDash \delta^*$ and $\mathcal{M}, 11 \not\vDash \delta^*$. Then $\mathcal{M}, 00 \not\vDash Kw\delta^*$. For each proper submodel of $\mathcal{M}$ generating by some ψ, there always exists an indistinguishable class which contains two distinct states. Such a updated model $\mathcal{M}_i = \langle S, \sim_i, V \rangle$ for each $i \in \{1, 2, 3\}$ must be one of the following cases:

- $\sim_1 = \{(00, 00), (10, 10), (11, 11), (00, 10), (10, 00)\}$,
- $\sim_2 = \{(00, 00), (10, 10), (11, 11), (00, 11), (11, 00)\}$,
- $\sim_3 = \{(00, 00), (10, 10), (11, 11), (10, 11), (11, 10)\}$.

Then,

- $\mathcal{M}_1, 00 \vDash \delta^*$ and $\mathcal{M}_1, 10 \not\vDash \delta^*$ (since $\mathcal{M}_1, 10 \not\vDash \neg p \wedge \neg q$ and $\mathcal{M}_1, 10 \not\vDash \hat{K}(p \wedge q)$)
- $\mathcal{M}_2, 00 \vDash \delta^*$ and $\mathcal{M}_2, 11 \not\vDash \delta^*$ (since $\mathcal{M}_2, 11 \not\vDash \neg p \wedge \neg q$ and $\mathcal{M}_2, 11 \not\vDash p \wedge \neg q$)
- $\mathcal{M}_3, 10 \vDash \delta^*$ and $\mathcal{M}_3, 11 \not\vDash \delta^*$ (since $\mathcal{M}_3, 11 \not\vDash \neg p \wedge \neg q$ and $\mathcal{M}_3, 11 \not\vDash p \wedge \neg q$)

Hence, for each $i \in \{1, 2, 3\}$, $\mathcal{M}_i \not\vDash Kw\delta^*$. Therefore, $\mathcal{M}, 00 \vDash \boxtimes \neg Kw\delta^*$.

The main idea of these formulas is that it is a combination of special Moorean sentences for each state "in the middle".

Can we have a version of wh-knowability if we allow multiple questions by using $\Diamond^n$? We will show this is also not possible. We first generalize the definition of δ^* if arbitrarily many propositional variables can be used.

Definition 6. *Given a countable set* **Prop** $= \{p_1, p_2, \ldots\}$ *of propositional variables, we can define δ_n for all $n > 0$ as follows:*

$$\delta_n := (0^n \vee (0^{n-1}1 \wedge \bigvee_{0^{n-1}1 < \sigma} \hat{K}\sigma) \vee \cdots \vee (01^{n-1} \wedge \hat{K}1^{n-1}0))$$

where σ is a binary string of length n, 0^n denotes an abbreviation for the formula $\neg p_1 \wedge \neg p_2 \wedge \cdots \wedge \neg p_n$, and $0^{n-1}1$ denotes an abbreviation for the formula $\neg p_1 \wedge \neg p_2 \wedge \cdots \wedge \neg p_{n-1} \wedge p_n$. Similarly, for other binary strings.

For example, we have

$\delta_3 := (\neg p_1 \wedge \neg p_2 \wedge \neg p_3) \vee (\neg p_1 \wedge \neg p_2 \wedge p_3 \wedge (\hat{K}(\neg p_1 \wedge p_2 \wedge \neg p_3) \vee \hat{K}(\neg p_1 \wedge p_2 \wedge p_3)$
$\vee \hat{K}(p_1 \wedge \neg p_2 \wedge \neg p_3))) \vee (\neg p_1 \wedge p_2 \wedge \neg p_3 \wedge (\hat{K}(\neg p_1 \wedge p_2 \wedge p_3) \vee \hat{K}(p_1 \wedge \neg p_2 \wedge \neg p_3)))$
$\vee (\neg p_1 \wedge p_2 \wedge p_3 \wedge \hat{K}(p_1 \wedge \neg p_2 \wedge \neg p_3))$

And δ^* in Proposition 2 is actually an instance of δ_2.

Proposition 3. *For each $n \in \mathbb{N}$, there exists an epistemic model $\mathcal{M}$ contains $2^n + 1$ states and a state s in $\mathcal{M}$ such that $\mathcal{M}, s \not\models \Diamond^n \mathsf{Kw}\delta_{n+1}$.*

On the other hand, for any φ, given enough chances to inquiry, an agent could always know whether φ holds, as the following proposition shows.

Proposition 4. *For each φ, there exists $n \in \mathbb{N}, n \leq |PV(\varphi)|$ such that $\models \Diamond^n \mathsf{Kw}\varphi$.*

We omit the proofs of Proposition 3 and Proposition 4 due to space limitation.

This suggests that the gap between the numbers of variables in δ_n and the layers of $\Diamond$ is exactly 1. Namely, there exists a series of "hard to clarify" formulas δ_n for each $n \in \mathbb{N}$ such that to settle the truth of δ_n we would have to use at least n announcements and clarify the truth values of all the propositional variables in δ_n.

4 Expressivity, Compactness and Interpolation

In this part, we first investigate the expressivity of **HAPAL**. It is well-known that in single-agent S5 **APAL** and most of its variants are as expressive as **EL** [2]. However, **HAPAL** is more expressive than **EL** in single-agent case.

Proposition 5. **HAPAL** *is strictly more expressive than* **EL** *in single-agent case.*

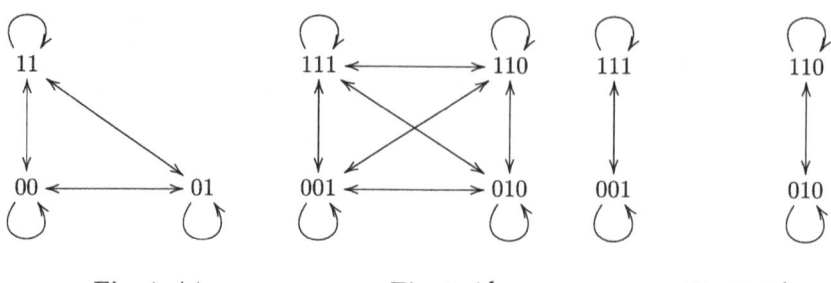

Fig. 4. $\mathcal{M}$ **Fig. 5.** $\mathcal{N}$ **Fig. 6.** $\mathcal{N}_r$

Proof. Let $\mathcal{M} = \langle S_\mathcal{M}, \sim_\mathcal{M}, V_\mathcal{M}\rangle$ be the model shown in Fig. 4 and $\mathcal{N} = \langle S_\mathcal{N}, \sim_\mathcal{N}, V_\mathcal{N}\rangle$ be the model shown in Fig. 5, where the first digit of each state represents the truth value of p, the second for q and the third (if exists) for r. Moreover, let $V_\mathcal{M}(r) = S_\mathcal{M}$ and for any variable $p_i \in \mathbf{Prop} \setminus \{p, q, r\}$, $V_\mathcal{M}(p_i) = S_\mathcal{M}$ and $V_\mathcal{N}(p_i) = S_\mathcal{N}$.

Let θ be $\hat{\mathsf{K}}(\neg p \wedge \neg q) \wedge \hat{\mathsf{K}}(\neg p \wedge q) \wedge \hat{\mathsf{K}}(p \wedge q)$. Now consider the formula $\varphi = \theta \wedge \Diamond(\hat{\mathsf{K}}p \wedge \neg\theta)$.

On one hand, we have $\mathcal{N}, 001 \vDash \theta \wedge \Diamond(\hat{\mathsf{K}}p \wedge \neg\theta)$ since $\mathcal{N}, 001 \vDash \theta \wedge \mathsf{K}[?r](\hat{\mathsf{K}}p \wedge \neg\theta)$. On the other hand, suppose that $\mathcal{M}, 00 \vDash \theta \wedge \Diamond(\hat{\mathsf{K}}p \wedge \neg\theta)$, then there exists $\psi \in \mathbf{EL}$ such that $\mathcal{M}, 00 \vDash \theta \wedge \mathsf{K}[?\psi](\hat{\mathsf{K}}p \wedge \neg\theta)$. For each properly updated model of $\mathcal{M}$ generating by some $\mathbf{EL}$ formula ψ, there always exists an indistinguishable class which contains two distinct states. Such a updated model must be one of the following cases:
$\mathcal{M}_i = \langle S_\mathcal{M}, \sim_i, V\rangle$ for each $i \in \{1, 2, 3\}$ where

- $\sim_1 = \{(00,00), (01,01), (11,11), (00,01), (01,00)\}$,
- $\sim_2 = \{(00,00), (01,01), (11,11), (00,11), (11,00)\}$,
- $\sim_3 = \{(00,00), (01,01), (11,11), (01,11), (11,01)\}$.

Since $\mathcal{M}_1, 00 \nvDash \hat{\mathsf{K}}p$, $\mathcal{M}_2, 01 \nvDash \hat{\mathsf{K}}p$, $\mathcal{M}_3, 00 \nvDash \hat{\mathsf{K}}p$, we have $\mathcal{M}, 00 \nvDash \theta \wedge \Diamond(\hat{\mathsf{K}}p \wedge \neg\theta)$. Meanwhile, it is easy to verify that $Z = \{(11, 111), (11, 110), (00, 001), (01, 010)\}$ is a bisimulation between $\mathcal{M}, 00$ and $\mathcal{N}, 001$ for variables other than r, then $\mathcal{M}, 00 \leftrightarroweq_{\mathbf{Prop}\setminus\{r\}} \mathcal{N}, 001$. Suppose there exists $\chi \in \mathbf{EL}$ which is equivalent to φ. Then let r be a variable which does not occur in χ. Then we have $\mathcal{M}, 00 \vDash \chi$ iff $\mathcal{N}, 001 \vDash \chi$ since $\mathcal{M}, 00 \leftrightarroweq_{\mathbf{Prop}\setminus r} \mathcal{N}, 001$. Contradiction! Therefore, $\mathbf{EL}$ is less expressive than $\mathbf{HAPAL}$, as it is a fragment of $\mathbf{HAPAL}$.

Then we immediately has the following corollary whose counterpart was proved in $\mathbf{APAL}$ with multiple agents [2].

Proposition 6. *Let $\mathcal{M}, s$ and $\mathcal{N}, t$ be arbitrary epistemic models. $\mathcal{M}, s \leftrightarroweq_Q \mathcal{N}, t$ does not imply $\mathcal{M}, s \equiv_Q \mathcal{N}, t$.*

Next we investigate the compactness of $\mathbf{HAPAL}$. It is worth noticing that as the previous proof has shown we can use formulas of the form $\Diamond \hat{\mathsf{K}}p$ to "count", e.g. distinguishing two "p states" from only one "p state". In contrast, no **EL** can express there are two different p states which without explicitly pointing out what the difference is. It could be shown more formally as the observation below:

Observation 1. *Note that for each $\varphi \in$ **HAPAL**, if $\mathcal{M}, s \models \varphi \wedge \Diamond(\neg\varphi \wedge \hat{\mathsf{K}}p)$ holds then there exists $\psi \in$ **EL** and there exist $s_1 \sim_\mathcal{M} s, s_2 \sim_\mathcal{M} s$ such that $\mathcal{M}, s_1 \models \psi$ and $\mathcal{M}_\psi, s_1 \models p$ as well as $\mathcal{M}, s_2 \models \neg\psi$ and $\mathcal{M}_\psi, s_2 \models p$. Since the truth of p is persevered under updating, we have $\mathcal{M}, s_1 \models \psi \wedge p$ and $\mathcal{M}, s_2 \models \neg\psi \wedge p$. And hence $\mathcal{M}, s \models \hat{\mathsf{K}}(\psi \wedge p) \wedge \hat{\mathsf{K}}(\neg\psi \wedge p)$.*

Based on the observation, it is not difficult to verify that infinite set of formulas $\Gamma = \{\theta \wedge (\mathsf{K}(p \to \lambda) \vee \mathsf{K}(p \to \neg\lambda)) \mid \lambda \in \mathbf{EL}\} \cup \{\theta \wedge \Diamond(\neg\theta \wedge \hat{\mathsf{K}}p)\}$ is finitely satisfiable but itself is not satisfiable. This provides a counterexample for compactness of **HAPAL**.

Proposition 7. ***HAPAL** is not compact.*

Furthermore, notice that in **HAPAL** the truth value of a variable that does not occur in some formula would have an impact on the truth of this formula. A natural but hardly being mentioned question follows immediately: does **HAPAL** have Craig Interpolation Property (CIP)? It turns out the answer is negative and we can find a counterexample for CIP. It suffices to show that there exists two pointed models $\mathcal{M}_1, s$ and $\mathcal{M}_2, t$ and two formulas φ and ψ such that (1) $\mathcal{M}_1, s \models \varphi$ and $\mathcal{M}_2, t \models \psi$, (2) $\models \varphi \to \neg\psi$ and (3) for any formula χ only contains propositional variables in $PV(\varphi) \cap PV(\psi)$, we have $\mathcal{M}_1, s \models \chi$ iff $\mathcal{M}_2, t \models \chi$. Recall that θ is the formula $\hat{\mathsf{K}}(\neg p \wedge \neg q) \wedge \hat{\mathsf{K}}(\neg p \wedge q) \wedge \hat{\mathsf{K}}(p \wedge q)$. And the valid formula in the following proposition is a proper instance for $\models \varphi \to \neg\psi$.

Proposition 8. $\models \theta \wedge \boxdot(\theta \vee \mathsf{K}\neg p) \to \neg(\hat{\mathsf{K}}(p \wedge r) \wedge \hat{\mathsf{K}}(p \wedge \neg r) \wedge \mathsf{K}((r \to (p \vee \neg q)) \wedge (\neg r \to (p \vee q))))$ *where r is a distinct variable from p and q.*

Proof. Suppose towards a contradiction that $\theta \wedge \boxdot(\theta \vee \mathsf{K}\neg p) \to \neg(\hat{\mathsf{K}}(p \wedge r) \wedge \hat{\mathsf{K}}(p \wedge \neg r) \wedge \mathsf{K}((r \to (p \vee \neg q)) \wedge (\neg r \to (p \vee q))))$ is not valid, then there exists a pointed model $\mathcal{M}, s$ such that $\mathcal{M}, s \models \theta \wedge \boxdot(\theta \vee \mathsf{K}\neg p)$ but $\mathcal{M}, s \models \hat{\mathsf{K}}(p \wedge r) \wedge \hat{\mathsf{K}}(p \wedge \neg r) \wedge \mathsf{K}((r \to (p \vee \neg q)) \wedge (\neg r \to (p \vee q)))$. Then on one hand, we have for each $\psi \in \mathbf{EL}$, there exists a state $s' \sim s$ such that $\mathcal{M}, s' \models [?\psi](\theta \vee \mathsf{K}\neg p)$ $(\star)$. On the other hand, since $\mathcal{M}, s \models \mathsf{K}((r \to (p \vee \neg q)) \wedge (\neg r \to (p \vee q)))$, we have that for each $s^* \sim s$, if $\mathcal{M}_r, s^* \models r$, then $\mathcal{M}_r, s^* \models \mathsf{K}\neg(p \wedge \neg q)$ and if $\mathcal{M}_r, s^* \models \neg r$, then $\mathcal{M}_r, s^* \models \mathsf{K}\neg(\neg p \wedge \neg q)$. Hence, for each $s^* \sim s$, $\mathcal{M}_r, s^* \models \neg\theta$. Moreover, $\mathcal{M}, s \models \hat{\mathsf{K}}(p \wedge r) \wedge \hat{\mathsf{K}}(p \wedge \neg r)$, then for each $s^* \sim s$, if $\mathcal{M}, s^* \models r$ then $\mathcal{M}_r, s^* \models \hat{\mathsf{K}}p$ and if $\mathcal{M}, s^* \models \neg r$ then we also have $\mathcal{M}_r, s^* \models \hat{\mathsf{K}}p$. Therefore, for each $s^* \sim s$, $\mathcal{M}_r, s^* \models \hat{\mathsf{K}}p$. It follows that for each $s^* \sim s$, $\mathcal{M}, s^* \models [?r](\neg\theta \wedge \hat{\mathsf{K}}p)$, which is contradictory to $(\star)$. Therefore, the formula $\theta \wedge \boxdot(\theta \vee \mathsf{K}\neg p) \to \neg(\hat{\mathsf{K}}(p \wedge r) \wedge \hat{\mathsf{K}}(p \wedge \neg r) \wedge \mathsf{K}((r \to (p \vee \neg q)) \wedge (\neg r \to (p \vee q))))$ is valid.

By taking $\mathcal{M}$ in Fig. 4 as $\mathcal{M}_1$ and $\mathcal{N}$ in Fig. 5 as $\mathcal{M}_2$, we have a counterexample for CIP.

Proposition 9. ***HAPAL** does not have Craig Interpolation Property.*

5 Axiomatization

Following [11], we first give a proof system with an infinitary rule, then strengthen it to a finite one. We first define necessity forms similarly as in [2]:

Definition 7 (Necessity forms). *Let $\psi \in \mathbf{HAPAL}$. Then, we recursively define necessity forms as follows:*

$$\tau(\sharp) : \sharp \mid (\psi \to \tau(\sharp)) \mid \mathsf{K}\tau(\sharp) \mid [?\psi]\tau(\sharp)$$

where $\sharp$ is not a formula, but a placeholder. We use $\tau(\varphi)$ to denote the result from replacing $\sharp$ in a necessity form $\tau(\sharp)$ by a formula φ. It is defined as follows:

$$\begin{aligned}
\sharp(\varphi) &= \varphi \\
(\psi \to \tau(\sharp))(\varphi) &= \psi \to \tau(\varphi) \\
(\mathsf{K}\tau(\sharp))(\varphi) &= \mathsf{K}\tau(\varphi) \\
([?\psi]\tau(\sharp))(\varphi) &= [?\psi]\tau(\varphi)
\end{aligned}$$

Definition 8 ($\mathbb{HAPAL}$ and $\mathbb{HAPAL}^+$). *The axiomatization $\mathbb{HAPAL}$ consists of the following axiom schemata and inference rules.*

Axiom Schemata

TAUT	all instances of propositional tautologies
DISTK	$\mathsf{K}(\varphi \to \psi) \to (\mathsf{K}\varphi \to \mathsf{K}\psi)$
T	$\mathsf{K}\varphi \to \varphi$
4	$\mathsf{K}\varphi \to \mathsf{K}\mathsf{K}\varphi$
5	$\neg\mathsf{K}\varphi \to \mathsf{K}\neg\mathsf{K}\varphi$
DUALK	$\hat{\mathsf{K}}\varphi \leftrightarrow \neg\mathsf{K}\neg\varphi$
?TOP	$[?\psi]\top \leftrightarrow \top$
?ATOM	$[?\psi]p \leftrightarrow p$
?NEG	$[?\psi]\neg\varphi \leftrightarrow \neg[?\psi]\varphi$
?COM	$[?\chi](\varphi \wedge \psi) \leftrightarrow [?\chi]\varphi \wedge [?\chi]\psi$
?K	$[?\psi]\mathsf{K}\varphi \leftrightarrow (\psi \to \mathsf{K}(\psi \to [?\psi]\varphi)) \wedge (\neg\psi \to \mathsf{K}(\neg\psi \to [?\psi]\varphi))$
Instantiation	$\mathsf{K}[?\psi]\varphi \to \Diamond\varphi$ where $\psi \in \mathbf{EL}$
DUAL$\Diamond$	$\Diamond\varphi \leftrightarrow \neg\Box\neg\varphi$

Rules

MP	φ and $\varphi \to \psi$ imply ψ
NECK	φ implies $\mathsf{K}\varphi$
RE	$\varphi \leftrightarrow \psi$ implies $\chi \leftrightarrow \chi[\psi/\varphi]$
MONO$\Diamond$	$\varphi \to \psi$ implies $\Diamond\varphi \to \Diamond\psi$
RK$\Box$	$\tau(\hat{\mathsf{K}}[?r]\varphi)$ where $r \notin PV(\tau) \cup PV(\varphi)$ implies $\tau(\Box\varphi)$

where $\tau(\sharp)$ is a necessity form.

Let RK$^+\Box$ be the rule

$$\frac{\tau(\hat{\mathsf{K}}[?\psi]\varphi) \text{ for each } \psi \in \mathbf{EL}}{\tau(\Box\varphi)}$$

then by replacing the rule RK◻ with the rule RK$^+$◻ we get the system $\mathbb{HAPAL}^+$.

If from a set of premises Γ a formula φ is provable in system $\mathbb{HAPAL}$, we write $\Gamma \vdash_{\mathbb{HAPAL}} \varphi$. If this holds for $\Gamma = \emptyset$ we also write $\vdash_{\mathbb{HAPAL}} \varphi$.

A derivation in system $\mathbb{HAPAL}^+$ is a well-founded (possibly) infinitely-branching tree. Given such a tree, we turn it into a finite derivation in $\mathbb{HAPAL}$ by replacing each application of the infinitary rule with the finitary rule and accordingly pruning the (infinitely many) branches (for the premises) except one. That, is for each RK$^+$◻ application with a conclusion $\tau(◻\varphi)$, one of its premises must be $\tau(\hat{\mathsf{K}}[?r]\varphi)$ for some fresh variable $r \notin PV(\tau) \cup PV(\varphi)$. Then we have a well-founded derivation tree of $\mathbb{HAPAL}$ without infinite branching. This is a counterpart of results in [2,11]. Then we have the following corollary.

Proposition 10. *For each $\varphi \in \mathbf{HAPAL}$, if $\vdash_{\mathbb{HAPAL}^+} \varphi$ then $\vdash_{\mathbb{HAPAL}} \varphi$.*

Note that validity of axioms and validity preservation of most rules including RK$^+$◻ are either routine or either directly followed by the semantics. To show the soundness of $\mathbb{HAPAL}$, it suffices to show that the derivation rule RK◻ is validity preserving. Note that in single agent S5, quantifying over **EL** formulas is actually equivalent to quantifying over **PL** formulas. The soundness of $\mathbb{HAPAL}$ can be shown almost exactly as in [11].

Proposition 11. $\mathbb{HAPAL}$ *and* $\mathbb{HAPAL}^+$ *are sound.*

Here we adapt the proof strategy in [11], using the detour proof through $\mathbb{HAPAL}^+$ to show the completeness of $\mathbb{HAPAL}$. That is, we firstly show $\mathbb{HAPAL}^+$ is weakly complete with respect to epistemic models, and by Proposition 10 each $\mathbb{HAPAL}^+$ derivation can be transformed into a $\mathbb{HAPAL}$ derivation. Then by the soundness of $\mathbb{HAPAL}$, we have (weak) completeness of system $\mathbb{HAPAL}$. Details are omitted due to page limitation.

Theorem 1 (Completeness). *For each $\varphi \in \mathbf{HAPAL}$, if $\vDash \varphi$ then $\vdash_{\mathbb{HAPAL}} \varphi$.*

6 Discussion and Future Work

In this paper, we proposed **HAPAL** as a framework to capture internalist knowability based on the idea of knowing how to know by inquiry. We showed that the resulting logic behaves quite differently compared to **APAL** and its variants in the single-agent S5 case.

There are various technical questions to be studied, such as the decidability of the satisfiability problem and whether **HAPAL** has finite model property. Moreover, we leave for a future occasion for a detailed comparison between our logic and the logics of knowing how. It is also interesting to see what the the logic without the dynamic question operator.

Furthermore, we can extend this work in two directions: Instead of asking a bounded number of questions, we could consider changing the semantics of ◇ w.r.t. a questioning strategy without the upper bound of the steps. Moreover, we can also consider different types of inquiries. That is, instead of considering a simple question of "whether or not", we can extend it into a general case $\exists x\, \mathsf{K}[?x]\psi$ where x is a general question that has possibly more than two answers. This also has a natural connection with the existing logics of knowing value [14,17]. Finally, it is well known that the most interesting part of **APAL** lies in its multi-agent case. So a natural attempt is to compare the expressivity between **HAPAL** and **APAL** in multi-agent case.

References

1. Ågotnes, T., Balbiani, P., van Ditmarsch, H., Seban, P.: Group announcement logic. J. Appl. Log. **8**(1), 62–81 (2010). https://doi.org/10.1016/j.jal.2008.12.002
2. Balbiani, P., Baltag, A., van Ditmarsch, H.V., Herzig, A., Hoshi, T., Lima, T.D.: Knowable 'as' known after an announcement. Rev. Symbolic Logic **1**(3), 305–334 (2008). https://doi.org/10.1017/S1755020308080210
3. Fitch, F.B.: A logical analysis of some value concepts. J. Symbolic Logic **28**(2), 135–142 (1963). https://doi.org/10.2307/2271594
4. French, T., van Ditmarsch, H.: Undecidability for arbitrary public announcement logic. Adv. Modal Logic **7**, 23–42 (2008)
5. Li, Y., Wang, Y.: Multi-agent knowing how via multi-step plans: a dynamic epistemic planning based approach. In: Blackburn, P., Lorini, E., Guo, M. (eds.) LORI 2019. LNCS, vol. 11813, pp. 126–139. Springer, Heidelberg (2019). https://doi.org/10.1007/978-3-662-60292-8_10
6. Liu, M., Fan, J., van Ditmarsch, H., Kuijer, L.B.: Logics for knowability. Logic Logical Philos. **31**(3), 385–426 (2022). https://doi.org/10.12775/LLP.2021.018
7. Plaza, J.: Logics of public communications. Synthese **158**(2), 165–179 (2007). https://doi.org/10.1007/s11229-007-9168-7
8. van Benthem, J.: What one may come to know. Analysis **64**(2), 95–105 (2004). https://doi.org/10.1093/analys/64.2.95
9. van Ditmarsch, H., van Der Hoek, W., Iliev, P.: Everything is knowable - how to get to know whether a proposition is true. Theoria **78**(2), 93–114 (2012). https://doi.org/10.1111/j.1755-2567.2011.01119.x
10. van Ditmarsch, H., Fan, J.: Propositional quantification in logics of contingency. J. Appl. Non-Class. Logics **26**(1), 81–102 (2016). https://doi.org/10.1080/11663081.2016.1184931
11. van Ditmarsch, H., French, T.: Quantifying over Boolean announcements. Logical Methods Comput. Sci. **18**(1), 4147 (2022). https://doi.org/10.46298/lmcs-18(1:20)2022
12. van Ditmarsch, H., French, T., Galimullin, R.: Satisfiability of quantified Boolean announcements (2022). https://doi.org/10.48550/arXiv.2206.00903
13. van Ditmarsch, H., French, T., Hales, J.: Positive announcements. Stud. Logica. **109**(3), 639–681 (2020). https://doi.org/10.1007/s11225-020-09922-1
14. van Eijck, J., Gattinger, M., Wang, Y.: Knowing values and public inspection. In: Ghosh, S., Prasad, S. (eds.) ICLA 2017. LNCS, vol. 10119, pp. 77–90. Springer, Heidelberg (2017). https://doi.org/10.1007/978-3-662-54069-5_7

15. Wang, Y.: Beyond knowing that: a new generation of epistemic logics. In: van Ditmarsch, H., Sandu, G. (eds.) Jaakko Hintikka on Knowledge and Game-Theoretical Semantics. OCL, vol. 12, pp. 499–533. Springer, Cham (2018). https://doi.org/10.1007/978-3-319-62864-6_21
16. Wang, Y.: A logic of goal-directed knowing how. Synthese **195**(10), 4419–4439 (2016). https://doi.org/10.1007/s11229-016-1272-0
17. Wang, Y., Fan, J.: Knowing that, knowing what, and public communication: public announcement logic with Kv operators. In: IJCAI, vol. 13, pp. 1147–1154 (2013)

Author Index

A
Adsul, Bharat 69

B
Basu, Sankha S. 84

D
De, Abhishek 97
Ding, Yiwen 111

E
Etedadialiabadi, Mahmood 3

G
Gaba, Pranshu 126
Gao, Su 3

J
Jain, Nehul 69
Jockwich, Santiago 138

K
Khan, Md. Aquil 151

L
Löwe, Benedikt 164

M
Makowsky, Johann A. 34
Manoorkar, Krishna 111
Mathew, Prince 176

P
Palmigiano, Alessandra 111
Paul, Bornali 190
Paul, Sandip 190
Penelle, Vincent 176
Pinchinat, Sophie 38

R
Ramsey, Nicholas 50
Ranjan, 151
Roy, Sayantan 84

S
Saivasan, Prakash 176
Sano, Katsuhiko 233
Sreejith, A. V. 176
Srinivasan, Bama 204
Sur, Arnab 126
Suzuki, Satoru 218

U
Udatsu, Kosuke 233

V
Vijayan, Mohan Raj 204

W
Wang, Ruoding 111
Wang, Yanjing 247
Wang, Yiting 247

X
Xiao, Han 164

© The Editor(s) (if applicable) and The Author(s), under exclusive license
to Springer Nature Switzerland AG 2025
C. Aiswarya et al. (Eds.): ICLA 2025, LNCS 15402, p. 263, 2025.
https://doi.org/10.1007/978-3-031-89610-1

The manufacturer's authorised representative in the EU is Springer Nature Customer Service Centre GmbH, Europaplatz 3, 69115 Heidelberg, Germany. If you have any concerns regarding our products, please contact ProductSafety@springernature.com

Printed and bound by CPI Group (UK) Ltd, Croydon, CR0 4YY

26/03/2026

02078962-0008